MICRO- AND NANOSTRUCTURED POLYMER SYSTEMS

From Synthesis to Applications

MICRO- AND NANOSTRUCTURED POLYMER SYSTEMS

From Synthesis to Applications

Edited by

Sabu Thomas, PhD, Robert A. Shanks, PhD, and Jithin Joy

AAP | APPLE ACADEMIC PRESS

Apple Academic Press Inc.
3333 Mistwell Crescent
Oakville, ON L6L 0A2
Canada

Apple Academic Press Inc.
9 Spinnaker Way
Waretown, NJ 08758
USA

Exclusive worldwide distribution by CRC Press, a member of Taylor & Francis Group

International Standard Book Number-13: 978-1-77188-100-5 (Hardcover)

International Standard Book Number-13: 978-1-4987-2159-2 (eBook)

Library and Archives Canada Cataloguing in Publication

Micro- and nanostructured polymer systems : from synthesis to applications / edited by Sabu Thomas, PhD, Robert A. Shanks, PhD, and Jithin Joy.

Includes bibliographical references and index.
Issued in print and electronic formats.
ISBN 978-1-77188-100-5 (hardcover).--ISBN 978-1-4987-2159-2 (pdf)
1. Polymers--Microstructure. 2. Nanostructured materials. 3. Biopolymers. I. Thomas, Sabu, author, editor II. Shanks, Robert (Robert A.), editor III. Joy, Jithin, author, editor

QC173.4.P65M52 2015 668.9 C2015-906173-3 C2015-906174-1

Library of Congress Cataloging-in-Publication Data

Names: Thomas, Sabu, editor. | Shanks, Robert (Robert A.) editor. | Joy, Jithin, editor.
Title: Micro- and nanostructured polymer systems : from synthesis to applications / editors, Sabu Thomas, PhD, Robert A. Shanks, PhD, and Jithin Joy.

Description: Toronto: Apple Academic Press, 2015
Includes bibliographical references and index.
Identifiers: LCCN 2015036429 | ISBN 9781771881005 (alk. paper)
Subjects: LCSH: Polymers. | Polymerization. | Green chemistry. |
Microstructure. | Nanostructured materials.
Classification: LCC QD381 .M47 2015 | DDC 668.9--dc23
LC record available at http://lccn.loc.gov/2015036429

Apple Academic Press also publishes its books in a variety of electronic formats. Some content that appears in print may not be available in electronic format. For information about Apple Academic Press products, visit our website at **www.appleacademicpress.com** and the CRC Press website at **www.crcpress.com**

ABOUT THE EDITORS

Prof. Sabu Thomas

Prof. Sabu Thomas is a Professor of Polymer Science and Engineering at the School of Chemical Sciences and Director of the International and Inter University Centre for Nanoscience and Nanotechnology at Mahatma Gandhi University, Kottayam, Kerala, India. He received his BSc degree (1980) in Chemistry from the University of Kerala, B.Tech. (1983) in Polymer Science and Rubber Technology from the Cochin University of Science and Technology, and PhD (1987) in Polymer Engineering from the Indian Institute of Technology, Kharagpur. The research activities of Professor Thomas include surfaces and interfaces in multiphase polymer blend and composite systems, phase separation in polymer blends, compatibilization of immiscible polymer blends, thermoplastic elastomers, phase transitions in polymers, nanostructured polymer blends, macro-, micro- and nanocomposites, polymer rheology, recycling, reactive extrusion, processing–morphology–property relationships in multiphase polymer systems, double networking of elastomers, natural fibers and green composites, rubber vulcanization, interpenetrating polymer networks, diffusion and transport and polymer scaffolds for tissue engineering. He has supervised 66 PhD thesis, 40 MPhil thesis, and 45 Masters theses. He has three patents to his credit. He also received the coveted Sukumar Maithy Award as the best polymer researcher in the country for the year 2008. Very recently, Professor Thomas received the MRSI and CRSI medals for his excellent work. With over 600 publications to his credit and over 17,000 citations, with an h-index of 67, Dr. Thomas has been ranked fifth in India as one of the most productive scientists.

Prof. Robert A. Shanks

Dr. Robert Shanks is an Emeritus Professor of Polymer Science at the Royal Melbourne Institute of Technology (RMIT University) in Melbourne, Australia. He is the author of over 230 refereed journal papers, 22 book chapters, eight patents, and 378 conference papers on or related to polymer science. He is the recipient of many awards, including the BHERT (Business Higher Education Round Table) Award for academic-industry collaboration, 2005; RACI Polymer Division Citation, 1990; the Cooperative Research Centres' (CRC) Association Award and CRC Award for

Polymers Chairman Award 2005; RMIT University Supervisor of the Year 1997, and RMIT University Most Published Awards 1995, 1996. Dr. Shanks has extensive consulting experience with several major industrial projects with many major companies and small businesses. This work included problem solving related to product failures, production problems and selection of materials. His research interests include synthesis and modification of nano-structured materials and biopolymers with enhanced physical, chemical and thermal properties; development of functional nano-materials; and the investigation of processes of controlling the self-assembly of nano-structures, with emphasis on thermochemical and thermophysical properties. His qualifications are DAppSc (RMIT) 2003, Grad Dip Ed (Tertiary, Hawthorn Inst EDUC) 1982, FRMIT (Management) 1978, PhD (La Trobe University) 1972, FRMIT (Applied Chemistry) 1971, Msc (La Trobe) 1969, ARMIT (Applied Chemistry) 1966, and FRACI (Royal Australian Chemical Institute) 1986–2010.

Jithin Joy

Mr. Jithin Joy is a Research Scholar at the International and Inter University Centre for Nanoscience and Nanotechnology at Mahatma Gandhi University, Kottayam, Kerala, India. He is engaged in doctoral studies in the area of nanocellulose-based polymer nanocomposites. He has also conducted research work at Clemson University, South Carolina in USA. He received his MSc degree in Chemistry from Mahatma Gandhi University. He is a co-editor of the book entitled *Natural Rubber Materials: Volume 2: Composites and Nanocomposites*, published by the Royal Society of Chemistry. In 2010, Mr. Jithin Joy received a prestigious research fellowship administered jointly by the Council of Scientific and Industrial Research and University Grants Commission of the Government of India. He has publications in international journals and conference proceedings to his credit. He is a co-author of several books chapters, peer-reviewed publications, and invited presentations in international forums.

CONTENTS

LIST OF CONTRIBUTORS

Rosalin Abraham
Department of Physics, St. Dominic's College, Kanjirapally, Kottayam-685612, Kerala, India.

Olufemi O. Adeyemi
Department of Chemical Sciences, Olabisi Onabanjo University, PO Box 364, Ago-Iwoye, Ogun state, Nigeria.

S. G. Adoor
Department of Chemistry, GSS College, Belgaum, Karnataka, India.

Rupa Bhattacharyya
Department of Chemistry, Narula Institute of Technology, 81, Nilgunj Road, Kolkata-700109, West Bengal, India.

S. Bhavani
Department of Physics, Sri Venkateswara University, Tirupati-517 502, Andhra Pradesh, India.

C. Anand Chairman
Department of Metallurgical and Materials Engineering, National Institute of Technology, Tiruchirappalli, Tamilnadu, India.

Debabrata Chakrabarty
Department of Polymer Science & Technology, 92, A. P. C. Road, Kolkata-700009, West Bengal, India.
Department of Materials Science, Mangalore University, Mangalagangothri-574199, Karnataka, India.

P. Parvathy Chandran
Division of Crop Utilization, Central Tuber Crops Research Institute, Sreekariyam, Thiruvananthapuram, Kerala, India.

Jai Prakash Chaudhary
Scale-Up & Process Engineering Unit, CSIR-Central Salt & Marine Chemical Research Institute, G.B.Marg, Bhavnagar-364002, Gujarat, India.

K. Chowdojirao
Department of Polymer Science & Technology, SKU, Anantapur, Andhra Pradesh, India.

Mahuya Das
Department of Chemistry, JIS College of Engineering, Kalyani, Nadia, Pin-741235, West Bengal, India.

Johnsy George
Food Engineering and Packaging, Defence Food Research laboratory, Siddarthanagar, Mysore-570 011, Karnataka, India.

Jayakumari Isac
Department of Chemistry, CMS College, Kottayam, Kerala, India.

Jithin Joy
International and Inter University Centre for Nanoscience and Nanotechnology, Mahatma Gandhi University, Kottayam, Kerala, India.

A. N. Jyothi
Division of Crop Utilization, Central Tuber Crops Research Institute, Sreekariyam, Thiruvananthapuram, Kerala, India.

V. Rajinee Kant
School of Mining, Metallurgy and Chemical Engineering, Doornfontein Campus, University of Johannesburg, South Africa.

K. Kiran Kumar
Department of Physics, Sri Venkateswara University, Tirupati-517 502, Andhra Pradesh, India.

S. P. Kumaresh Babu
Department of Metallurgical and Materials Engineering, National Institute of Technology, Tiruchirappalli, Tamilnadu, India.

Edyta Kusiak
Department of Chemistry, Institute of Polymer & Dye Technology Technical University of Lodz, Poland,116 Zeromskiego Str., 90-924 Lodz, Poland.

P. Lovely Mathew
Department of Chemistry, Newman College, Thodupuzha, Kerala, India.

R. K. Mandal
Department of Metallurgical Engineering, Institute of Technology, Banaras Hindu University, Varanasi-221005, Uttar Pradesh, India.

Anna Marzec
Institute of Polymer and Dye Technology, Technical University of Lodz, ul. Stefanowskiego 12/16 90-924 Lodz, Poland.

Ramavatar Meena
Scale-Up & Process Engineering Unit, CSIR-Central Salt & Marine Chemical Research Institute,G.B.Marg, Bhavnagar-364002, Gujarat, India.

Tanki Mochochoko
Department of Chemistry and Chemical Technology, Walter Sisulu University, Private Bag X1, WSU 5117, Mthatha, Eastern Cape, South Africa.

Antoine F. Mulaba-Bafubiandi
School of Mining, Metallurgical and Chemical Engineering, Doornfontein Campus, University of Johannesburg, Johannesburg, South Africa.

K. Nandakumar
International and Inter University Centre for Nanoscience and Nanotechnology, Mahatma Gandhi University, Kottayam, Kerala, India.

R. Nandini
Department of Chemistry, MITE, Moodbidri, Karnataka, India.

Soumya B. Nair
Division of Crop Utilization, Central Tuber Crops Research Institute, Sreekariyam, Thiruvananthapuram, Kerala, India.

Amod A. Ogale
Department of Chemical Engineering and Center for Advanced Engineering Fibers and Films, Clemson University, Clemson, SC 29634-0909, USA.

Oluwatobi S. Oluwafemi
Department of Chemistry and Chemical Technology, Walter Sisulu University, Private Bag X1, WSU 5117, Mthatha, Eastern Cape, South Africa.

Manish P. Patel
Department of Chemistry, Sardar Patel University, Vallabh Vidyanagar-388120, Gujarat, India.

Yatin N. Patel
Department of Chemistry, Sardar Patel University, Vallabh Vidyanagar-388120, Gujarat, India.

Y. Pavani
Department of Physics, Sri Venkateswara University, Tirupati-517 502, India.

V. V. R. N. Rao
Department of Physics, Sri Venkateswara University, Tirupati-517 502, Andhra Pradesh, India.

M. Ravi
Department of Physics, Sri Venkateswara University, Tirupati-517 502, Andhra Pradesh, India.

S. Ravindra
School of Mining, Metallurgy and Chemical Engineering, Doornfontein Campus, University of Johannesburg, Johannesburg, South Africa.

S. N. Sabapathy
Food Engineering and Packaging, Defence Food Research Laboratory, Siddarthanagar, Mysore-570 011, Karnataka, India.

A. K. Sharma
Department of Physics, Sri Venkateswara University, Tirupati-517 502, Andhra Pradesh, India.

Siddaramaiah
Sri Jayachamarajendra College of Engineering, Mysore-570 006, Karnataka, India.

Manjeet Singh
Centre for Applied Chemistry, Central University of Jharkhand, Brambe, Ranchi-835 205, Jharkhand, India.

I. Sinha
Department of Applied Chemistry, Institute of Technology, Banaras Hindu University, Varanasi-221005, Uttar Pradesh, India.

Sandile P. Songca
Faculty of Science, Engineering and Technology, Walter Sisulu University, PO Box-19712, Tecoma, East London, South Africa.

P. Souda
Soft Materials Research Laboratory, Department of Chemistry, National Institute of Technology, Calicut, Kerala, India.

Lisa Sreejith
Soft Materials Research Laboratory, Department of Chemistry, National Institute of Technology, Calicut, India.

G. Suja
Division of Crop Utilization, Central Tuber Crops Research Institute, Sreekariyam, Thiruvananthapuram, Kerala, India.

P. C. Thapliyal
Organic Building Materials Group, CSIR-Central Building Research Institute, Roorkee-247667, Uttarakhand, India.

Sabu Thomas
International and Inter University Centre for Nanoscience and Nanotechnology, Mahatma Gandhi University, Kottayam, Kerala, India.

Byron S. Villacorta
Department of Chemical Engineering and Center for Advanced Engineering Fibers and Films, Clemson University, Clemson, SC 29634-0909, USA.

B. Vishalakshi
Department of Chemistry, MITE, Moodbidri, Karnataka, India.

Runcy Wilson
School of Chemical Science, Mahatma Gandhi University, Kottayam, Kerala, India.

Marian Zaborski
Institute of Polymer and Dye Technology, Technical University of Lodz, ul. Stefanowskiego 12/16 90-924 Lodz, Poland.

LIST OF ABBREVIATIONS

AA	acrylic acid
AFM	atomic force microscope
AgNPs	silver nanoparticles
APS	ammonium per sulfate
BC	bacterial cellulose
BCNC	bacterial cellulose nanocrystals
BER	bulk electrical resistivity
BFE	basalt fabric reinforced epoxy
BFP	basalt fabric reinforced polyester composite
BM	batch mixer
BNN-PS	barium sodium niobate-polystyrene
CF	carbon fibers
CNCs	cellulose nanocrystals
CNFs	carbon nanofibers
CNTs	carbon nanotubes
COD	chemical oxygen demand
CP/MASNMR	crossed polarization and magic angle spinning nuclear magnetic resonance
CPE	chlorinated polyethylene
CT	computer tomography
CVD	chemical vapor deposition
DIDP	di-isodecyl phthalate
DMM	digital multi-meter
DSC	differential scanning calorimeter
DSM	twin-screw micro-compounder
EAPaP	"smart material" or "electroactive paper"
EM SE	electromagnetic shielding effectiveness
EMC	electromagnetic compatibility
EMI	electromagnetic interference
ENR	epoxidized natural rubber
ESD	electrostatic discharge
ESD	electrostatic dissipation
ETFE	ethylene tetrafluoroethylene
FTIR	fourier transform infrared
FWHM	full width at half maximum
GC	gas chromatography

Gd_2O_3	gadolinium (III) oxide
GPC	gel permeation chromatography
HALS	hindered amine light stabilizer
HR-TEM	high resolution transmission electron Microscopy
HPLC	high performance liquid chromatography
HPMC	hydroxypropyl methylcellulose
HT	thermal treatment
IDT	initial decomposition temperature
i-PMMA	PMMA isotactic
IPN	inter penetrating network
KGM	starch-konjac glucomannan
KOH	potassium hydroxide
LLDPE	linear, low-density polyethylene
LSPR	localized surface plasmon resonance
MA	maleic acid
MB	methylene blue
MBTS	mercaptobenzothiazyl disulfide
mCPBA	*m*-chloroperoxybenzoic acid
MEKP	methyl ethyl ketone peroxide
MFA	multifunctional acrylates
MWNT	multi-walled carbon nanotubes
MWS	Maxwell-Wagner-Sillars polarization
nano-$CaCO_3$	nanoscale calcium carbonate
NBR	acrylonitrile-butadiene rubber
NMBA	N,N'-methylenebisacrylamide
NRF	National Research Foundation
OM	optical microscopy
PAAAM	poly(acrylate- acrylic acid-co maleic acid)
PAM	polyacrylamide
PAMA	poly(n-amyl methacrylate)
PBA	poly(butyl acrylate)
PBMA	poly(butyl methacrylate)
PCP	polychloroprene
PDMDPS	polydimethyldiphenylsiloxane
PDPS	polydiphenylsiloxane
PE	polyethylene
PEA	poly(ethyl acrylate)
PEG	polyethylene glycol
PEMA	poly(ethyl methacrylate)
PEO	polyethylene oxide
PHO	poly(β-hydrox-yoctanoate)
PIB	poly(iso butylene)

PLA	polylactic acid
PMMA	poly(methyl methacrylate)
POE	polyolefin elastomer
POM	polyoxymethylene
PPy	polypyrrole
PVA	poly(vinyl alcohol)
PVC	poly(vinyl chloride)
PVP	poly (vinyl pyrrolidone)
RGA	residual gas analyzer
SAP	superabsorbent polymers
SAXS	small-angle x-ray scattering
SCE	saturated calomel electrode
SEI	secondary electron image
SEM	scanning electron micrographs
SEM-EDX	scanning electron microscopy-energy dispersive X-ray
SERS	surface enhancing Raman scattering
SICART	sophisticated instrument center for applied research and testing
SMA	styrene maleic anhydride
SP/s	seaweed polysaccharide/s
SPHs	SUPERPOROUS hydrogels
s-PMMA	syndiotactic
TEM	transmission electron microscopy
TG	thermogravimetry
TGA	thermo gravimetric analyzer
TMTD	tetramethylthiuram disulfide
ToF-SIMS	time of flight secondary ion mass spectrometry
TPS	thermoplastic starch
TPU	thermoplastic polyurethane
UHV	ultra-high vacuum
UPVC	unplasticized poly(vinyl chloride)
USEPA	US Environmental Protection Agency
UV/Vis	UV/visible spectroscopy
WHO	World Health Organization
WVTR	water vapor transmission rate
XRD	X-ray diffraction
X-RD	X-ray diffractometry

PREFACE

This book, **Micro- and Nanostructured Polymer Systems: From Synthesis to Applications**, describes the recent advances in the development and characterization of multicomponent polymer blends and composites. It covers occurrence, synthesis, isolation and production, properties and applications, modification, and also the relevant analysis techniques to reveal the structures and properties of polymer systems. Bio-based polymer blends and composites occupy a unique position in the dynamic world of new biomaterials. Natural polymers have attained their cutting-edge technology through various platforms; yet, there is a lot of novel information about them, that is discussed in this book.

This book covers topics such as biopolymer-synthetic systems, nanomaterial-polymer structures, multi-characterization techniques, polymer blends, composites, polymer gels, polyelectrolytes and many other interesting aspects. It is written in a systematic and comprehensive manner. The content of the present book is unique. It covers an up-to-date record on the major findings and observations in the field of micro- and nanostructured polymer systems.

This book will be a very valuable reference source for university and college faculties, professionals, post-doctoral research fellows, senior graduate students, polymer technologists, and researchers from R&D laboratories working in the area of nanoscience, nanotechnology, and polymer technology.

Finally, the editors would like to express their sincere gratitude to all the contributors of this book, who extended excellent support to the successful completion of this venture. We are grateful to them for the commitment and the sincerity they have shown toward their contribution in the book. Without their enthusiasm and support, the compilation of this volume could not have been possible. We would like to thank all the reviewers who have taken their valuable time to make critical comments on each chapter. We also thank the publisher Apple Academic Press for recognizing the demand for such a book and for realizing the increasing importance of the area of micro- and nanostructured polymer systems and for starting such a new project.

—Sabu Thomas, Robert A. Shanks, and Jithin Joy

CHAPTER 1

NATURAL POLYMER BLENDS AND THEIR COMPOSITES: MICRO AND NANO STRUCTURED POLYMER SYSTEMS

JITHIN JOY[1,2], RUNCY WILSON[2], P. LOVELY MATHEW[3], and SABU THOMAS[1,2]

[1] International and Inter University Centre for Nanoscience and Nanotechnology, Mahatma Gandhi University, Kottayam, Kerala, India

[2]School of Chemical Science, Mahatma Gandhi University, Kottayam, Kerala, India

[3]Department of Chemistry, Newman College, Thodupuzha, Kerala, India

E-mail: jithinjoyjrf@gmail.com, runcy21@gmail.com, lovely.mathew@gmail.com, sabuchathukulam@yahoo.co.uk

CONTENTS

ABSTRACT

There is a global interest in replacing synthetic polymer composites for daily applications with environmentally friendly alternatives from renewable resources. The large quantities of the petroleum-based polymeric materials raised a negative effect on the environment. Nanotechnology has a great potential in the development of high-quality biopolymer-based products. Biopolymers are attracting considerable attention as a potential replacement for petroleum-based plastics due to an increased consciousness for sustainable development and high price of crude oil. Biopolymers ideally maintain the carbon dioxide balance after their degradation or incineration. By using biodegradable grades, they will also save energy on waste disposal. Bio-reinforced materials are attractive and an alternative low-cost substitute and widely available. This chapter deals with the preparation, characterization, and applications of natural polymer blends and their composites in a systematic and detailed way.

1.1 INTRODUCTION

Natural polymers have attracted an increasing attention over the past two decades, mainly due to their abundance and low cost in addition to environmental concerns, and the anticipated depletion of petroleum resources. This has led to a growing interest in developing chemical and biochemical processes to acquire and modify natural polymers, and to utilize their useful inherent properties in a wide range of applications of industrial interests in different fields (Zhao, Jin, Cong, Liu, & Fu, 2013). The development of commodities derived from petrochemical polymers has brought many benefits to mankind. However, it is becoming more evident that the ecosystem is considerably disturbed and damaged as a result of the non-degradable plastic materials used in disposable items. Therefore, the interest in polymers from renewable resources has recently gained exponential momentum and the use of biodegradable and renewable materials to replace conventional petroleum plastics for disposable applications is becoming popular (Yu, Dean, & Li, 2006). Within the broad family of renewable polymers, starch is one of the most attractive and promising sources for biodegradable plastics because of the abundant supply, low cost, renewability, biodegradability, and ease of chemical modifications (Mathew & Dufresne, 2002; Mohanty, Misra, & Hinrichsen, 2000).

Compared to polymeric resources from petroleum, natural polymers from renewable resources have the advantages of biodegradability, biocompatibility, non-toxicity, high reactivity, low cost, ease of availability, and so forth (Mecking, 2004; Smith, 2005; Wool, 2005). They also possess great reinforcing function in composites, similar to traditional inorganic nanofillers (Angellier, Molina-Boisseau, & Dufresne, 2005a; Angellier, Molina-Boisseau, Belgacem, & Dufresne, 2005b; Angellier, Molina-Boisseau, Dole, & Dufresne, 2006; Labet, Thielemans, & Dufresne, 2007; Thielemans & Wool, 2005; Yuan, Nishiyama, Wada, & Kuga, 2006), by virtue of their rigidity to form strong physical interactions. Due to the similarity with living tissues, natural polymers are biocompatible and non-cytotoxic (Druchok & Vlachy, 2010; Gentile, Chiono, Carmagnola, & Hatton, 2014; Rutledge, Cheng, Pryzhkova, Harris, & Jabbarzadeh, 2014). A great variety of materials derived from natural sources have been studied and proposed for different biomedical uses (Hietala, Mathew, & Oksman, 2013; Mohanty et. al., 2000). Starch-based polymers present enormous potential for wide use in the biomedical field ranging from bone replacement to engineering of tissue scaffolds and drug-delivery systems (Crépy, Petit, Wirquin, Martin, & Joly, 2014; Maran, Sivakumar, Sridhar, & Immanuel, 2013; Xie, Pollet, Halley, & Avérous, 2013).

1.2 CELLULOSE-BASED BLENDS, COMPOSITES AND NANOCOMPOSITES

Consumers, industry, and government are increasingly demanding products made from renewable and sustainable resources that are biodegradable, non-petroleum based, carbon neutral, and have low environmental, animal/human health and safety risks. Natural cellulose-based materials (wood, hemp, cotton, linen, etc.) have been used by our society as engineering materials for thousands of years and their use continues today as verified by the enormity of the worldwide industries in forest products, paper, textiles, etc. Natural materials develop functionality, flexibility and high mechanical strength/weight performance by exploiting hierarchical structure design that spans nanoscale to macroscopic dimensions.

Cellulose is a linear chain of ringed glucose molecules and has a flat ribbon-like conformation. The intra- and inter-chain hydrogen bonding network makes cellulose a relatively stable polymer. These cellulose fibrils contain both crystalline and amorphous phases, (Fig. 1.1a–c).

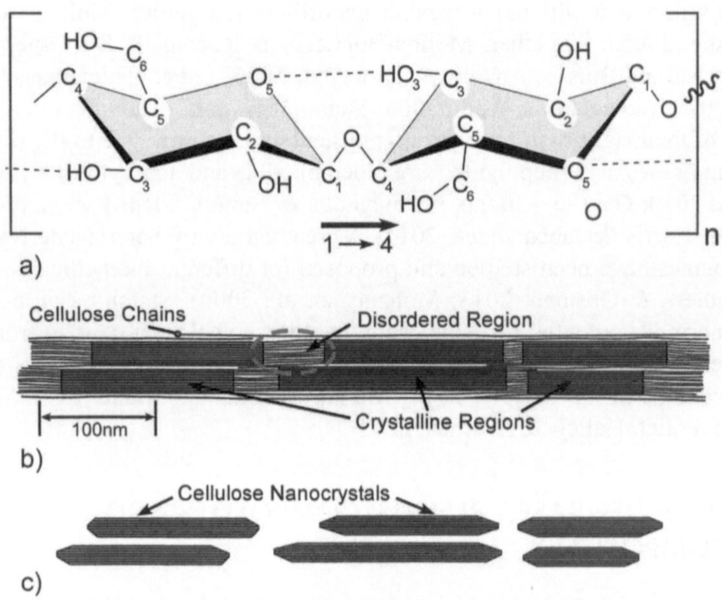

FIGURE 1.1 (a) Structure of cellulose monomeric unit; (b) Schematic representations of crystalline and amorphous regions in the cellulose micro fibril; (c) cellulose nanocrystals after acid hydrolysis (Moon, Martini, Nairn, Simonsen, & Youngblood, 2011).

The nanocellulose was prepared mainly by three different techniques, namely mechanical treatment (Hubbe, Rojas, Lucia, & Sain, 2008; Karimi, Tahir, Karimi, Dufresne, & Abdulkhani, 2014; Siró & Plackett, 2010), acid hydrolysis (Habibi, Lucia, & Rojas, 2010; Satyamurthy & Vigneshwaran, 2013; Syverud & Stenius, 2009; Tang, Yang, Zhang, & Zhang, 2014), and enzymatic hydrolysis (Penttilä et al., 2013; Siqueira, Tapin-Lingua, Bras, da Silva Perez, & Dufresne, 2011). Mechanical processes, such as high-pressure homogenizers (Cervin, Aulin, Larsson, & Wågberg, 2012; Li et al., 2012), grinders/refiners (Baheti, Abbasi, & Militky, 2012; Jin et al., 2011; Vartiainen et al., 2011), cryocrushing (Chakraborty, Sain, & Kortschot, 2005), high-intensity ultrasonic treatments (Mishra, Manent, Chabot, & Daneault, 2011; Mishra, Thirree, Manent, Chabot, & Daneault, 2010), and microfluidization (Tanaka, Hjelt, Sneck, & Korpela, 2012; Zhu, Sabo, & Luo, 2011), have been used for removing cellulose fibrils from different source materials. Repeated mechanical treatment leads to the reduction in percentage of cellulose crystallinity due to increased mechanical damage to the crystalline cellulose. For the extraction of nano crystalline cellulose particles from a variety of cellulose sources, acid hydrolysis technique was mainly used. By this technique, the amorphous regions within the cellulose microfibrils were removed. Sulfuric acid is the most typically used acid reagent, as it provides a net negative surface charge on the particles, leading to more stable suspensions (Mandal & Chakrabarty, 2011; Tang et al., 2014; Jiang & Hsieh, 2013), but other acids have also been used, such as hydrochloric acid (Jiang& Hsieh, 2013; Ludueña, Fasce, Alvarez, & Stefani, 2011), phosphoric acid (Camarero, Kuhnt, Foster, & Weder, 2013), oxalic acid (Abraham et al., 2011; Abraham et al., 2013; Chirayil et al., 2014) etc.

The reason for the use of nanocellulose in reinforced polymer matrix composites processing is that it is desirable to have both, a fine nanocellulose dispersion while also forming a nanocellulose network structure. The research has focused on producing composites with maximized and reproducible properties by developing processing routes that maximize fine nanocellulose dispersion. Four processing techniques have been used to make nanocellulose reinforced polymer matrix composites: solution casting (Azizi, Alloin, & Dufresne, 2005; Habibi et al., 2010; Liu, Sui, & Bhattacharyya, 2014), melt-compounding (Cobut, Sehaqui, & Berglund, 2014; Eichhorn et al., 2010; Hietala et al., Kamel, 2007; 2013; Reddy & Rhim, 2014), partial dissolution and electrospinning (Gabr et al., 2014; Liu, Yuan, & Bhattacharyya, 2012; Peresin et al., 2010; Rojas, Montero, & Habibi, 2009; Zoppe, Peresin, Habibi, Venditti, & Rojas, 2009). In addition to the low cost of the raw material, cellulose has other benefits like low energy consumption; high specific properties, low density; renewable nature, biodegradability; relatively reactive surface, great availability (Espert, Vilaplana, & Karlsson, 2004; Faruk, Bledzki, Fink, & Sain, 2014; Gandini & Belgacem, 2008). High moisture absorption, incompatibility with most polymeric matrices, low wettability, limitations of processing temperature are some limitations of nanocellulose over other fillers (Han, Zhou, Wu, Liu, & Wu, 2013; Wambua, Ivens, & Verpoest, 2003).

Jonoobi et al., has reported a comparative study of modified and unmodified cellulose nanofiber reinforced polylactic acid (PLA) prepared by twin screw extrusion (Jonoobi, Mathew, Abdi, Makinejad, & Oksman, 2012), Liu et al., also reported the characterization of solution cast cellulose nanofiber reinforced poly (lactic acid) (Liu, Yuan, Bhattacharyya, & Easteal, 2010). Fortunati et al. described the effects of modified cellulose nanocrystals on the barrier and migration properties of PLA nano-biocomposites (Fortunati et al., 2012). Song et al. reported a work on lowering water vapor transmission rate (WVTR) of paper by the incorporation of hydrophobic modified nanocellulose fiber on PLA biodegradable composites (Song, Xiao, & Zhao, 2014). Correction to transparent nanocellulosic multilayer thin films on polylactic acid with tunable gas barrier properties is described by Aulin, Karabulut, Tran, Wågberg, and Lindström (2013). Nanocellulose is a promising reinforcement for PLA composites.

The nanocellulose from various sources, such as cotton, tunicate, algae, bacteria, ramie, and wood for preparation of high-performance composite materials, have been investigated extensively (Azizi, et. al., 2005). Both natural and synthetic polymers were explored as the matrixes. Natural polymers such as poly (β-hydroxyoctanoate) (PHO) (Dubief, Samain, & Dufresne, 1999), soy protein (Wang, Cao,

& Zhang, 2006), silk fibroin (Noishiki, Nishiyama, Wada, Kuga, & Magoshi, 2002) reinforced with cellulose whiskers were reported. Meanwhile, Poly-(styrene-co-butylacrylate) (poly(S-co-BuA)) (Helbert, Cavaille, & Dufresne, 1996), poly (vinyl chlo-ride) (PVC) (Chazeau, Cavaille, Canova, Dendievel, & Boutherin, 1999), polypropylene (Ljungberg et al., 2005), waterborne polyurethane (Cao, Dong, & Li, 2007), were also used as synthetic matrixes.

1.3 STARCH-BASED BLENDS, COMPOSITES AND NANOCOMPOSITES

Starch from a variety of crops such as corn, wheat, rice and a potato is a source of biodegradable plastics, which are readily available at low cost when compared with most synthetic plastics (Ma, Chang, & Yu, 2008a). Starch is comprised of amylose, a linear polymer with molecular weight between 103 and 106 and amylopectin, a branched polymer with a-(1-6)-linked branch points (Fig. 1.2a–b) (Dufresne & Vignon, 1998).

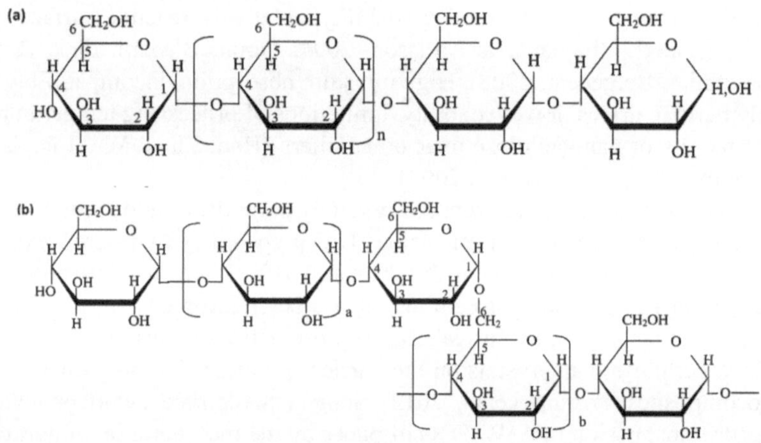

FIGURE 1.2 (a–b) Structure of amylose and amylopectin

In the glassy state, starch tends to be brittle and is very sensitive to moisture. In order to extrude or mold an object from starch, it is often converted into a thermoplastic starch (TPS) (Dufresne, Dupeyre, & Vignon, 2000). Both native starch and TPS can suffer from poor mechanical properties and high-water uptake compared to conventional polymers. Moreover, these properties may change after processing (Averous, 2004; Ma, Chang, Wang, & Yu, 2008b). Micro cellulose fiber and bacterial cellulose have also been reported as promising candidates for starch reinforcement. Dufresne et al. (2000) aimed to improve the thermomechanical properties and reduce the water sensitivity of potato starch-based nanocomposites, while preserving the biodegradability of the material through addition of Micro cellulose. This significantly reinforced the starch matrix, regardless of the plasticizer content, and the increase in tensile modulus as a function of filler content was almost linear.

1.4 CHITIN AND CHITOSAN-BASED BLENDS, COMPOSITES AND NANOCOMPOSITES

Chitin, poly (β-(1-4)-N-acetyl-d-glucosamine), is a natural polysaccharide of major importance, first identified in 1884. Chitin occurs in nature as ordered crystalline microfibrils forming structural components in the exoskeleton of arthropods or in the cell walls of fungi and yeast. It is also produced by a number of other living organisms in the lower plant and animal kingdoms, serving in many functions where reinforcement and strength are required. The most important derivative of chitin is chitosan (Fig. 1.3), obtained by (partial) deacetylation of chitin in the solid state under alkaline conditions (concentrated NaOH) or by enzymatic hydrolysis in the presence of chitin deacetylase. Because of the semicrystalline morphology of chitin, chitosan obtained by solid-state reaction has a heterogeneous distribution of acetyl groups along the chains.

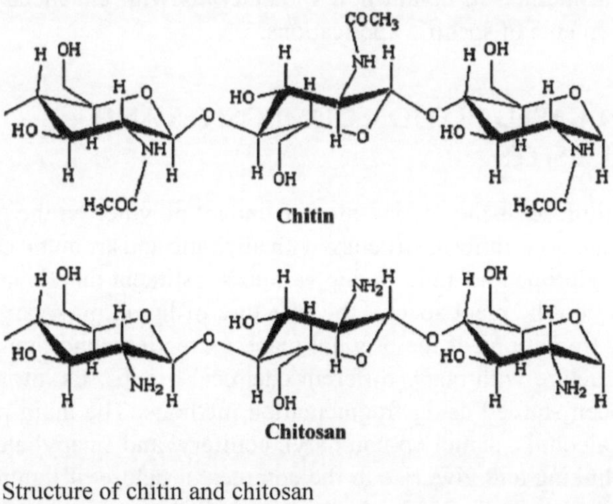

FIGURE 1.3 Structure of chitin and chitosan

Among the natural polymers, chitosan occupies a special position due to its abundance, versatility, facile modification, and unique properties, including biodegradability (Kean & Thanou, 2010), biocompatibility (Costa-Pinto et al., 2014; Hsu et al., 2011; Shi et al., 2012), non-toxicity (Stephen-Haynes, Gibson, & Greenwood, 2014), and anti-bacterial (Adila, Suyatma, Firlieyanti, & Bujang, 2013; Champer et al., 2013; Kong, Chen, Xing, & Park, 2010), as well as, hydrophilicity (Xiao, Gao, & Gao, 2010). Chitin and chitosan are biocompatible, biodegradable, and non-toxic polymers (Anitha et al., 2014). This has made chitosan a very useful compound in a wide range of applications in medical, pharmaceutical, chemical, agricultural, and environmental fields (Croisier & Jérôme, 2013). Due to its non-toxicity, it can be applied in the food industry. Chitosan is also biocompatible and antibacterial, with wound-healing effects. Therefore, it can play a role in various medical applications (Croisier & Jérôme, 2013; Dash, Chiellini, Ottenbrite, & Chiellini, 2011; Huang, 2011; Jayakumar, Prabaharan, Sudheesh Kumar, Nair, & Tamura, 2011; Jayakumar, Menon, Manzoor, Nair, & Tamura, 2010a; Jayakumar, Prabaharan, Nair, & Tamura, 2010b; Larsson et al., 2013; Silva et al., 2013; Stephen-Haynes et al., 2014). This is due to its capability of binding a high amount of fats (about 4 to 5 times its weight) compared to other fibers. Chitosan is safely used in the field of agriculture for controlled agrochemical release, seed coating and making fertilizer due to its biodegradability and natural origin (Dzung, Khanh, & Dzung, 2011; El-Sawy, Abd El-Rehim, Elbarbary, & Hegazy, 2010; Zeng & Luo, 2012). In environmental applications, chitosan plays an important role in wastewater treatment and industrial toxic pollution management as it has the ability to adsorb dyes, pesticides, and toxic metals from water and waste water (Altaher, 2012; Chen, Chen, Bai, & Li, 2013; DelaiáSun, 2011). Blending of chitosan with other biopolymers has been proposed as an interesting method to obtain new biomaterials with enhanced properties to meet the requirements of specific applications.

1.5 LIGNIN-BASED BLENDS, COMPOSITES AND NANOCOMPOSITES

Lignin, after cellulose, is the second-most abundant polymer on the planet. Lignin has a complex and non-uniform structure with aliphatic and aromatic constituents. It consists of phenylpropanoid units having various substituent functional groups (Fig. 1.4). Depending on the plant species the structure of lignin may vary by botanical origin, growing conditions of the original plant, extraction conditions, etc. It has a cross-linked structure with many different chemical groups. Its intricate chemical structure has been studied using fragmentation methods. The main precursors are three aromatic alcohols, namely p-coumaryl, coniferyl and sinapyl alcohols, which undergo cross linking and give rise to the complex structure of lignin (Buranov & Mazza, 2008; Floudas et al., 2012; Martínez et al., 2010).

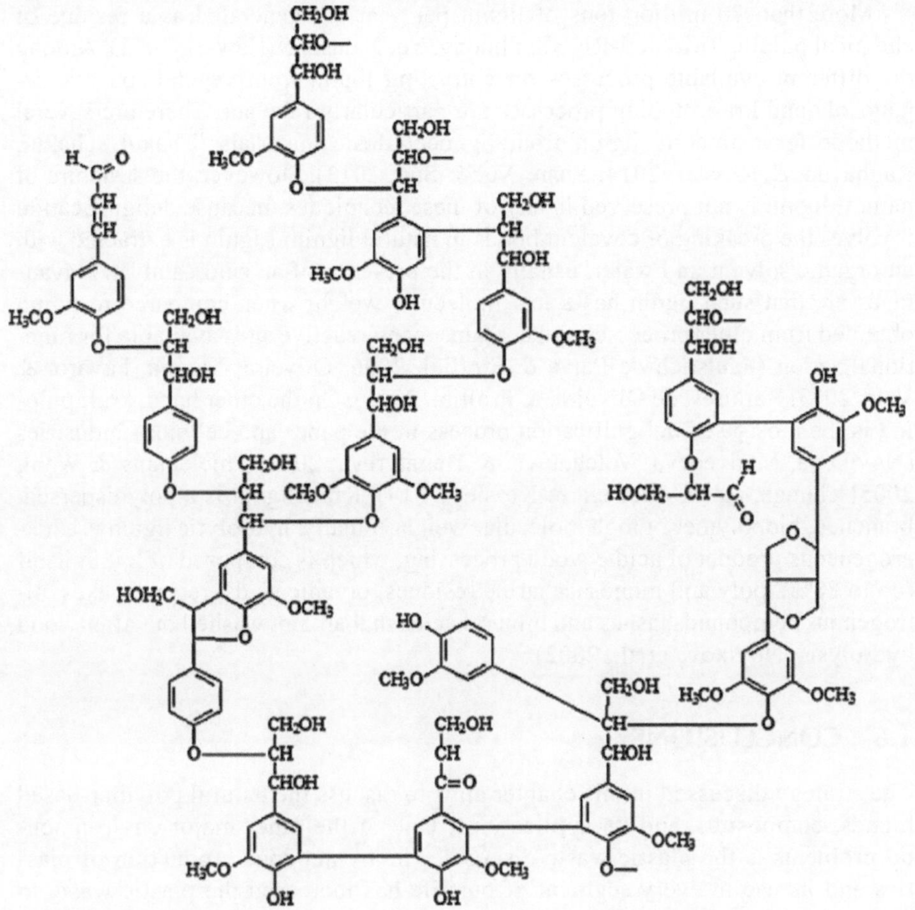

FIGURE 1.4 Structure of lignin

Lignin is more reactive than cellulose and other natural polymers due to the presence of specific functional groups. Studies conducted on the reactivity of functional groups of the biopolymer, as well as its modification, have facilitated a succession of new methods of application in various industries. Due to its structure, lignin has been investigated as compatibilizer and as an antioxidant. Its utilization as compatibilizer is justified by the presence of both aliphatic and polar groups, which may provide compatibility between nonpolar polymers and lignocellulosic fibers (Fig. 1.4) (Gregorová, Cibulková, Kosiková, & Simon, 2005; Guigo, Vincent, Mija, Naegele, & Sbirrazzuoli, 2009; Morandim-Giannetti et al., 2012).

More than 70 million tons of lignin per year are generated as a residue of chemical pulping (Monteil-Rivera, Phuong, Ye, Halasz, & Hawari, 2013). Among the different available processes for extracting lignin from vegetal sources, organosolv and kraft pulping processes are particularly relevant. There are several methods for extracting lignin from lignocellulosic materials (Thakur, Thakur, Raghavan, & Kessler, 2014; Yuan, Xu, & Sun, 2013). However, the structure of natural lignin is not preserved in any of these techniques, because delignification involves the breaking of covalent bonds in natural lignin. Lignin is extracted with an organic solvent and water, usually in the presence of an acid catalyst. Advantages are that such lignin has a low molecular weight when compared to lignin obtained from other processes and contains many reactive sites available for functionalization (Faulstich de Paiva & Frollini, 2006; Oliveira, Marcia, Favaro, & Vale, 2013; Ramires, de Oliveira, & Frollini, 2013). On the other hand, kraft pulping is the most-used delignification process in the paper and cellulose industries (Novikova, Medvedeva, Volchatova, & Bogatyreva, 2002; Thielemans & Wool, 2005; Thunga, Chen, Grewell, & Kessler, 2014). Kraft lignin is a polydispersed, branched, biopolymer of high molecular weight. Finally, hydrolytic lignin is a heterogeneous product of acidic wood processing, which is composed of lignin itself (up to 88%), poly and monosaccharide residues, organic acids, resins, waxes, nitrogenous compounds, ashes and mineral acids that are not washed out after wood hydrolysis (Novikova et al., 2002).

1.6 CONCLUSIONS

The strategy discussed in this chapter aims to discuss the natural polymer-based blends, composites, and its applications. One of the other major environmental problems is the plastic waste disposal. The tremendous production of plastics and its use in every segment of our life has increased the plastic waste to huge quantities. The growing global environmental and social concern, high rate of depletion of petroleum resources, and new environmental regulations have forced the search for new composites and green materials, compatible with the environment. The research field of biodegradable polymers is still in its early stages, but is growing in popularity every day. Currently, a lot of polymer biodegradable matrices have appeared as commercial products, offered by various producers. The future outlook for development in the field of biopolymers material is promising because of its eco friendly behavior. Bio-composites often lead to the fabrication of light weight and low-cost materials having potential applications in the fields of environmental protection and the maintenance of physical health.

KEYWORDS

- **polymer blend**
- **polymer composites**
- **cellulose**
- **chitosan**
- **chitin**
- **starch**
- **lignin**

REFERENCES

Abraham, E., Deepa, B., Pothan, L. A., Jacob, M., Thomas, S., Cvelbar, U., & Anandjiwala, R. (2011). Extraction of nanocellulose fibrils from lignocellulosic fibres: A novel approach. (4), 1468–1475.

Abraham, E., Deepa, B., Pothen, L. A., Cintil, J., Thomas, S., John, M. J., & Narine, S. S. (2013). Environmental friendly method for the extraction of coir fibre and isolation of nanofiber. (2), 1477–1483.

Adila, S. N., Suyatma, N. E., Firlieyanti, A. S., & Bujang, A. (2013). Antimicrobial and physical properties of Chitosan film as affected by solvent types and glycerol as plasticizer. , 155–159.

Altaher, H. (2012). The use of chitosan as a coagulant in the pre-treatment of turbid sea water. , 97–102.

Angellier, H., Molina-Boisseau, S., & Dufresne, A. (2005a). Mechanical properties of waxy maize starch nanocrystal reinforced natural rubber. (22), 9161–9170.

Angellier, H., Molina-Boisseau, S., Belgacem, M. N., & Dufresne, A. (2005b). Surface chemical modification of waxy maize starch nanocrystals. (6), 2425–2433.

Angellier, H., Molina-Boisseau, S., Dole, P., & Dufresne, A. (2006). Thermoplastic starch-waxy maize starch nanocrystals nanocomposites. (2), 531–539.

Anitha, A., Sowmya, S., Kumar, P. T., Deepthi, S., Chennazhi, K. P., Ehrlich, H., ... & Jayakumar, R. (2014). Chitin and chitosan in selected biomedical applications. .

Aral, C., & Akbuğa, J. (1998). Alternative approach to the preparation of chitosan beads. (1), 9–15.

Aulin, C., Karabulut, E., Tran, A., Wågberg, L., & Lindström, T. (2013). Correction to transparent nanocellulosic multilayer thin films on polylactic acid with tunable gas barrier properties. ACS (20), 10395–10396.

Averous, L. (2004). Biodegradable multiphase systems based on plasticized starch: A review. , 231–274.

Azizi Samir, M. A. S., Alloin, F., & Dufresne, A. (2005). Review of recent research into cellulosic whiskers, their properties and their application in nanocomposite field. (2), 612–626.

Baheti, V. K., Abbasi, R., & Militky, J. (2012). Ball milling of jute fibre wastes to prepare nanocellulose. 9(1), 45–50.

Buranov, A. U., & Mazza, G. (2008). Lignin in straw of herbaceous crops. (3), 237–259.

Camarero, E. S., Kuhnt, T., Foster, E. J., & Weder, C. (2013). Isolation of thermally stable cellulose nanocrystals by phosphoric acid hydrolysis. 14(4), 1223–1230.

Cao, X., Dong, H., & Li, C. M. (2007). New nanocomposite materials reinforced with flax cellulose nanocrystals in waterborne polyurethane. (3), 899–904.

Cervin, N. T., Aulin, C., Larsson, P. T., & Wågberg, L. (2012). Ultra porous nanocellulose aerogels as separation medium for mixtures of oil/water liquids. 19(2), 401–410.

Chakraborty, A., Sain, M., & Kortschot, M. (2005). Cellulose microfibrils: A novel method of preparation using high shear refining and cryocrushing. 59(1), 102–107.

Champer, J., Patel, J., Fernando, N., Salehi, E., Wong, V., & Kim, J. (2013). Chitosan against cutaneous pathogens. (1), 37.

Chazeau, L., Cavaille, J. Y., Canova, G., Dendievel, R., & Boutherin, B. (1999). Viscoelastic properties of plasticized PVC reinforced with cellulose whiskers. (11), 1797–1808.

Chen, Y., Chen, L., Bai, H., & Li, L. (2013). Graphene oxide–chitosan composite hydrogels as broad-spectrum adsorbents for water purification. (6), 1992–2001.

Chirayil, C. J., Joy, J., Mathew, L., Mozetic, M., Koetz, J., & Thomas, S. (2014). Isolation and characterization of cellulose nanofibrils from plant. , 27–34.

Cobut, A., Sehaqui, H., & Berglund, L. A. (2014). Cellulose nanocomposites by melt compounding of TEMPO-treated wood fibers in thermoplastic starch matrix. 9(2), 3276–3289.

Costa-Pinto, A. R., Martins, A. M., Castelhano-Carlos, M. J., Correlo, V. M., Sol, P. C., Longatto-Filho, A., & Neves, N. M. (2014). In vitro degradation and in vivo biocompatibility of chitosan–poly (butylene succinate) fiber mesh scaffolds. (2), 137–151.

Crépy, L., Petit, J.-Y., Wirquin, E., Martin, P., & Joly, N. (2014). Synthesis and evaluation of starch-based polymers as potential dispersants in cement pastes and self leveling compounds. , 29–38.

Croisier, F., & Jérôme, C. (2013). Chitosan-based biomaterials for tissue engineering. (4), 780–792.

Dash, M., Chiellini, F., Ottenbrite, R. M., & Chiellini, E. (2011). Chitosan—A versatile semi-synthetic polymer in biomedical applications. (8), 981–1014.

de Oliveira Taipina, M., Ferrarezi, M. M. F., Yoshida, I. V. P., & do Carmo Gonçalves, M. (2013). Surface modification of cotton nanocrystals with a silane agent. (1), 217–226.

DelaiáSun, D. (2011). Facile fabrication of porous chitosan/TiO 2/Fe 3 O 4 microspheres with multifunction for water purifications. 35(1), 137–140.

Druchok, M., & Vlachy, V. (2010). Molecular dynamics study of the hydrophobic 6,6-ionene oligocation in aqueous solution with sodium halides. (10), 1943–1955. 13p.

Dubief, D., Samain, E., & Dufresne, A. (1999). Polysaccharide microcrystals reinforced amorphous poly (β-hydroxyoctanoate) nanocomposite materials. (18), 5765–5771.

Dufresne, A. & Vignon, M. R. (1998). Improvement of starch film performances using cellulose microfibrils. , 2693–2696.

Dufresne, A., Dupeyre, D. & Vignon, M. R. (2000). Cellulose microfibrils from potato tuber cells: Processing and characterization of starch-cellulose microfibril composites. (14), 2080–2092.

Dzung, N. A., Khanh, V. T. P., & Dzung, T. T. (2011). Research on impact of chitosan oligomers on biophysical characteristics, growth, development and drought resistance of coffee. (2), 751–755.

Eichhorn, S. J., Dufresne, A., Aranguren, M., Marcovich, N. E., Capadona, J. R., Rowan, S. J., ... & Peijs, T. (2010). Review: Current international research into cellulose nanofibres and nanocomposites. (1), 1–33.

El-Sawy, N. M., Abd El-Rehim, H. A., Elbarbary, A. M., & Hegazy, E. S. A. (2010). Radiation-induced degradation of chitosan for possible use as a growth promoter in agricultural purposes. (3), 555–562.

Espert, A., Vilaplana, F., & Karlsson, S. (2004). Comparison of water absorption in natural cellulosic fibres from wood and one-year crops in polypropylene composites and its influence on their mechanical properties. (11), 1267–1276.

Faruk, O., Bledzki, A. K., Fink, H. P., & Sain, M. (2014). Progress Report on Natural Fiber Reinforced Composites. (1), 9–26.

Faulstich de Paiva, J. M., & Frollini, E. (2006). Unmodified and modified surface Sisal fibers as reinforcement of phenolic and lignophenolic matrices composites: Thermal analyses of fibers and composites. (4), 405–417.

Floudas, D., Binder, M., Riley, R., Barry, K., Blanchette, R. A., Henrissat, B., ... & Patyshakuliyeva, A. (2012). The Paleozoic origin of enzymatic lignin decomposition reconstructed from 31 fungal genomes. (6089), 1715–1719.

Fortunati, E., Peltzer, M., Armentano, I., Torre, L., Jiménez, A., & Kenny, J. M. (2012). Effects of modified cellulose nanocrystals on the barrier and migration properties of PLA nano-biocomposites. (2), 948–956.

Gabr, M. H., Phong, N. T., Okubo, K., Uzawa, K., Kimpara, I., & Fujii, T. (2014). Thermal and mechanical properties of electrospun nano-celullose reinforced epoxy nanocomposites. , 51–58.

Gandini, A., & Belgacem, M. N. (2008). Chemical Modification of Wood. In Monomers, Polymers and Composites from Renewable Resources, 1st ed.; Gandini, A., Belgacem, M. N., Eds.; Elsevier: Oxford, UK, 419–432.

Gentile, P., Chiono, V., Carmagnola, I., & Hatton, P. V. (2014). An overview of poly(lactic-co-glycolic) acid (PLGA)-based biomaterials for bone tissue engineering. (3), 3640–59.

Gregorová, A., Cibulková, Z., Kosiková, B., & Simon, P. (2005). Stabilization effect of lignin in polypropylene and recycled polypropylene. , 553–558.

Guigo, N., Vincent, L., Mija, A., Naegele, H., & Sbirrazzuoli, N. (2009). Innovative green nanocomposites based on silicate clays/lignin/natural fibres. (11), 1979–1984.

Habibi, Y., Lucia, L. A., & Rojas, O. J. (2010). Cellulose nanocrystals: Chemistry, self-assembly, and applications. (6), 3479–3500.

Han, J., Zhou, C., Wu, Y., Liu, F., & Wu, Q. (2013). Self-assembling behavior of cellulose nanoparticles during freeze-drying: Effect of suspension concentration, particle size, crystal structure, and surface charge. (5), 1529–1540.

Helbert, W., Cavaille, J. Y., & Dufresne, A. (1996). Thermoplastic nanocomposites filled with wheat straw cellulose whiskers. Part I: Processing and mechanical behavior. (4), 604–611.

Hietala, M., Mathew, A. P., & Oksman, K. (2013). Bionanocomposites of thermoplastic starch and cellulose nanofibers manufactured using twin-screw extrusion. , (4), 950–956.

Hsu, S. H., Chang, Y. B., Tsai, C. L., Fu, K. Y., Wang, S. H., & Tseng, H. J. (2011). Characterization and biocompatibility of chitosan nanocomposites. (2), 198–206.

Huang, J. (2011). New waterborne polyurethane-based nanocomposites reinforced with low loading levels of chitin whisker. (4), 362–373.

Hubbe, M. A., Rojas, O. J., Lucia, L. A., & Sain, M. (2008). Cellulosic nanocomposites: A review. (3), 929–980.

Jayakumar, R., Menon, D., Manzoor, K., Nair, S. V., & Tamura, H. (2010a). Biomedical applications of chitin and chitosan-based nanomaterials—A short review (2), 227–232.

Jayakumar, R., Prabaharan, M., Nair, S. V., & Tamura, H. (2010b). Novel chitin and chitosan nanofibers in biomedical applications. (1), 142–150.

Jayakumar, R., Prabaharan, M., Sudheesh Kumar, P. T., Nair, S. V., & Tamura, H. (2011). Biomaterials based on chitin and chitosan in wound dressing applications. (3), 322–337.

Jiang, F., & Hsieh, Y. L. (2013). Chemically and mechanically isolated nanocellulose and their self-assembled structures. (1), 32–40.

Jin, H., Kettunen, M., Laiho, A., Pynnönen, H., Paltakari, J., Marmur, A., & Ras, R. H. (2011). Superhydrophobic and superoleophobic nanocellulose aerogel membranes as bioinspired cargo carriers on water and oil. (5), 1930–1934.

Jonoobi, M., Harun, J., Mathew, A. P., & Oksman, K. (2010). Mechanical properties of cellulose nanofiber (CNF) reinforced polylactic acid (PLA) prepared by twin screw extrusion. (12), 1742–1747.

Jonoobi, M., Mathew, A. P., Abdi, M. M., Makinejad, M. D., & Oksman, K. (2012). A comparison of modified and unmodified cellulose nanofiber reinforced polylactic acid (PLA) prepared by twin screw extrusion. (4), 991–997.

Kamel, S. (2007). Nanotechnology and its applications in lignocellulosic composites, a mini review. , (9), 546–575.

Karimi, S., Tahir, P. M., Karimi, A., Dufresne, A., & Abdulkhani, A. (2014). Kenaf bast cellulosic fibers hierarchy: A comprehensive approach from micro to nano. , , 878–885.

Kean, T., & Thanou, M. (2010). Biodegradation, biodistribution and toxicity of chitosan. (1), 3–11.

Kong, M., Chen, X. G., Xing, K., & Park, H. J. (2010). Antimicrobial properties of chitosan and mode of action: A state of the art review. (1), 51–63.

Labet, M., Thielemans, W., & Dufresne, A. (2007). Polymer grafting onto starch nanocrystals. , (9), 2916–2927.

Larsson, M., Huang, W. C., Hsiao, M. H., Wang, Y. J., Nydén, M., Chiou, S. H., & Liu, D. M. (2013). Biomedical applications and colloidal properties of amphiphilically modified chitosan hybrids. e, 38(9), 1307–1328.

Li, J., Wei, X., Wang, Q., Chen, J., Chang, G., Kong, L., & Liu, Y. (2012). Homogeneous isolation of nanocellulose from sugarcane bagasse by high pressure homogenization. , (4), 1609–1613.

Liu, D. Y., Sui, G. X., & Bhattacharyya, D. (2014). Synthesis and characterisation of nanocellulose-based polyaniline conducting films. , 31–36.

Liu, D. Y., Yuan, X. W., Bhattacharyya, D., & Easteal, A. J. (2010). Characterisation of solution cast cellulose nanofiber—reinforced poly (lactic acid). (1), 26–31.

Liu, D., Yuan, X., & Bhattacharyya, D. (2012). The effects of cellulose nanowhiskers on electrospun poly (lactic acid) nanofibers (7), 3159–3165.

Ljungberg, N., Bonini, C., Bortolussi, F., Boisson, C., Heux, L., & Cavaillé, J. Y. (2005). New nanocomposite materials reinforced with cellulose whiskers in atactic polypropylene: Affect of surface and dispersion characteristics. , (5), 2732–2739.

Ludueña, L., Fasce, D., Alvarez, V. A., & Stefani, P. M. (2011). Nanocellulose from rice husk following alkaline treatment to remove silica. (2), 1440–1453.

Ma, X. F., Chang, P. R. & Yu, J. G. (2008a). Properties of biodegradable thermoplastic pea starch/carboxymethyl cellulose and pea starch/microcrystalline cellulose composites. , 369–375.

Ma, X. F., Chang, P. R., Yu, J. G. & Wang, N. (2008b). Preparation and properties of biodegradable poly(propylene carbonate)/thermoplastic dried starch composites. , 229–234.

Mandal, A., & Chakrabarty, D. (2011). Isolation of nanocellulose from waste sugarcane bagasse (SCB) and its characterization. (3), 1291–1299.

Maran, J. P., Sivakumar, V., Sridhar, R., & Immanuel, V. P. (2013). Development of model for mechanical properties of tapioca starch-based edible films. , (Complete), 159–168.

Martínez, Á. T., Speranza, M., Ruiz-Dueñas, F. J., Ferreira, P., Camarero, S., Guillén, F., ... & del Río, J. C. (2010). Biodegradation of lignocellulosics: Microbial, chemical, and enzymatic aspects of the fungal attack of lignin. (3), 195–204.

Mathew, A. P., & Dufresne, A. (2002). Morphological investigation of nanocomposites from sorbitol plasticized starch and tunicin whiskers. , (3), 609–617.

Mecking, S. (2004). Nature or petrochemistry?—biologically degradable materials. , (9), 1078–1085.

Mishra, S. P., Manent, A. S., Chabot, B., & Daneault, C. (2011). Production of nanocellulose from native cellulose–various options utilizing ultrasound. (1), 0422–0436.

Mishra, S. P., Thirree, J., Manent, A. S., Chabot, B., & Daneault, C. (2010). Ultrasound-catalyzed TEMPO-mediated oxidation of native cellulose for the production of nanocellulose: Effect of process variables. (1), 121–143.

Mohanty, A. K., Misra, M., & Hinrichsen, G. (2000). Biofibres, biodegradable polymers and biocomposites: An overview. , (1), 1–24.

Monteil-Rivera, F., Phuong, M., Ye, M., Halasz, A., & Hawari, J. (2013). Isolation and characterization of herbaceous lignins for applications in biomaterials. , , 356–364.

Moon, R. J., Martini, A., Nairn, J., Simonsen, J., & Youngblood, J. (2011). Cellulose nanomaterials review: Structure, properties and nanocomposites. (7), 3941–3994.

Morandim-Giannetti, A. A., Agnelli, J. A. M., Lanças, B. Z., Magnabosco, R., Casarin, S. A., & Bettini, S. H. (2012). Lignin as additive in polypropylene/coir composites: Thermal, mechanical and morphological properties. (4), 2563–2568.

Muzzarelli, R. A., Boudrant, J., Meyer, D., Manno, N., DeMarchis, M., & Paoletti, M. G. (2012). Current views on fungal chitin/chitosan, human chitinases, food preservation, glucans, pectins and inulin: A tribute to Henri Braconnot, precursor of the carbohydrate polymers science, on the chitin bicentennial. , (2), 995–1012.

Noishiki, Y., Nishiyama, Y., Wada, M., Kuga, S., & Magoshi, J. (2002). Mechanical properties of silk fibroin–microcrystalline cellulose composite films. , (13), 3425–3429.

Novikova, L. N., Medvedeva, S. A., Volchatova, I. V., & Bogatyreva, S. A. (2002). Changes in macromolecular characteristics and biological activity of hydrolytic lignin in the course of composting. , (2), 181–185.

Penttilä, P. A., Várnai, A., Pere, J., Tammelin, T., Salmén, L., Siika-aho, M., & Serimaa, R. (2013). Xylan as limiting factor in enzymatic hydrolysis of nanocellulose. , 135–141.

Peresin, M. S., Habibi, Y., Vesterinen, A. H., Rojas, O. J., Pawlak, J. J., & Seppälä, J. V. (2010). Effect of moisture on electrospun nanofiber composites of poly (vinyl alcohol) and cellulose nanocrystals. (9), 2471–2477.

Ramires, E. C., de Oliveira, F., & Frollini, E. (2013). Composites based on renewable materials: Polyurethane-type matrices from forest byproduct/vegetable oil and reinforced with lignocellulosic fibers. , (4), 2224–2233.

Reddy, J. P., & Rhim, J. W. (2014). Characterization of bionanocomposite films prepared with agar and paper-mulberry pulp nanocellulose. , 480–488.

Rojas, O. J., Montero, G. A., & Habibi, Y. (2009). Electrospun nanocomposites from polystyrene loaded with cellulose nanowhiskers. (2), 927–935.

Rutledge, K. E., Cheng, Q., Pryzhkova, M., Harris, G., & Jabbarzadeh, E. (2014). Enhanced differentiation of human embryonic stem cells on ECM-containing osteomimetic scaffolds for bone tissue engineering. .

Satyamurthy, P., & Vigneshwaran, N. (2013). A novel process for synthesis of spherical nanocellulose by controlled hydrolysis of microcrystalline cellulose using anaerobic microbial consortium. (1), 20–25.

Shi, S. F., Jia, J. F., Guo, X. K., Zhao, Y. P., Chen, D. S., Guo, Y. Y., ... & Zhang, X. L. (2012). Biocompatibility of chitosan-coated iron oxide nanoparticles with osteoblast cells , 5593.

Silva, S. S., Popa, E. G., Gomes, M. E., Cerqueira, M., Marques, A. P., Caridade, S. G., ... & Reis, R. L. (2013). An investigation of the potential application of chitosan/aloe-based membranes for regenerative medicine. (6), 6790–6797.

Siqueira, G., Tapin-Lingua, S., Bras, J., da Silva Perez, D., & Dufresne, A. (2011). Mechanical properties of natural rubber nanocomposites reinforced with cellulosic nanoparticles obtained from combined mechanical shearing, and enzymatic and acid hydrolysis of sisal fibers. (1), 57–65.

Siró, I., & Plackett, D. (2010). Microfibrillated cellulose and new nanocomposite materials: A review. (3), 459–494.

Smith, R. (Ed.). (2005). . Boca Raton: CRC Press.

Song, Z., Xiao, H., & Zhao, Y. (2014). Hydrophobic-modified nano-cellulose fiber/PLA biodegradable composites for lowering water vapor transmission rate (WVTR) of paper. , 442–448.

Stephen-Haynes, J., Gibson, E., & Greenwood, M. (2014). Chitosan: A natural solution for wound healing (1), 48–53.

Syverud, K., & Stenius, P. (2009). Strength and barrier properties of MFC films. , (1), 75–85.

Tanaka, A., Hjelt, T., Sneck, A., & Korpela, A. (2012). Fractionation of nanocellulose by foam filter. (12), 1771–1776.

Tang, Y., Yang, S., Zhang, N., & Zhang, J. (2014). Preparation and characterization of nanocrystalline cellulose via low-intensity ultrasonic-assisted sulfuric acid hydrolysis. (1), 335–346.

Thakur, V. K., Thakur, M. K., Raghavan, P., & Kessler, M. R. (2014). Progress in Green Polymer Composites from Lignin for Multifunctional Applications: A Review. ,

Thielemans, W., & Wool, R. P. (2005). Lignin esters for use in unsaturated thermosets: Lignin modification and solubility modeling. , (4), 1895–1905.

Thielemans, W., Belgacem, M. N., & Dufresne, A. (2006). Starch nanocrystals with large chain surface modifications. , (10), 4804–4810.

Thunga, M., Chen, K., Grewell, D., & Kessler, M. R. (2014). Bio-renewable precursor fibers from lignin/polylactide blends for conversion to carbon fibers. , , 159–166.

Vartiainen, J., Pöhler, T., Sirola, K., Pylkkänen, L., Alenius, H., Hokkinen, J., … & Laukkanen, A. (2011). Health and environmental safety aspects of friction grinding and spray drying of microfibrillated cellulose. (3), 775–786.

Wambua, P., Ivens, J., & Verpoest, I. (2003). Natural fibres: Can they replace glass in fibre reinforced plastics?. , (9), 1259–1264.

Wang, Y., Cao, X., & Zhang, L. (2006). Effects of cellulose whiskers on properties of soy protein thermoplastics. , (7), 524–531.

Wool, R., & Sun, X. S. (2011). . Boston: Academic Press.

Xiao, C., Gao, F., & Gao, Y. (2010). Controlled preparation of physically crosslinked chitosan-g-poly (vinyl alcohol) hydrogel. (5), 2946–2950.

Xie, F., Pollet, E., Halley, P. J., & Avérous, L. (2013). Starch-based nano-biocomposites. (10–11), 1590–1628.

Yu, L., Dean, K., & Li, L. (2006). Polymer blends and composites from renewable resources. (6), 576–602

Yuan, H., Nishiyama, Y., Wada, M., & Kuga, S. (2006). Surface acylation of cellulose whiskers by drying aqueous emulsion. , (3), 696–700.

Yuan, T.-Q., Xu, F., & Sun, R.-C. (2013). Role of lignin in a biorefinery: Separation characterization and valorization. (3), 346–352.

Zeng, D., & Luo, X. (2012). Physiological effects of Chitosan coating on wheat growth and activities of protective enzyme with drought tolerance. (03), 282.

Zhao, W., Jin, X., Cong, Y., Liu, Y., & Fu, J. (2013). Degradable natural polymer hydrogels for articular cartilage tissue engineering. (3), 327–339.

Zhu, J. Y., Sabo, R., & Luo, X. (2011). Integrated production of nano-fibrillated cellulose and cellulosic biofuel (ethanol) by enzymatic fractionation of wood fibers. (5), 1339–134.

Zoppe, J. O., Peresin, M. S., Habibi, Y., Venditti, R. A., & Rojas, O. J. (2009). Reinforcing poly (ε-caprolactone) nanofibers with cellulose nanocrystals. (9), 1996–2004.

CHAPTER 2

PERFORMANCE ENHANCEMENT OF POLYMERIC MATERIALS THROUGH NANOTECHNOLOGY

P.C. THAPLIYAL*

Organic Building Materials Group, CSIR-Central Building Research Institute, Roorkee-247667, Uttarakhand, India

*E-mail: pct866@yahoo.com

CONTENTS

2.1 INTRODUCTION

Nanotechnology is gaining widespread attention and being applied in many fields to formulate materials with novel functions due to their unique physical and chemical properties. Major nanotechnology applications are identified as energy, agricultural productivity, water treatment, disease diagnosis, drug delivery system, food processing, air pollution and control, construction, health monitoring, etc. In the construction sector, nanotechnology is being used in a variety of ways to produce innovative materials. Using nanotechnology as a tool, it is possible to modify the nano/basic structure of the materials to improve the material's bulk properties such as mechanical performance, volume stability, durability, and sustainability. The applications of nano materials in construction improve the essential properties of building materials such as strength, durability bond strength, corrosion resistance, abrasion resistance, and novel collateral functions such as energy saving, self-healing, anti-fogging and super hydrophobic.

Newer applications in the field of advanced materials are related to matter for which the surface-to-volume ratio is very high. Nanotechnology significantly improves and enhances the performance of these materials. In fact, nanotechnology-based polymeric materials can be developed into multifunctional materials. Therefore, the combination at the nano size level of inorganic/organic components into a single material may lead to an immense new area of materials science leading to development of multifunctional polymeric materials (Chao et al., 2001; Grinthal & Aizenberg, 2014; Kowalczyk & Spychaj, 2009; Lee et al., 2010; Thapliyal, 2011; Wang et al., 2013; Zhao et al., 2012).

2.2 ROLE OF NANOTECHNOLOGY IN POLYMERIC MATERIALS

Today's buildings contain many polymeric materials, including neoprene, silicone, poly(vinyl chloride) (PVC), ethylene tetrafluoroethylene (ETFE), laminated glass using polyvinylbutyral and fiber-reinforced polymer composites. Many of these polymeric materials were discovered and used successfully in industry decades before their application in buildings. Polymeric materials are also important components of paints and coating systems. These polymeric materials are expected to have characteristics such as (a) excellent weatherability (exterior durability), (b) film integrity, (c) tunable mechanical performance, (d) processability, and (e) amenable for environmentally friendly coating formulations, among others.

Using nanotechnology, polymeric materials, including advanced coatings systems can improve energy efficiency, durability, aesthetics, and other functionalities of buildings and superstructures. For example, cool-roof coatings (high solar reflection and thermal emission) have been very effective in increasing building efficiency and thereby reducing energy consumption for cooling. Solar heat-absorbing polymeric materials are becoming essential components of solar collectors used in solar-energy harvesting. Super-durable coatings with self-cleaning properties are in many demands for applications on super-structures, monuments and areas where re-painting is very costly.

2.3 CURRENT STATUS

Polymeric materials such as coating systems are reported for corrosion prevention based on alkyds, acrylics, polyurethanes, polyesters, and epoxies. Among them, epoxies have a number of advantages such as better physico-mechanical properties and improved chemical resistance. Its low UV resistance and higher cost led to development of innovative epoxies by blending with low cost renewable natural resins. The epoxy resin and modified epoxy cardanol resin-based coatings form a kind of inter penetrating network (IPN) on the surface of steel and concrete, thus providing a barrier to the attack by moisture. IPNs possess several interesting characteristics in comparison to normal polyblends, because varied synthetic techniques yield IPNs of such diverse properties that their engineering potential spans a broad gamut of modern technology (Sperling, 1981; Thapliyal, 2010).

In the Indian scenario, in ongoing research efforts on polymeric materials at IIT Bombay, researchers are considering the basic issues like homogeneous dispersion of CNT in polymer matrix and adequate interfacial adhesion between the phases and a novel CNT material, that is, SMA-g-MWNT is being developed by grafting acid functionalized MWNT with styrene maleic anhydride (SMA) dissolved in THF solvent. The R&D work on development of heat reflecting coating on flat glass is being done at CSIR-CGCRI. CSIR-CBRI has the expertise in the area of polymeric materials, especially adhesives, sealants, and coatings. In the past, CSIR-CBRI scientists have done work in the field of synthesis, formulation and testing of different types of polymeric materials. As a result, CSIR-CBRI had published a number of research publications, and several technologies were transferred to private organizations. For example, CSIR-CBRI has developed natural cardanol resin-based epoxy coating systems for corrosion protection (Aggarwal et al., 2007; Thapliyal, 2010).

2.4 A NEW ERA OF POLYMERIC MATERIAL INNOVATIONS FOR BUILDINGS

Recent developments in the field of fabrication and characterization of objects at the nano-scale make it possible to design and realize new materials with special functional properties. For example, materials can be strengthened or, conversely, made more flexible, or materials can be given greater electrical resistance and lower thermal resistance. The possibilities are virtually endless, particularly in relation to the coupling between living cells and specific functional nanoparticles, nanosurfaces or nanostructures. Artificially inserted organic particles or surfaces can influence a cell to the extent that it takes on an entirely new functionality, such as fluorescence or magnetism. Insertion of these particles or surfaces in cells may even result in the production of fresh biomaterials. These couplings open up many new scientific and commercial avenues.

New material—polyamide, or nylon—has emerged in applications as a "smart" vapor barrier in exterior envelopes. Their water vapor permeability increases ten times even in conditions of very high humidity and is particularly useful when moisture is trapped within a wall assembly. The vapor barrier becomes more permeable and allows moisture to escape, reducing the risk of corrosion, rot, and the growth of mould and mildew. Although nylon was discovered in 1931, its properties as a vapor barrier were described after 1999, and it was recently commercialized as vapour barrier. Both of these examples illustrate opportunities that arise from addressing the needs of the built environment with polymeric materials science and engineering. The first resulted from an unintended consequence of an aesthetic choice, the second from an overlooked property of a common polymeric material. Both examples raise the question of why our built environment has been so resistant to change when new polymeric materials may offer better performance and more satisfying aesthetic results (Munirasu et al., 2009; Singh et al., 2010; Thapliyal, 2010).

2.5 CONCLUSIONS

Building new polymeric materials at the atomic and nano scale and structuring or combining existing materials, resulting in entirely new characteristics of these materials, make the application areas virtually limitless. The international interest in this area is demonstrated clearly by the growing number of major research programmes being funded in Europe, Japan, and the USA, as well as in Australia, Canada, China, S. Korea, Singapore, and Taiwan, etc. However, the introduction of unfamiliar polymeric materials in buildings is difficult because of life safety concerns, first-cost constraints, and the reluctance of builders to adopt new practices in the field. In addition, the very long life of buildings that serve as host to unproven polymeric materials compounds the risk of legal exposure for all involved, from researchers to builders. However, it is likely that latent opportunities for achieving a substantially improved built environment await the attention of building experts, and the polymeric/materials science community united in common research goals.

KEYWORDS

- **nanotechnology**
- **building materials**
- **self-healing**
- **anti fogging**
- **polymeric materials**

REFERENCES

Aggarwal, L.K.; Thapliyal P.C.; Karade, S.R. Progress in Organic Coating. 2007, 59, 76.

Chao, T.P.; Chandrasekaran, C.; Limmer, S.J.; Seraji, S.; Wu, Y.; Forbess, M.J.; Neguen, C.; Cao, G.Z. Journal of Non-Crystalline Solids. 2001, 290, 153.

Grinthal, A.; Aizenberg, J. Chemistry of Materials, 2014, 26, 698.

Kowalczyk, K.; Spychaj, T. Surface & Coatings Technology. 2009, 204, 635.

Lee, J.; Mahendra S.; Alvarez, P.J.J. ACS Nano. 2010, 4, 3580.

Munirasu, S.; Aggarwal R.; Baskaran, D. Chemical Communication. 2009, 30, 4518.

Singh, L.P.; Thapliyal P.C.; Bhattacharyya, S.K. Nanodigest. 2010, 2, 45.

Sperling, L.H. Advances in Interpenetrating Polymer Networks, Lancaster: Technomic. 1981, 2, 284.

Thapliyal, P.C. Composite Interfaces. 2010a, 17, 85.

Thapliyal, P.C. Nanodigest. 2011, 3, 46.

Thapliyal, P.C. Proceedings of GTGE 2010b. 2010, 29.

Thapliyal, P.C. Proceeding of International Workshop on Nanotechnology in the Science of Concrete. 2010c, 69–74.

Wang, H., Zhao, Y.-L.; Nie, G.-J. Frontiers of Materials Science. 2013, 7, 118.

Zhao, Y.; Xu, Z.; Wang X.; Lin, T. Langmuir. 2012, 28, 6328.

CHAPTER 3

NATURAL CELLULOSE FIBERS: SOURCES, ISOLATION, PROPERTIES AND APPLICATIONS

TANKI MOCHOCHOKO[1], OLUWATOBI S. OLUWAFEMI[2,3], OLUFEMI O. ADEYEMI[4], DENIS N. JUMBAM[1], and SANDILE P. SONGCA[1]

[1]Department of Chemistry, Walter Sisulu University, Private Bag X1, WSU 5117, Mthatha, Eastern Cape, South Africa

[2]Department of Applied Chemistry, University of Johannesburg, P.O. Box 17011, Doornfontein 2028, Johannesburg, South Africa

[3]Centre for Nanomaterials Science Research, University of Johannesburg, Johannesburg, South Africa

[4]Department of Chemical Sciences, Olabisi Onabanjo University, P.O Box 364, Ago-Iwoye, Ogun state, Nigeria

CONTENTS

3.1 INTRODUCTION

In recent years, the focus of research has shifted to the use of natural fibers for various applications. Petrochemical based polymers such as polyolefins, polyesters, polyamides, etc. are mostly used in packaging materials because of their abundance, low cost, and good functionality characteristics such as good tensile and tear strength, good barrier properties to O and aroma compounds, heat seal ability and low water vapor transmission rate. However, they are completely non-biodegradable causing environmental pollution (Tharanathan, 2003). Although the total abolition of petrochemical-based polymers is very difficult, their utilization should be minimized to lower the environmental pollution and emission of green house gases. This could be achieved through use of natural biopolymers such as chitin, chitosan, starch, cellulose, and pullulan (Fig 3.1).

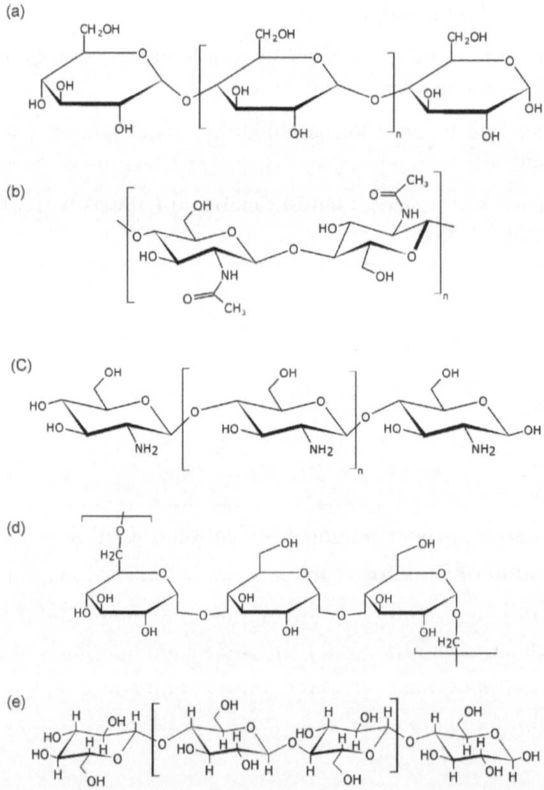

FIGURE 3.1 Typical structures of (a) starch (amylase), (b) chitin, (c) chitosan, (d) pullulan, and (e) cellulose

Among many of these natural fibers, cellulose and its derivatives have attracted a lot of attention due to their high flexibility and elasticity properties, which enable them to uphold their high aspect ratio (Frone et al., 2011). Cellulose was first discovered and named by a French Chemist called Anselme Payen in 1838, when he suggested that the cell walls of almost all plants are made of similar material. In 1921, Haworth discovered the dimer of cellulose known as cellobiose. Five years later (1926), Sponsler and Dore became the first group to propose its molecular structure (Kontturi, 2005).

As against the use of inorganic fiber counterparts, natural cellulose fibers have numerous advantages such as source renewability, abundance, biodegradability (Teixeira et al., 2011), non-food agriculture based economy, low cost, low density, environmental friendliness, low abraisivity, high sound attenuation of lignocellulosic based composite materials, relatively reactive surfaces that can be used to graft specific groups and high specific modulus (Samir et al., 2005; Pandey et al., 2009; Zhao et al., 2009; Cabiac et al., 2011; Peng et al., 2011). These properties have enabled large scale applications of cellulose fibers in nanocomposites, packagings, pharmaceuticals (Mansouri et al., 2012), coatings, and dispersion technology (Stenstad et al., 2008; Johnson et al., 2009; Qua et al., 2009). In the past, cellulose had also been used for ropes, twines (Siqueira et al., 2010), rugs, mats, mattresses (Methacanon et al., 2010), sails, papers, and timbers (Eichhorn et al., 2010). The lignocellulosic fibers also find application as alternative sources of energy in order to minimize the dependence of the world on petroleum-based energy sources (Obi et al., 2012), which emit green house gases and consequently result in global warming (Balat, 2011; Sarkar et al., 2012). Moreover, the continued increase in prices and decrease in the reserves base of non-renewable mineral deposit (energy sources) have posed a huge energy sustainability and economic challenge for the world leaders (Zhao et al., 2012).

Native fibers are complex materials, which consist largely of cellulose, hemicellulose lignin (Samir et al., 2005; Jargalsaikhan and Saraçoğlu, 2008; Balat, 2011; Cabiac et al., 2011; Ibbet et al., 2011; Li et al., 2012; Ludueña et al., 2012), and small fractions of extractives (Chattopadhyay and Sarkar, 1946) bonded together through covalent bonding, various intermolecular bridges, and van der Waals forces (Kumar et al., 2010) (Fig 3.2).

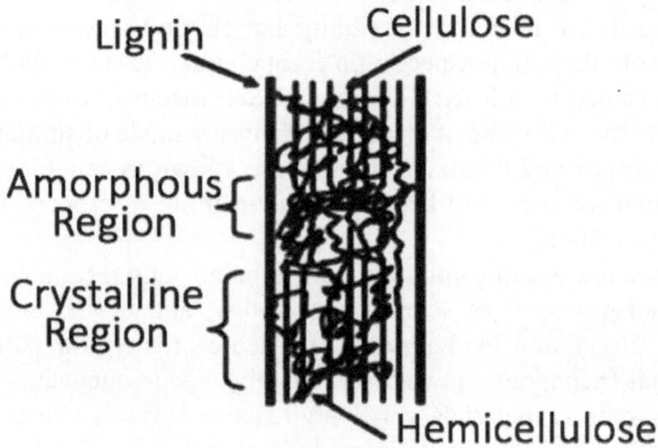

FIGURE 3.2 The schematic diagram for a complex composite of cellulose, lignin, and hemicellulose (Zhang and Shahbazi, 2011).

Cellulose being the most abundant polymer in nature, constituting 40–50% of the biomass by weight (McKendry, 2002), forms the principal ingredient of woody plants, which diversifies its applications. It contains about 44.2% carbon, 49.5% oxygen, and 6.3% hydrogen and it forms a major component of plant cell walls. It is a linear high molecular weight homopolymer made from β-D-glucopyranose units linked together by β-(1→4)-glycosidic bonds. The hydroxyl groups present in these units help cellulose undergo several reactions such as esterification, etherification, oxidation, halogenations, and ceric salt-initiated grafting (Hon and Shiraishi, 1991).

Hemicelluloses are branched amorphous polymers of shorter chain lengths (about 500 to 3000 units) (Achyuthan et al., 2010) constituting 20–40% of the biomass by weight (McKendry, 2002). They are linked to the crystalline cellulose part through hydrogen bonding and they consist of pentoses (xylose and arabinose), hexoses (mannose, glucose, galactose), and some sugar acids such as 4-O-methyl-D-glucuronic and galaturonic acids (Fig 3.3) (Saha, 2003; Moràn et al., 2008).

Lignin is a group of amorphous, branched, and cross-linked polymeric compounds (Saarinen et al., 2009) that consists of phenylpropanoid units such as p-coumaryl, coniferyl, and sinapyl alcohols, (Fig 3.4) (Achyuthan et al., 2010; FitzPatrick, 2011; Li et al., 2012; Ludueña et al., 2012). These units may have zero, one, or two methoxyl groups bonded to the rings giving rise to the above mentioned structures usually termed I, II, and III respectively. The ratios of each one of the structures is the function of different sources of polymers with I being prevalent in plants such as grasses, II in the wood of conifers, and III in deciduous wood (McKendry, 2002). Lignin binds strongly to cellulose and hemicellulose through chemical bonds such as ether and ester linkages and it makes plant cells to be impermeable to water (Achyuthan et al., 2010). It also protects trees from chemical and biological attacks, transports water in plant stems, as well as providing structural integrity (Vinardell, 2008; Pinkert et al., 2011).

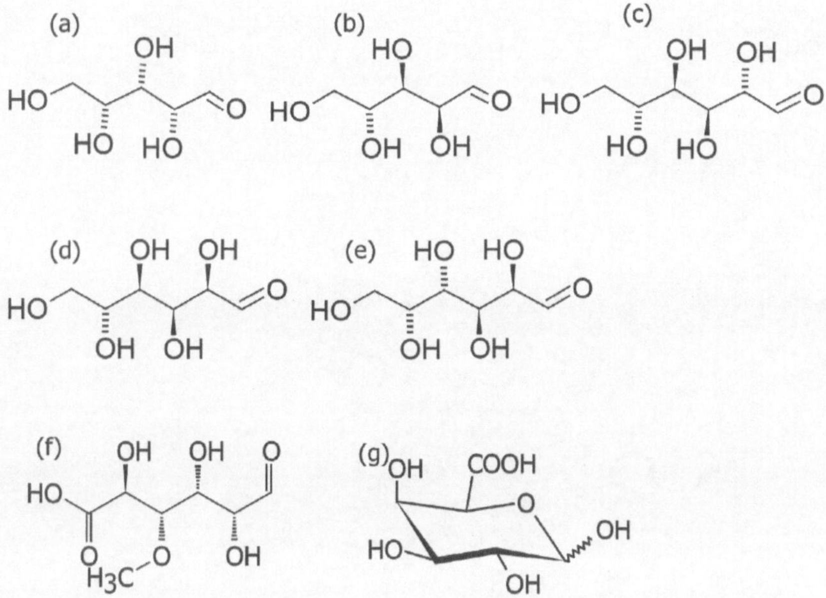

FIGURE 3.3 Primary monomeric structures of sugars constituting hemicelluloses: (a) xylose, (b) arabinose (pentoses), (c) mannose, (d) glucose, (e) galactose (hexoses), (f) 4-O-methyl-D-glucuronic acid, and (g) galaturonic acid

Extractives on the other hand, refer to heterogeneous mixture of small molecular weight individual compounds such as terpenoids, steroids, fats, waxes, and phenolic constituents (Fig 3.5) (Yokoyama et al., 2002), which are usually removed from the cellulose matrix using the solvent extraction technique. The commonly deployed solvents and/or mixture of solvents include acetone, ethanol, toluene/ethanol (Taylor et al., 2008), benzene/ethanol (Bhimte and Tayade, 2007), and chloroform/ethanol (Liu, 2006) in different proportions.

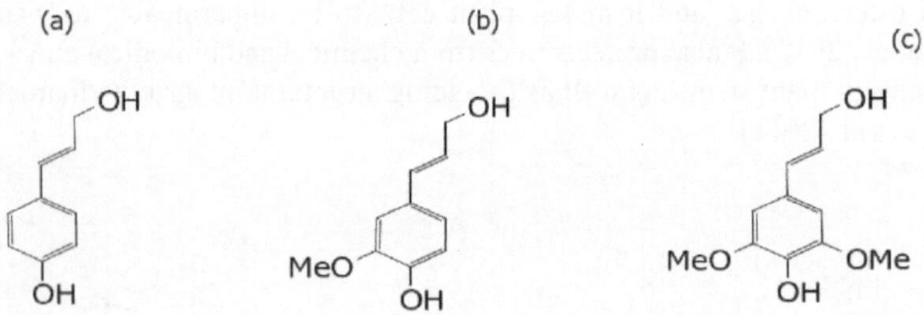

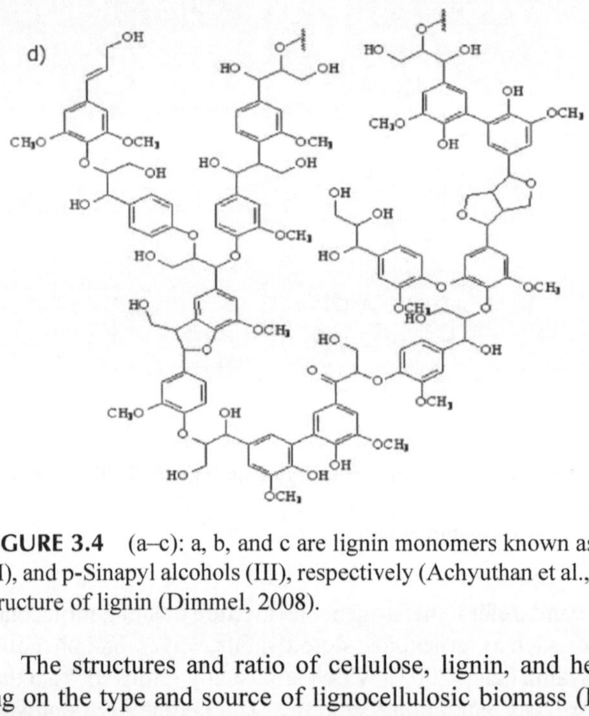

FIGURE 3.4 (a–c): a, b, and c are lignin monomers known as p-Coumaryl (I), p-Coniferyl (II), and p-Sinapyl alcohols (III), respectively (Achyuthan et al., 2010); (d) shows a polymeric structure of lignin (Dimmel, 2008).

The structures and ratio of cellulose, lignin, and hemicellulose vary depending on the type and source of lignocellulosic biomass (Pauly and Keegstra, 2010; Cabiac et al., 2011; Li et al., 2012). However, the cellulose content in biomass is relatively higher than lignin and hemicellulose and forms a major component of a cell wall predominantly in secondary wall (Khalil et al., 2012). In the cell wall, the cellulose microfibrils are formed from parallel cellulose chains and are bound to the hemicellulose and lignin matrix. These microfibrils form alternating crystalline and amorphous regions (Saha et. al., 2012).

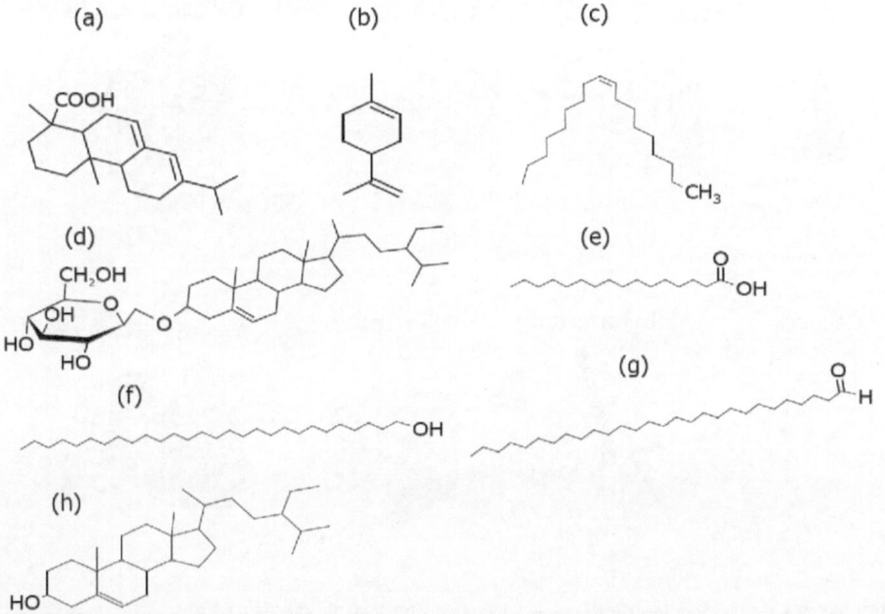

FIGURE 3.5 Some examples of extractives from wood and non-wood natural fibers: (a) Abietic acid (diterpenoids), (b) Limonene (monoterpene), (c) Oleic acid (fatty acids), (d) Sitosteryl 3β-D-glucopyranoside, (e) palmitic acid, (f) octacosanol, (g) octacosanal, and (h) Sitosterol (Marques et. al., 2010).

The cell wall of woody fibers consists of divisions of several layers namely: middle lamella (ML), primary cell wall (P), and secondary cell wall which is also divided into three parts (S1, S2, and S3), and warty layer (Fig 3.6). The difference in these layers lies in their structures and chemical composition. The middle lamella serves as a boundary between cells and is chiefly composed of lignin, while the primary cell wall consists of hemicelluloses, lignin, some pectin, and a little cellulose. The secondary cell wall consists largely of cellulose (Ahola, 2008; Khalil et al., 2012). Despite the divisions in their layers, the biochemical functions of cell walls are to help in the reorganization of signal transduction as a result of pathogen attack, environmental stress, and during different stages of development. They also provide tensile strength to the plant body (Sarkar et. al., 2009).

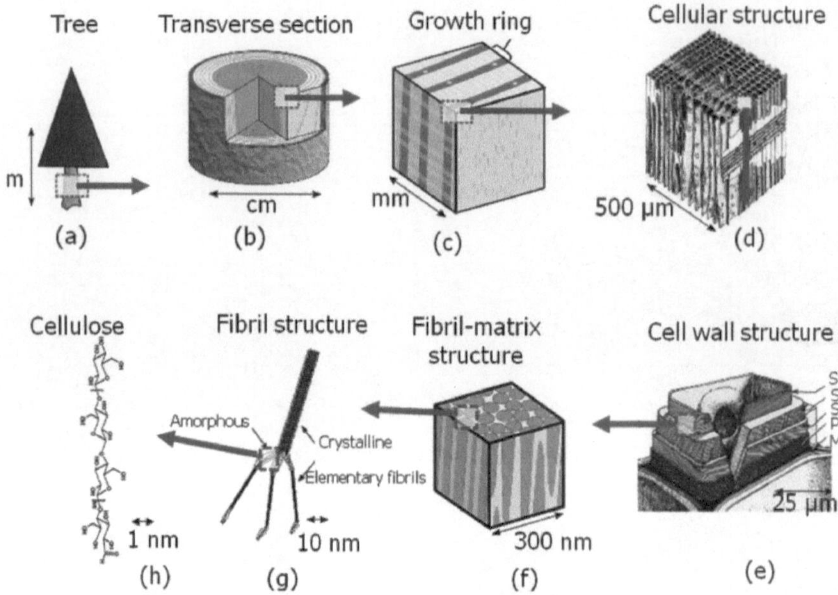

FIGURE 3.6 The schematic diagram showing different layers of wood fibers (Wegner and Jones, 2009).

Composite materials derived from natural fibers have diverse applications such as in railway, aircraft, irrigation system, furniture industries, sports and leisure items (Samir et al., 2005). Cellulose fibers in particular can also be used as binders and fillers in medical tablets, supporters for various biologically active substances, stabilizers in various suspensions, controller of flow and fat replacer in the food (Fleming et. al., 2001), and in cosmetics (Terinte et al., 2011) industries. However, the utilization of cellulose fibers is restricted due to their insolubility in many solvents and stability during thermal degradation because of intra- and inter-hydrogen bonding (Zhang and Xu, 2012).

As a result of their diverse architecture, different descriptors such as whiskers, nanowhiskers, cellulose nanocrystals, nano-crystalline cellulose, monocrystals or microcrystallites are used in literature to describe crystalline rod-like nanoparticles (Eichhorn et al., 2010; Frone et al., 2011; Khalil et al., 2012), while terms such as microfibrils, microfibrillated cellulose or nanofibrillated cellulose are used to describe cellulosic nanoparticles that are obtained as a result of mechanical shearing disintegration process (Siqueira et al., 2010; Siqueira et al., 2011). In this paper, sources, isolation, property and applications of natural cellulose fibers are reviewed.

3.2 SOURCES, ISOLATION, PROPERTIES AND APPLICATION OF CELLULOSE FIBERS

3.2.1 SOURCES

All plants consist largely of cellulose materials which account for about one-third to half of the plant tissues. This is usually produced through the process of photosynthesis. The total annual cellulose production is approximated at about 10 tonnes (Goodger, 1976; Sun et al., 2004; Samir et al., 2005), from various wood and non-wood biomass sources such as bamboo, wood (Bhimte and Tayade, 2007), milkweed (Reddy and Yang, 2009), grass, bagasse, reeds (Liu et al., 2006), rice husks and wheat straws (Binod et al., 2010; Jiang et al., 2011), hemp (Siqueira et al., 2010), kenaf (Wang et al., 2010), cotton (Siro and Plackett, 2010), sisal and jute (Moràn et al., 2008), flax (Qua et al., 2009), cassava waste, water hyacinth (Isarankura-Na-Ayudhya et al., 2007), groundnut shells (Lee et al., 2000; Adel et al., 2011), peel of prickly pear fruit (Habibi et al., 2009), grass, reeds (Cheng et al., 2010), sorghum, barley, pineapple and banana plant wastes. Cellulose fibers are densely distributed in higher plants such as wood (Liu et al., 2006), several marine animals and to a lesser extent in algae, fungi, bacteria (Reddy and Yang, 2009), invertebrates, and amoeba (Sjöström, 1993 ; Habibi et al., 2010).

3.2.2 STRUCTURE

Cellulose is a linear high molecular weight homopolymer made of repeating monomer units of β-D-glucopyranose linked together by β-(1→4)-glycosidic bonds (Brännvall, 2007; Leppänen et al., 2009; Abbott and Bismarck, 2010; Habibi et al., 2010; Siqueira et al., 2010; Panaitescu et al., 2011; Saha et al., 2012 ;) (Fig 3.7).

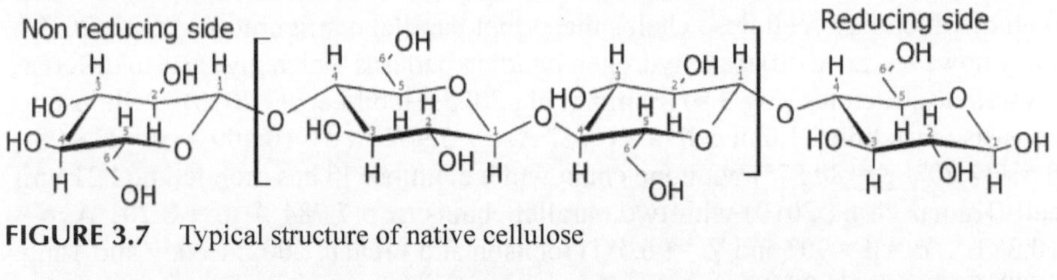

FIGURE 3.7 Typical structure of native cellulose

Each unit of β-D-glucopyranose is connected to its neighbour at an angle of 180 (Hubbe et al., 2008; Johnson and French, 2009) to form a homopolymer with chemically reducing functionality at one end and nominal non reducing functionality at the other end. The dimer of cellulose is termed celloboise (Brännvall, 2007; Achyuthan et al., 2010; Habibi et al., 2010). All the β-D-glucopyranose molecules adopt a C chair conformation (Nishiyama et al., 2002) which causes the hydroxyl groups at C, C, and C to be equatorially positioned and hydrogen atoms axially positioned.

The stability of this cellulose structure results from both an inter and intramolecular hydrogen bonding network (Zhao et al., 2009; Eichhorn et al., 2010) which protracts from the O-H hydroxyl of one unit to the O ring oxygen of another unit across the glycosidic bond and also from the O-H hydroxyl to O-H hydroxyl of another unit (Habibi et al., 2010). This network results in the formation of a highly crystalline structure of cellulose (Brännvall, 2007) which alternates with amorphous regions (Fig 3.8) (Abbott and Bismarck, 2010; Eichhorn et al., 2010).

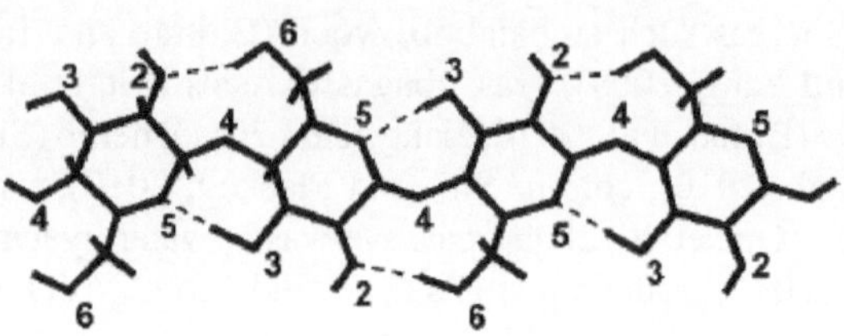

FIGURE 3.8 The hydrogen bonding network in native cellulose (Habibi et al., 2010)

Native cellulose has a degree of polymerization (DP) of about 10,000 and 15,000 glucopyranose units in wood and cotton plants, respectively (Samir et al., 2005; Habibi et al., 2010). It is very difficult to measure the chain lengths of insoluble, high molecular weight polymers like cellulose as they present possible enzymatic and mechanical degradation during analysis (Samir et al., 2005), and are dependent on cellulose sources and types. Wood sources give cellulose fibers with typical chain lengths of 100–300 nm and chain width of 3–10 nm (Goetz et al., 2009). Cellulose has six different interconvertible polymorphs, namely, cellulose I, II, III, III, IV and IV (Samir et al., 2005; Moràn et al., 2008; Habibi et al., 2010; Kumar et al., 2010; Ludueña et al., 2012). Native cellulose (cellulose I) has allomorphs Iα and Iβ both of which have cellulose chains that adopt parallel configurations (Oke, 2010). They however, have different hydrogen bonding patterns which give rise to different crystalline structures (Fig 3.9) (Samir et al., 2005; Habibi et al., 2010). Cellulose Iα consists of triclinic P1 unit cell (a = 6.717 Å, b = 5.962 Å, c = 10.400 Å, α = 118.08°, β = 114.80°, ɣ = 80.37°) with one chain while cellulose Iβ has monoclinic P21 unit cell (Kumar et al., 2010) with two parallel chains (a = 7.784 Å, b = 8.201 Å, c = 10.384 Å, α = β = 90° and ɣ = 96.5°) (Johnson and French, 2002; Reddy and Yang, 2009; Sarkar et al., 2009).

The ratio of these two allomorphs depends on the different sources of cellulose. Cellulose Iα is usually known to be predominant in bacteria and algae, while Iβ allomorph is prevalent in higher plants (Brännvall, 2007; Siqueira et al., 2010). However, Atalla and VanderHart have demonstrated some peculiarities in the NMR spectra of higher plants cellulose, which seemed to suggest that higher plants contain only cellulose Iβ (Atalla and VanderHart, 1999). The unit cell dimensions for other cellulose polymorphs are summarized in Table 1.

TABLE 3.1 The dimensions of cellulose polymorph unit cells (Krässig, 1996)

-axis /Å	-axis/Å	-axis/Å	Lattice angle(γ)/°	Polymorphs
7.85	8.17	10.34	96.4	Cellulose I
9.08	7.92	10.34	117.3	Cellulose II
9.9	7.74	10.3	122	Cellulose III
4.450	7.850	10.31	105.10	Cellulose III
4.45	7.64	10.36	106.96	Cellulose III
7.9	8.11	10.3	90	Cellulose IV

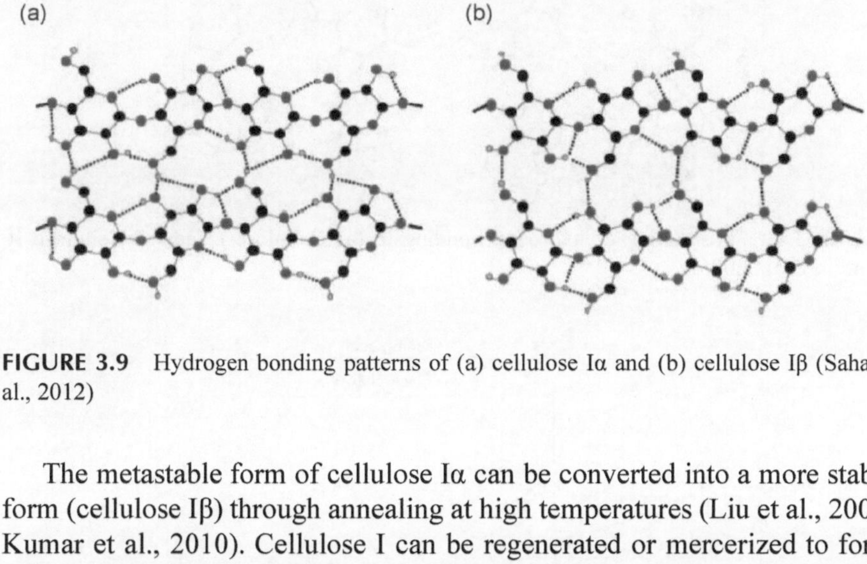

FIGURE 3.9 Hydrogen bonding patterns of (a) cellulose Iα and (b) cellulose Iβ (Saha et al., 2012)

The metastable form of cellulose Iα can be converted into a more stable form (cellulose Iβ) through annealing at high temperatures (Liu et al., 2006; Kumar et al., 2010). Cellulose I can be regenerated or mercerized to form cellulose II which in turn, forms cellulose III upon treatment with ammonia and subsequent removal of swelling agent. When cellulose Iβ is treated with ammonia at −80° C, it gives rise to cellulose III (Hon and Shiraishi, 1991). The regeneration step involves the dissolution of cellulose I in suitable solvents such as solutions of heavy metal- amine complexes for example cupric ethylene diamine while mercerization involves the swelling of native cellulose in solutions of concentrated sodium hydroxide to yield cellulose II, which is thermodynamically more stable than cellulose I (Sarkar et al., 2009;

Khalil et al., 2012; Saha et al., 2012). Cellulose I and II are the most studied polymorphs and they show the structural difference brought about by the hydrogen bonding patterns as shown in (Fig 3.10). The hydrogen bond in cellulose I is between the O–H and the O ring oxygen of another unit, while that of cellulose II is between the O–H hydroxyl to O ring oxygen of another unit. The chains in cellulose I run in parallel direction, while in cellulose II are antiparallel (Sarkar et al., 2009; Habibi et al., 2010). The treatment of cellulose III and III with glycerol at 260 °C produces respective cellulose IV and IV (Brännvall, 2007). The inter conversion of cellulose polymorphs is presented in scheme 3.1.

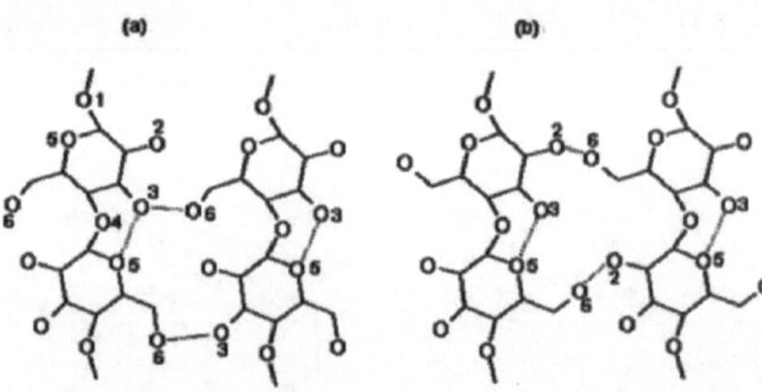

FIGURE 3.10 Positioning of hydrogen bonding in (a) Cellulose I and (b) Cellulose II (Kuntturi et al., 2006)

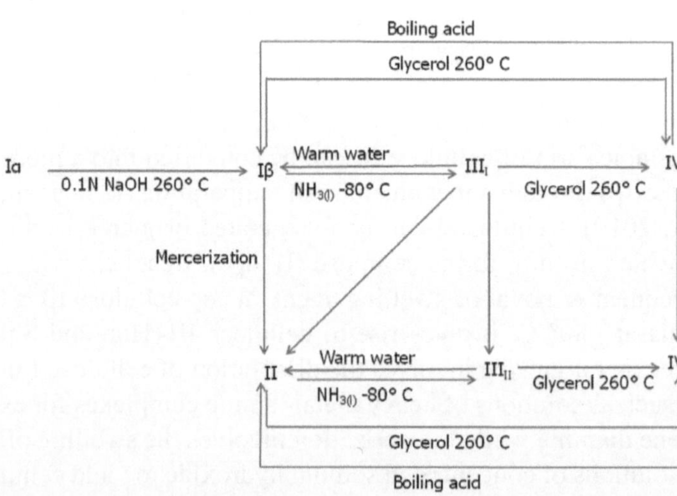

SCHEME 3.1 The typical representation of the inter conversion of cellulose polymophs (Kroom-Batenburg et al., 1996)

3.2.3 PRE-TREATMENT OF CELLULOSE

The isolation and purification of cellulose from different sources has been carried out following different treatment methods such as mechanical, chemical, physico-chemical, and biological. The commonly utilized physical pre-treatment methods for the isolation of cellulose include steaming, homogenization, grinding and milling, extrusion, irradiation (electron beam, ultrasounds, and microwaves) (Wang et al., 2010) temperature and pressure (Jiang et al., 2011), and cryo-crushing (Isarankura-Na-Ayudhya et al., 2007). The use of these methods is aimed at reducing particle size and crystallinity, increasing pore size and surface area of cellulose (Jiang et al., 2011; Harun et al., 2011) that result in an increase in the total hydrolysis yield of cellulose. Although these physical pre-treatment methods usually remove the whole lignocellulosic material (Abdul Khalil et al., 2012); they do not however, produce satisfactory results if used individually. They are therefore, used in combination with chemical treatment methods to improve the efficiency of the process. Besides, they are energy consuming, thereby making large industrial applications unviable (Hendriks et al., 2009).

Solvent extraction is an important method deployed to remove the extractable fractions such as fats, waxes and resins from the cellulosic fibers (van der Werf et al., 1994) using solvents and/or mixture of solvents such as acetone (Yokoyama et al., 2002), ethanol, toluene/ethanol (Taylor et al., 2008), benzene/ethanol (Bhimte and Tayade, 2007), and chloroform/ethanol (Liu et al., 2006).

Alkaline treatment of cellulosic material is normally carried out using the hydroxides of sodium, potassium, calcium, and ammonium (Brodeur et al., 2011). This method of treatment is responsible for the degradation of ester and glycosidic side chains, which result in the alteration of lignin structure, cellulose swelling, partial reduction in the crystallinity of cellulose, increase in internal surface area, as well as partial solvation of hemicelluloses (Sjöström, 1993). An extensive study on the use of sodium hydroxide in lignocellulosic biomass has shown that, it is able to disrupt the lignin structure of the biomass, thus increasing the accessibility of cellulose and hemicelluloses to enzymatic hydrolysis (MacDonald et al., 1983). Failure to remove lignin from the lignocellulosic biomass during alkaline treatment prevents enzymatic hydrolysis as the enzymes will bind on to the surface of lignin and not on cellulose chains (Brodeur et al., 2011). In their study of determining the effectiveness of different alkaline solutions during the treatment of lignocellulosic material, Sun and co-workers found that the optimal condition was achieved by using 1.5% sodium hydroxide at 20 °C for 144 h which released 60% and 80% of lignin and hemicelluloses, respectively. They also found that xylose was the major fraction of the hemicelluloses present, while glucose and galactose appeared to be the least fractions (Sun et al., 1995).

Calcium hydroxide (lime) is another alkali also used more often in the treatment of lignocellulosic feedstock such as corn stover, switchgrass, bagasse, wheat, rice straw, etc (Brodeur et al., 2011). Pre-treatment of lignocellulosic material with calcium hydroxide solution offers the advantage of lowest cost of lime amongst the alkaline treatment methods. In 2005, the cost of hydrated lime was estimated at $70/ton compared to $270/ton for ammonia (fertilizer grade hydrate), and $320/ton for sodium hydroxide (50 wt%) and potassium hydroxide (45 wt%) (Sun and Cheng, 2002). Ammonia pre-treatment also finds application in the chemical treatment of lignocellulosic materials where it causes biomass swelling, disruption of lignin-carbohydrate linkage, hemicelluloses hydrolysis, ammonolysis of glucuronic cross-linked bonds, and decrystallization of cellulose structure making it more accessible to enzymatic hydrolysis (Brodeur et al., 2011). The advantage of this treatment is the recovery of ammonia for reuse.

In comparison to other methods of lignocellulosic treatments, alkali treatments offer advantages of less severity of conditions. It can be operated at lower temperatures and pressures of up to even room conditions. However, long pre-treatment times and conversion of alkali into irrecoverable salts and the incorporation of salts into the lignocellulosic material still remain as challenges (Brodeur et al., 2011). The general equations for the alkaline treatment reactions are shown in scheme 3.2 where M represents sodium or calcium metals.

$$(C\,H\,O(OH))+nMOH\quad(C\,H\,O(OH)OM)+nH\,O \tag{3.1}$$

$$(C\,H\,O(OH))+nKOH\quad(C\,H\,O(OH)OK)+nH\,O \tag{3.2}$$

SCHEME 3.2 Reactions showing alkaline pre-treatment process during cellulose isolation

This is another form of chemical pre-treatment that has also gained popularity owing to its capacity to whiten the products and improve their ageing resistance (Saha et al., 2012). It further ensures the removal of lignin and some hemicelluloses through the use of oxidants such as hydrogen peroxide (H O), ozone (O), carbon dioxide (CO), peracids (Alvira et al., 2010). However, in order to reduce the high energy cost during chemical hydrolysis, pre-treatment with microorganisms such as bacteria (Acetobacter xylinum) and fungi (and are better alternatives and are also environmentally friendly (Hamelinck et al., 2005; Sari et al., 2011). However, a major drawback in the use of microorganisms is the low hydrolysis rate of most biological materials (Alvira et al., 2010).

This is a chemical pre-treatment of great importance in the treatment of lignocellulosic materials, which involves the use of dilute and concentrated mineral acids such as sulphuric acid (H SO), hydrochloric acid (HCl), nitric acid (HNO), and phosphoric acid (H PO) (Jiang et al., 2011). Organic acids, such as maleic and oxalic acids have also been tested. They are responsible for reducing the degree of polymerization of cellulose and removing the amorphous parts (hemicelluloses) through hydrolytic cleavage of the glycosidic bonds (Das et al., 2009). In particular, hydrolysis of cellulose with sulphuric acid (H SO) was found to produce more stable cellulose crystals than other acids. It reduces the degree of polymerization of cellulose and concurrently–under controlled conditions (hydrolysis time, temperature, and concentration)–introduces negatively charged sulphate moieties on the surface of cellulose microfibrils, which create an electrostatic repulsion between cellulose fibers (Kamel, 2007; Qua et al., 2009; Cheng et al., 2010 ; Elanthikkal et al., 2010; Lu and Hsieh, 2010; Adebayo et al., 2011; Nadanathangam and Satyamurthy, 2011). The whiskers produced therefore, neither precipitate nor flocculate (Elanthikkal et al., 2010). Although it forms stable cellulose whiskers, sulphuric acid-hydrolyzed cellulosic materials tend to have reduced thermal stability (Rosa et al., 2010). Besides, acid-hydrolysis treatment is environmentally hostile, energy intensive, and it chemically modifies the surface of the nanocellulose (Nadanathangam and Satyamurthy, 2011). Time and temperature, acid type and acid concentration need optimization during the hydrolysis process as they have been shown to influence the degree of hydrolysis and therefore, the type of products produced (Elanthikkal et al., 2010; Oke, 2010). The reaction and reaction mechanism for sulphuric acid hydrolysis and esterification are shown in scheme 3.3.

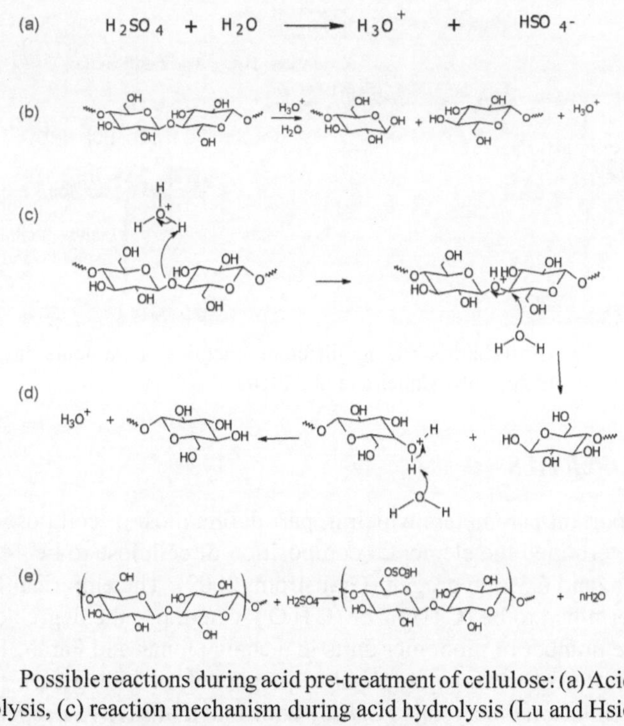

SCHEME 3.3 Possible reactions during acid pre-treatment of cellulose: (a) Acid dissociation, (b) acid hydrolysis, (c) reaction mechanism during acid hydrolysis (Lu and Hsieh, 2010), and (d) esterification with sulphuric acid (Adebayo et al., 2011)

There are undoubtedly numerous chemical pre-treatment methods used in the literature for the isolation of cellulose fibers from different lignocellulosic sources. The choice of these chemical treatment methods is influenced by the properties of lignocellulosic materials such as their chemical composition, internal fiber structure, microfibril structure, microfibril angle, cell dimensions and the defects which are in return influenced by the type and the sources of the lignocellulosic materials (Siqueira et al., 2010). The intended use of the cellulosic fiber product could also have an influence on the choice of chemical treatment method (Dufresne et al., 2008). The schematic diagram showing different pre-treatment methods during cellulose isolation are depicted in (Fig 3.11).

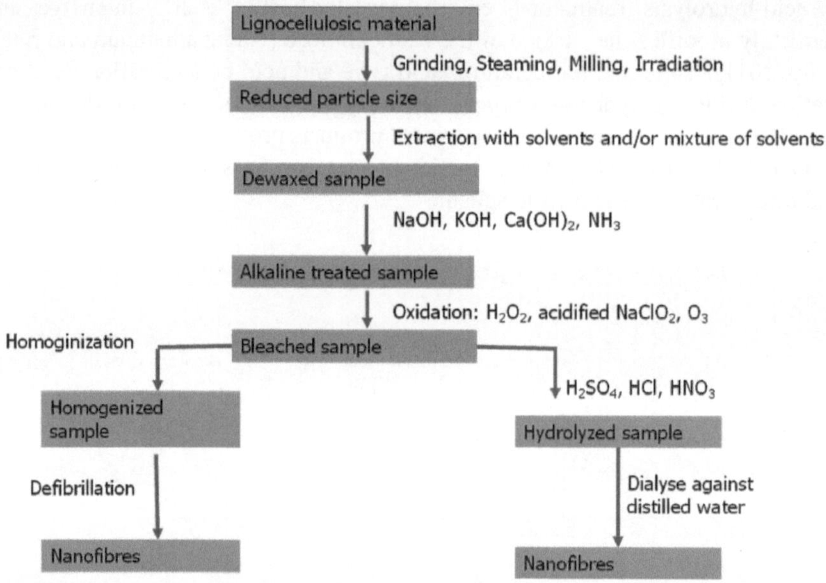

FIGURE 3.11 The flowchart showing different chemical treatments involved during cellulose nanofibers preparation (Siqueira et al., 2010)

3.2.4 PROPERTIES

These are important parameters which in part define most of cellulose characteristics. Payen determined the elemental composition of cellulose to be 44.2% carbon, 49.5% oxygen, and 6.3% hydrogen (Granström, 2009). The empirical formula was therefore determined to be (C H O) or (C H O) where p = the degree of polymerization; n = the number of monomer units in a chain (Jonas and Farah, 1998). When subjected to different treatments (physical, chemical, or biological), native cellulose forms cellulose nanocrystals (CNCs) with much attractive properties such as high surface-to-volume ratio, low density (1.566 g/cm), and high aspect ratio (L/D, where L = length and D = diameter). This makes it a suitable candidate for improving the mechanical properties of polymer composites (Brännvall, 2007). The transverse mechanical properties, size, and surface of CNCs dictate polymer properties of composites interphase (Pakzad et al., 2012). CNCs theoretically, have Young's modulus stronger than that of steel and similar to that of aramid fibers (Kevlar). Using the Raman spectroscopy technique, elastic modulus of native cellulose crystals from tunicate and cotton were measured and found to be to 143 GPa and 105 GPa respectively (Brännvall, 2007). Nishino and co-workers found that the tensile strength of uniaxially reinforced all-cellulose composite was 480 MPa at 25 °C, and 20 GPa at 300 °C for dynamic storage modulus. These were relatively higher than those of conventional glass-fiber-reinforced composites. Moreover, the linear thermal expansion coefficient was about 10 K . All-cellulose composite therefore, showed excellent mechanical and thermal performance during use (Nishino et al., 2004). However, the use of CNCs has a great deal of limitations, which include its low bulk density, high lubricant sensitivity, poor flow characteristics and the influence of moisture on the compression characteristics (El-Sakhawy and Hassan, 2007).

These give information about the thermal stability and purity of cellulose by looking at its degradation patterns. It is normally carried out using thermogravimetry (TG) or differential scanning calorimeter (DSC). The TGA analysis shows the weight loss of absorbed water in cellulose between 50 °C and 150 °C. The presence of hemicelluloses, lignin, and pectin in an untreated cellulosic fiber results in low decomposition temperature at around 202 °C while that of purified (free from lignin, hemicelluloses, and pectin) cellulose occurs at 253 °C. The increase in temperature to 500 °C results in minimum carbonaceous residue in purified cellulose (Sundari and Ramesh, 2012).

The presence of the hydroxyl groups in the structure of cellulose make it hygroscopic, absorbing about 8–14% water under normal atmospheric conditions. This property is not desirable in composites. It leads to the degradation of fiber-matrix interfacial region creating poor stress transfer efficiencies, which consequently results in a reduction of mechanical and dimensional properties (Dhakal et al., 2007). In their investigation of the relationship between the moisture absorption of pineapple-leaf-reinforced low density polyethylene composites and fiber loadings (10 wt%, 20 wt%, and 30 wt%), George and co-workers found that moisture absorption increased almost linearly with the fiber loading (George et al., 1998). The sensitivity of some mechanical properties of natural fibers reinforced polymeric composites to moisture uptake can be reduced by using coupling agents and fiber surface treatment (Mulinari et al., 2010).

The hydroxyl groups of cellulose also allow its functionalization with other molecules in many different ways such as esterification, etherification, grafting, fluorescently labelled, silylation, cationisation, which modify it for further applications such as in drug delivery, protein immobilization, and metal reaction template (Fig 3.12) (Peng et al., 2011; Saha et al., 2012).

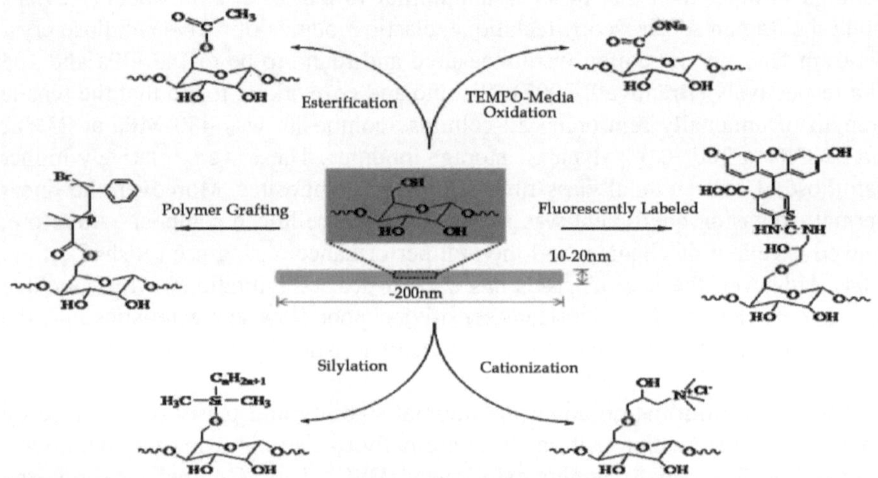

FIGURE 3.12 The schematic diagram showing the functionalization of cellulose fibers (Peng et al., 2011)

On the other hand, solvents and mixtures of solvents such as N, N'-dimethyl acetamide (DMAc)/LiCl, NaOH/Urea (Hashaikeh and Abushammala, 2011), dimethyl sulfoxide (DMSO)/tetrabutylammonium fluoride (Hadenström et al., 2009), N, N-dimethylformamide (DMF)/NO (Turbak et al., 1978), N-methyl morpholine-N-oxide (Vitz et al., 2010), trifluoroacetic acid (TFA) (Sarkar et al., 2009) liquid ammonia–ammonium thiocyanate–water, and LiCl, 3-dimethyl-2-imidazolidinone (Xu et al., 2009) have been found quite capable of dissolving cellulose (Zhao et al., 2009). However, most of these solvents suffer great limitations of high toxicity, cost, difficulty in solvent recovery and lack of stability during processing (Xu et al., 2009, 112; Zhao et al., 2012), which compromise the principles of green chemistry (Pinkert et al., 2011) – a concept based on the safe synthesis of products that are environmentally benign and recyclable, with minimal production of waste and hazardous by-products (Fleming et al., 2001).

As a result of the above shortcomings, a class of solvents known as ionic liquids has been found suitable for the dissolution of a wide range of inorganic and organic compounds (Martin, 2002) including cellulose. Ionic liquids are organic salts which consist of ions (Earle and Seddon, 2000) and are liquids at low temperatures (below 100 °C) (FitzPatrick, 2011; Frone et al., 2011; Zhao et al., 2012). They are often referred to as "green" solvents (Martin, 2002) because they possess attractive properties, such as high thermal and chemical stability (Sun and Cheng, 2010), insignificant vapor pressure, low toxicity and recyclability (Zhao et al., 2009; Zhao et al., 2012). They also act as catalysts in reactions like Friedel-Crafts (Martin, 2002). Once used, ionic liquids can be recovered using different separation techniques, such as evaporation, ion exchange, and reverse osmosis after which they can be re-used (Frone et al., 2011). Some of these ionic liquids are shown in the Table 3.2.

TABLE 3.2 Some commonly used ionic liquids to solubilize cellulose fibers

Ionic Liquids	Authors
1-N-butyl-3-methylimidazolium chloride ((C mim) Cl))	(Zhao et al., 2009)
1-ethyl-3-methylimidazolium diethylphosphonate ((EMIM)DEP)	(Kuhlmann et al., 2007)
1-methyl-3-methyl imidazolium dimethyl phosphate ((MMIM)(DMP))	(Kuhlmann et al., 2007)
1-ethyl-3-methylimidazolium alkylbenzsulfonates ((EMIM)ABS)	(Pinkert et al., 2011)
1-butyl-3-methylimidazolium acesulfamate (BMIM)Ace	(Frade et al., 2009)
Ethylammonium nitrate	(McMurry, 2004)
1-allyl-3-methylimidazolium chloride (AMIMCl)	(Frone et al., 2011)

The inherent renewability, biodegradability, abundance, low cost, low density, improved safety over conventional polymers (Cheng et al., 2011) coupled with high surface area and high aspect ratios make lignocellulosic fibers a polymer of choice for industrial scale applications. One such application is in automotive industry, where they are incorporated into door panel trims, package trays, trunk trims, and other exterior and interior parts (Soykeabkaew, 2007). As native cellulose consists of fermentable sugars, it can be used for the production of biofuel as an alternative source of energy to reduce over dependence on fossil fuels (i.e., coals, natural gas, and petroleum). The latter have negative impacts on the environment due to the emission of greenhouse gases such as CO , CH , and CO, which cause pollution and global warming (Saratale and Oh, 2011; Hassan et al., 2012). Besides, if not conservatively used by finding alternative sources, the petroleum reserves might soon be depleted. This is evidenced by dramatic growth in oil demand in China, India, and other developing countries (Yang and Lu, 2007). This could potentially result in price hikes leading to a chaotic economy (Yang and Wyman, 2008). Cellulose is used to prepare porous nanopaper of remarkably high toughness (Henriksson et al., 2008). Furthermore, its utilization as reinforcing material in polymer or polymer composites gives rise to high performance nanocomposites (Peng et al., 2011). In recent years, bacterial cellulose produced by species has shown unique properties including high mechanical strength, high water absorption capacity, and ultra-fine and highly pure fiber network structure (free from lignin and hemicelluloses) which opened its application as food matrix, dietary (de Santa Maria et al., 2009), filter membrane, ultra-strength paper and as a reticulated fine fiber network with coating, binding, thickening, and suspending characteristics. A wet spinning process has been developed for the production of textile fibers from bacterial cellulose (Vandamme et al., 1998). All-cellulose composite in which both the matrix and the filler are cellulose has been prepared and its structural, mechanical, and thermal properties investigated using X-ray diffraction, a scanning electron microscope, a tensile test, and dynamic viscoelastic and thermomechanical analyses (Nishino et al., 2004). Natural cotton fibers and different regenerated cellulose fibers such as viscose (cellulose xanthate), modal, and lyocell find medical application in hygiene and health care (i.e., bed linen, mattress covers, incontinence care pads, nappies, tampons, etc) and in external textile materials (i.e., wound dressings, bandages, gauzes, etc). The vital characteristics of these fibers responsible for their medical applications are: moisture and liquids' adsorption, antistatic behavior, low content of dyes, high mechanical stability, and ease of laundering and stabilization (Strnad et al. 2010). Cellulose also finds applications for the treatment of renal failure, for scaffolds in tissue engineering, temporary skin substitute, as hemostatic agent, post operative adhesion barrier, as well as a culture material for hepatocytes. It can also be used as wound dressing material for skin injuries, such as basal cell carcinoma/skin graft, severe body burns, facial peeling, sutures, dermabrasions, skin lesions, chronic ulcers (Czaja et al., 2006) and therefore, replaces the long use of pig skins and human cadaver skins as dressing material. The latter are very costly and could only be used for a short period of time (Hoenich, 2006). Although cellulosic fibers themselves do not show any antimicrobial properties to prevent wound infection (de Santa Maria et al., 2009) their functionalization with metallic nanoparticles such as silver and various coating strategies at the finishing stages using chitin and its derivative chitosan, alginate-based products, triclosan, etc. give such cellulose composites strong antimicrobial properties (Strnad et al. 2010; Li et al., 2011). Cellulose fibers are used for making cellulose paper often referred to as "smart material" or "electroactive paper" (EAPaP) which is used in sensors and actuator materials due to its many advantages which include lightweight, dryness, large deformation, low actuation voltage, low power consumption, low cost, and biodegradability (Yun et al., 2008). Cellulose also finds application as tablet binders in pharmaceuticals (Habibi et al., 2009), hydrogels (Samir et al., 2005; Builders et al., 2010; Oksman et al., 2011), laminates, optical films (Kim and Yun, 2006) and cosmetics (Isarankura-Na-Ayudhya et al., 2007; Stenstad et al., 2008; Heydarzadeh et al., 2009; Cheng et al., 2010).

3.3 CHARACTERIZATION OF CELLULOSE FIBERS

Different techniques such as Fourier transform infrared (FT-IR), UV/visible spectroscopy (UV/Vis), high performance liquid chromatography (HPLC), gas chromatography (GC), transmission electron microscopy (TEM), high resolution transmission electron Microscopy (HR-TEM), and X-Ray Diffraction (XRD) have been employed for the characterization of cellulose crystals.

3.3.1 UV/VISIBLE SPECTROSCOPY

This characterization technique involves the use of electromagnetic radiation between 190 nm and 800 nm. It is divided into ultraviolet (190–400nm) and visible (400–800nm) regions. It is aimed at studying the changes in the energy levels within a molecule during the promotion of electrons from occupied lower energy levels to unoccupied high energy levels (either non-bonding or π-orbitals). Although pure cellulose is not a good light absorber, the ultraviolet absorption spectra of wood depict that cellulose does absorb in the range of 200–300 nm with a tail of absorption extending to 400 nm. This could be attributed to the presence of a carbonyl group that accidentally introduces into cellulose at the time of isolation and preparation or the acetal or ketonic carbonyl groups at C position of non reducing unit of glucose (Hon and Shiraishi, 1991).

3.3.2 FOURIER TRANSFORM INFRARED (FTIR) SPECTROSCOPY

FTIR is a spectroscopic technique that is normally used for the identification of functional groups in a molecule and monitoring of any changes which occur during the reaction (Kumar, 2006). The FTIR spectrum of native cellulose shows absorption peaks in the following ranges: 3600–3000 cm and 3000–2800 cm which are indicative of stretching modes of O-H and C-H bonds, respectively. The peak at around 1726 cm and 1736 cm are attributed to the C-O stretch of either acetyl and uronic ester groups of hemicellulose or to the ester linkages of carboxylic groups of the ferulic and p-coumeric acids of lignin and/or hemicelluloses (Chen et al., 2011; Li et al., 2012). The absorption bands at around 1723 cm and 1517 cm are ascribed to C=O stretch in lignin, while the band at 1610 cm is ascribed to the C=C stretch in lignin (Oksman et al., 2011). The absorption band around 1640 cm indicates the O-H bending mode of water (Moràn et al., 2008). The band in the range 1436–1430 cm is due to the symmetric bending of CH (Lu and Hsieh, 2010). The bands at around 1381 cm, 1323 cm, and 1261 cm are indicative of O-H and C-H bending, C-C and C-O stretching bands (Liu et al., 2006). The bands in the range 1140–1070 cm are characteristic of C-O-C stretch of the β-(1→4)-glycosidic bond. A weak band at around 1118 cm originates from the skeletal vibration, while the band at around 904 cm corresponds to the glycosidic C -H deformation (Liu et al., 2006).

Upon subjecting native cellulose to different pre-treatments, the bands at 1516 cm , 1610 cm , and 1723 cm diminish (Fig 3.13).

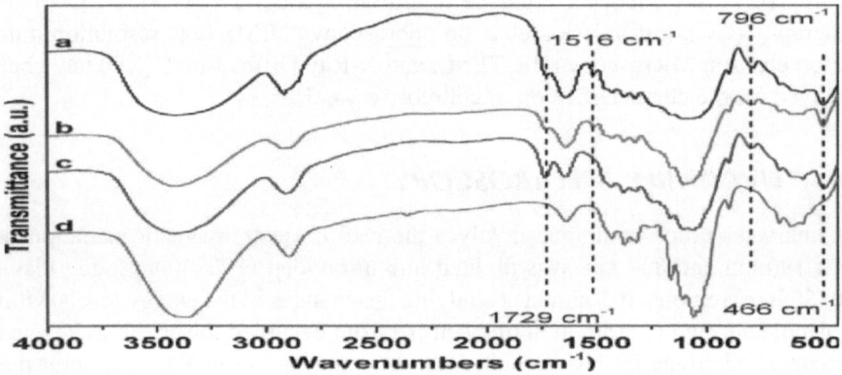

FIGURE 3.13 The FTIR spectra of rice straw subjected to the following treatments: (a) water washing, (b) de-waxing, (c) de-lignification, and (d) removal of hemicelluloses and silica (Lu and Hsieh, 2012)

3.3.3 CROSSED POLARIZATION AND MAGIC ANGLE SPINNING NUCLEAR MAGNETIC RESONANCE SPECTROSCOPY (CP/MAS NMR)

This is an essential analytical technique which helps in the study of molecular structures, relative configurations, relative and absolute concentrations, and intermolecular interactions. The analysis of ultrastructure of cellulose using NMR has been rife for three decades, after Attala and Vanderhert using high resolution solid state C CPMAS NMR, realized that it is possible to investigate native cellulose (Foston and Ragauskas, 2010). In their research, they noticed several resonances due to various cellulose carbon atoms. In addition, the position and appearance of these resonances changed with the sources of the cellulose. Based on this information coupled with the Raman spectroscopy analyses, they proposed that the changes observed were indicative of natural enrichment of particular crystalline allomorphs within the native cellulose (Atalla, 1999; Foston and Ragauskas, 2010). Investigation of the molecular order in cellulose can also be carried out by determining the appearance of the crystalline and non-crystalline regions of sub-spectra based on variations in the chemical shift and C relaxation (Newman and Hemmingston, 1994). Furthermore, apart from giving the relative amounts of cellulose allomorphs present, the information extracted from C CPMAS NMR can also be used for the determination of the average lateral dimensions of fibrils and fibril aggregates (Newman, 1999). Figure 3.14 shows a typical C CPMAS spectrum of cellulose isolated from untreated.

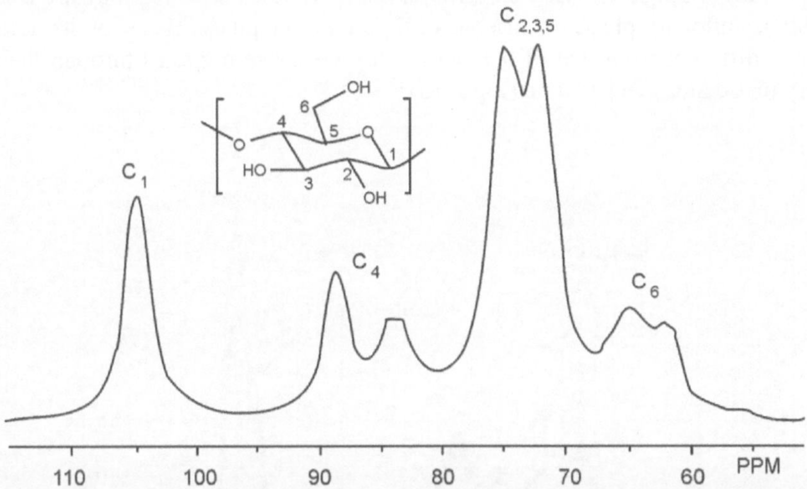

FIGURE 3.14 The 13C CPMAS spectrum of cellulose isolated from raw (Foston and Ragauskas, 2010)

The 13C NMR spectrum for cellulose depicts the characteristic signals of C-6 between 60 ppm and 70 ppm. Signals between 70 ppm and 80 ppm are ascribed to C-2, C-3, and C-5 (Hesse-Ertelt et al., 2008), while signals from 80 ppm to 90 ppm are due to C-4. The chemical shifts from 98 ppm to 110 ppm are indicative of C-1 (Liu et al., 2006). The signals between 87 ppm and 90 ppm are characteristics C-4 of highly crystalline cellulose fibers, while those signals between 79 ppm and 86 ppm correspond to C-4 of amorphous cellulose fibers. The signal at around 64.5 ppm is imputed to C-6 of crystalline cellulose chains and a shoulder between 60 ppm and 63 ppm usually corresponds to C-6 of amorphous and disordered cellulose fibers (Cheng et al., 2010). Upon subjecting native cellulose to various treatments, signals at around 56 ppm and 110–160 ppm – characteristic of methoxy and aromatic groups of lignins – diminish, indicating the removal of lignin (Liu et al., 2006).

3.3.4 HIGH PERFORMANCE LIQUID AND GAS CHROMATOGRAPHY (HPLC) AND (GC)

These techniques allow the identification, separation, and quantification of different components of a mixture through the use of eluants and the column packing materials. The elution of the analyte samples occurs at different migration rates resulting in different retention times (Jeffery et al., 1978; Skoog et al., 2004). These techniques could be used to purify extracted cellulose from the impurities such as phenols, oils, waxes, hemicelluloses, and resins. HPLC allows separation, purification, and identification of compounds based on differences in the rates at which they are carried through a stationary phase by gaseous or liquid mobile phase. Based on its polarity which is different from that of hemicelluloses, cellulose migrates through the stationary phase at a specific time (Fig 3.15).

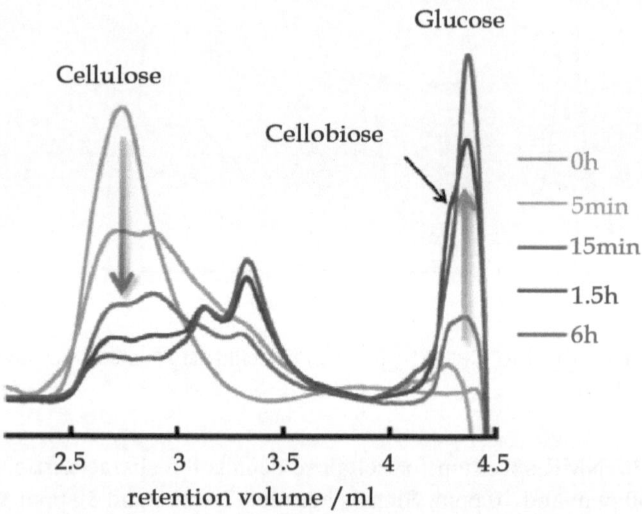

FIGURE 3.15 The high performance chromatogram of hydrolyzed cellulose in water with immobilized cellulose as stationary phase (adopted from Kuroda et al., 2012)

3.3.5 TRANSMISSION ELECTRON MICROSCOPY (TEM)

This is a very powerful tool for the analysis of microstructures at high resolutions (Rai and Subramanian, 2009). It gives information about the morphology, size, and crystallinity of the specimen under study through the use of a series of electromagnetic lenses to help them direct the electron beam produced at a high potential in an electrically heated filament. As this electron beam penetrates the sample, the diffraction patterns show the structure of the specimen (Tonejc, 1999). The TEM micrographs of cellulose from banana rachis shown in Fig 3.16A and B depict that the network of bundled and individualized fibers produced from different treatments have lengths of several micrometers with few nanometer sized diameter. This means therefore, that the produced fibers have high aspect ratio.

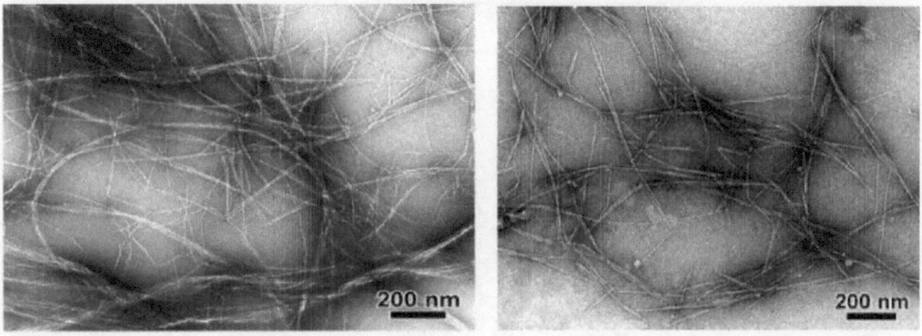

FIGURE 3.16 TEM micrographs of banana rachis cellulose treated with (a) peroxide alkaline, (b) peroxide alkaline-hydrochloric acid (Zuluaga et al., 2009)

3.3.6 X-RAY DIFFRACTION (XRD)

This is a non-sample destructive technique used in the determination of size, size distribution of materials, chemical composition and crystallography of the specimen. A representative X-ray diffractogram of native cellulose fiber is shown in Fig 3.17. The diffractogram show characteristic peaks at **2θ** degree of 14.7°, 16.4°, 22.6°, and 34.6° corresponding to 1 10, (110), (200), and (004) planes, respectively. Upon hydrolysis of native cellulose with concentrated sulphuric acid, the peak at **2θ**= 22.6° corresponding to (200) plane becomes sharper indicating improved crystallinity, while the peak at **2θ**= 14.7° becomes more intense and resolved from the peak at **2θ**= 16.4°. These results therefore show that controlled acid hydrolysis produces cellulose Iß, the more crystalline form of cellulose than native cellulose (Lu and Hsieh, 2010).

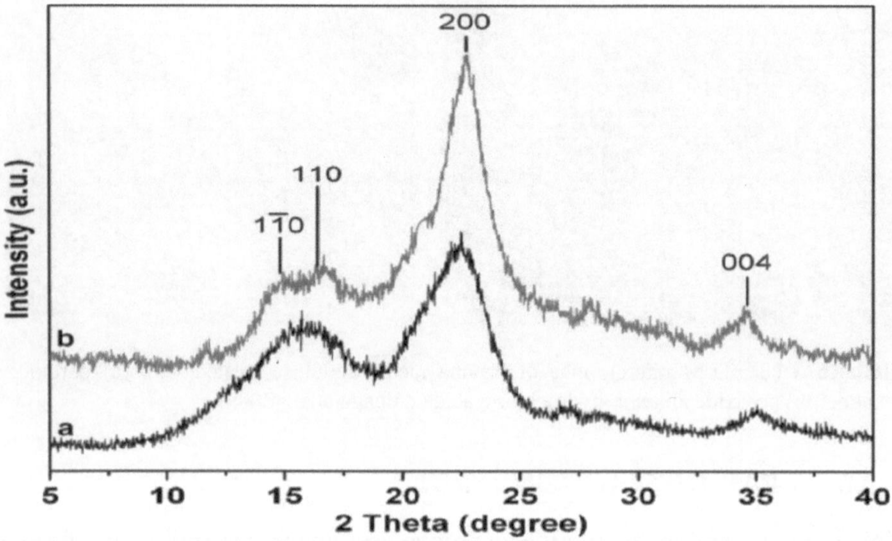

FIGURE 3.17 The XRD of: (a) cellulose extracted from grape skins (b) hydrolyzed cellulose nanocrystals adapted from (Lu and Hsieh, 2012)

3.3.7 GEL PERMEATION CHROMATOGRAPHY (GPC)

Gel permeation chromatography broadly known as size exclusion chromatography is a powerful separation technique in which the analyte is separated according to its hydrodynamic volume. A mobile phase is usually used for the transportation of the analyte through beads of porous polymeric material known as stationary phase (Anthony and Cooper, 2012). It is presently one of the most popular methods for the analysis and characterization of synthetic and biopolymers. However, its application in biopolymers is not as widespread as it is in synthetic polymers (Eremeeva, 2003). This is due to the highly ordered crystalline nature of cellulose structure, which does not allow easy penetration of the solvents into the cellulose fibers and to break the intermolecular hydrogen bonding network. GPC is used in cellulose analysis to provide in-depth information by allowing calculation and comparison of its molecular weight averages and distribution, branching and intrinsic viscosity. The molecular weight distribution of Avicel, α-cellulose, and sigmacell was analyzed. From the differential molecular weight distributions for Avicel, α-cellulose, and sigmacell, shown in (Fig 3.18a, b, and c), the weight average molecular weights (M) was determined to be 28,400 g/mol, 10,900 g/mol, and 76,100 g/mol, respectively (Engel et al., 2012).

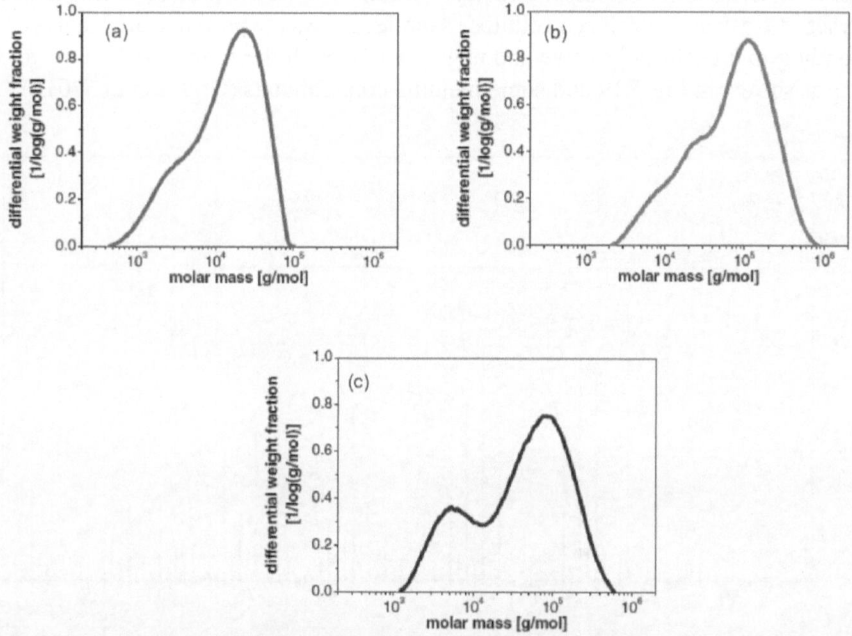

FIGURE 3.18 The differential weight distribution for (a) Avicel, (b) α-cellulose, and (c) sigmacell by Gel permeation chromatography (Engel et al., 2012)

3.3.8 TIME OF FLIGHT SECONDARY ION MASS SPECTROMETRY (TOF-SIMS)

Time of flight secondary ion mass spectroscopy (ToF-SIMS) is a very powerful technique representing the two most widely used ultra-high vacuum (UHV) surface analysis techniques for biomaterials (Belu et al., 2003). By using mass spectrometry to analyze the atomic and molecular secondary ions that are emitted from a solid surface when bombarded with ions, one obtains detailed information about the chemical composition of the surface. The principle behind ToF-SIMS is based upon the use of pulsed ion beam to remove molecules from the surface of the sample. The detached particles then accelerate at different speeds based on their mass into the flight tube through to the detector where their masses are recorded according to the exact time they have reached that detector. The sampling depths for ToF-SIMS are 0.2–1 nm. Thus, it gives a detailed characterization of the surface properties of the sample (Belu et al., 2003; Freire et al., 2006). The form of samples normally analyzed using ToF-SIMS include thermally unstable organic molecules on surfaces, synthetic polymers, and synthetically prepared molecular surface films and particles, (Benninghoven, 1994; Belu et al., 2003). Gram et al ., analyzed cellulose using ToF-

SIMS to determine its chemical structure and the composition of its surface. The analyses of fresh and pyrolyzed cellulose samples shows the presence of signals corresponding to a chain of positive and negative ions with the formulas C H O ± and C H ± as shown in Fig 3.19 and some metallic contaminants (Grams et al., 2012).

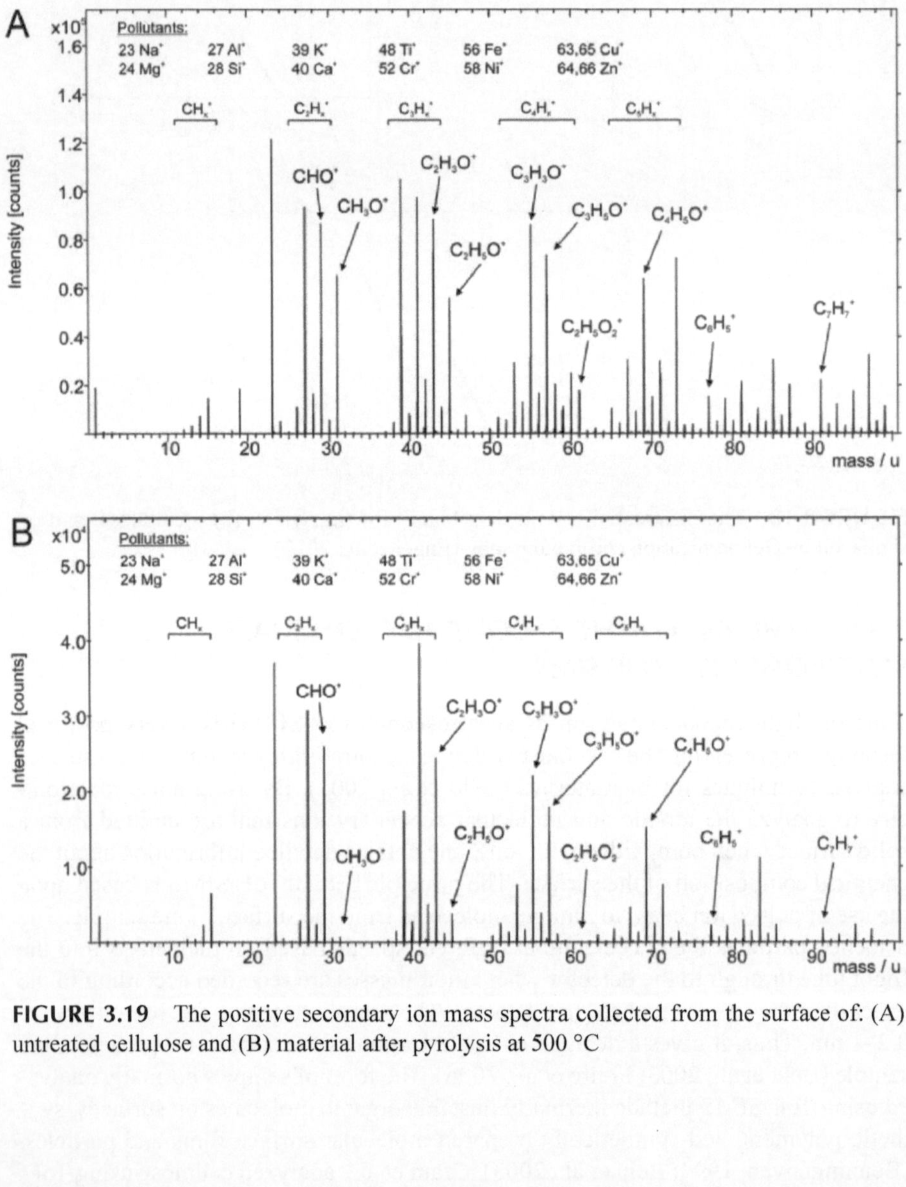

FIGURE 3.19 The positive secondary ion mass spectra collected from the surface of: (A) untreated cellulose and (B) material after pyrolysis at 500 °C

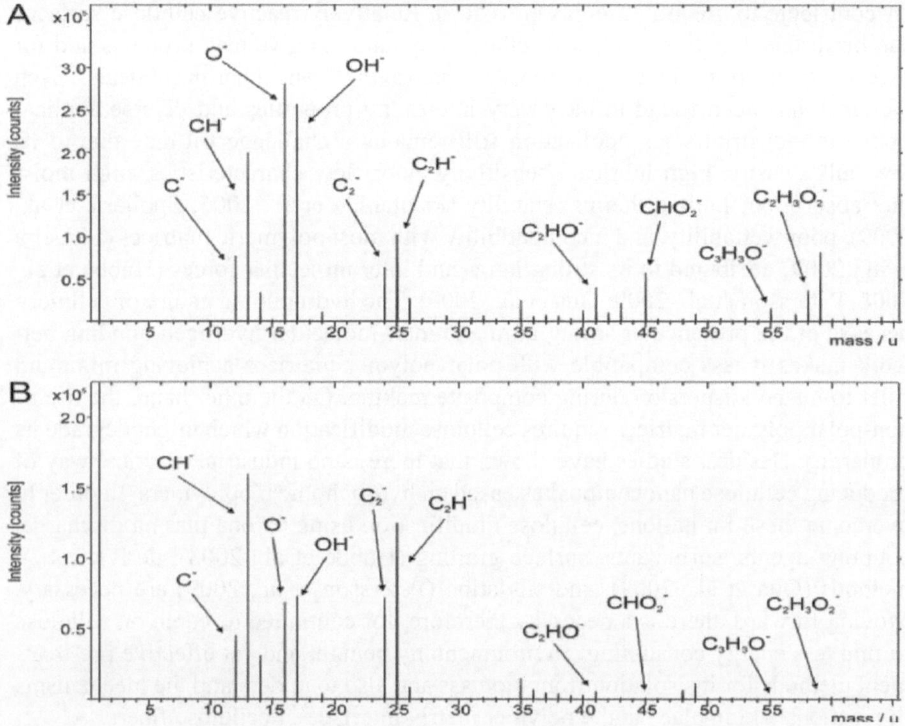

FIGURE 3.20 The negative secondary ion mass spectras collected from the surface of: (A) untreated cellulose and (B) material after pyrolysis at 500 °C

The spectra depict that there is no significant difference between the untreated cellulose and the residue left after pyrolysis in terms of qualitative composition of the analyzed surface. However, considerable difference was observed in the intensities of some ions between the untreated and pyrolyzed cellulose which could be due to migration of oxygen atoms from the catalyst support into the carbonaceous substance. The increase in pyrolysis temperature revealed that there is decrease in the hydrogen content of pyrolyzed residue and the lowest hydrogen was determined to be in the residue with the presence of nickel catalyst, which suggests that cellulose can undergo dehydration in the catalytic reaction (Basnar et al., 2001; Santos and Castanho, 2004; Dufrane, 2012; Grams et al., 2012; Hernández-Pedro et al., 2012).

3.4 CONCLUSIONS

Due to their renewability, abundance, biodegradability, non-food agricultural based economy, and low cost, biopolymers such as cellulose fibers remain the promising materials to mitigate the over dependence on petroleum-based sources as the latter contribute to green house gas emissions. Relatively reactive cellulose surfaces can be further functionalized with other compounds to give new products and for diverse applications in nanocomposites, packagings, and pharmaceuticals. Even though it has been found to have very interesting properties and diverse applications, its industrial scale application still remains a challenge till date due to its low bulk density, high lubricant sensitivity, poor flow characteristics, high moisture absorption, limited thermal stability (Piantanida et al., 2005; Spoljaric et al., 2009), poor wettability, and incompatibility with most polymeric matrices (Siqueira et al., 2009) attributed to its strong intra- and inter molecular forces (Hubbe et al., 2008; Petersson et al., 2009; Qua et al., 2009). The hydrophobic nature of cellulose because of the presence of strong intra and inter-molecular hydrogen bonding network makes it less compatible with polar polymer matrices achieving minimum filler to matrix dispersion during composite making. On the other hand, the use of non-polar polymer matrices requires cellulose modification which might degrade its properties. Besides, studies have shown that there is no industrial practical way of producing cellulose nanocomposites based on hydrophobic biopolymers. In order to overcome these limitations, cellulose modification using carona plasma discharge, coupling agents, surfactants, surface grafting (Hubbe et al., 2008), derivatization methods (Qua et al., 2009), and silylation (Petersson et al., 2009) are necessary. Moving forward, there is a necessity therefore, for continued research on cellulose to find less energy consuming, environmentally benign, and yet effective pre-treatment methods for its isolation from biomass and also to understand the mechanisms of reactions taking place at the polymer matrix interface of cellulose fibers.

ACKNOWLEDGEMENTS

The authors would like to thank the National Research Foundation (NRF), South Africa and directorate of research, Walter Sisulu University (WSU) for financial support.

KEYWORDS

- **Cellulose**
- **Lignin**
- **Fiber**
- **Isolation**
- **Characterization**

REFERENCES

Abbott, A. and A. Bismarck, Cellulose, 2010. **17**: p. 779–791.

Abdul Khalil, H.P.S., A.H. Bhat, and A.F. Ireana Yusra, Carbohydrate Polymers, 2012. **87**(2): p. 963–979.

Achyuthan, E.K., et al., Molecules, 2010.

Adebayo, A.A., et al., Aquatic Invasions, 2011. **6**(1): p. 91–96.

Adel, A.M., et al., Carbohydrate Polymers, 2011. **83**: p. 676–687.

Ahola, S., , in 2008, Helsinki University of Technology: Espoo.

Alvira, P., et al., Bioresource Technology, 2010. **101**: p. 4851–4861.

Anthony, R. and Â. Cooper, . 2012; Available from: http://www.accessscience.com.

Atalla, 1999, Amsterdam, Netherlands: Elsevier.

Atalla, R.H. and D.L. VanderHart, Solid State Nuclear Magnetic Resonance, 1999. **15**(1): p. 1–19.

Balat, M., Energy Conservation and Management, 2011. **52**: p. 858–875.

Basnar, B., et al., Applied Surface Science, 2001. **171**(3–4): p. 213–225.

Belu, A.M., D.J. Graham, and D.G. Castner, Biomaterials, 2003. **24**(21): p. 3635–3653.

Benninghoven, A., Angewandte Chemie International Edition in English, 1994. **33**(10): p. 1023–1043.

Bhimte, A.N. and T.P. Tayade, AAPS PharmSciTech, 2007. **8**(1): p. 8

Binod, P., et al., Bioresource Technology, 2010. **101**: p. 4767–4774.

Brännvall, E., , 2007, KTH, Royal Institute of Technology: Stockholm, Sweden.

Brodeur, G., et al., Enzyme Research, 2011.

Builders, F.P., et al., International Journal of Pharmaceutics, 2010. **388**: p. 159–167.

Cabiac, A., et al., Applied Catalysis A: General, 2011. **402**: p. 1–10.

Chattopadhyay, H. and P.B. Sarkar, , 1946.

Chen, W., et al., Carbohydrate Polymers, 2011. **83**: p. 1804–1811.

Cheng, Q., et al., , in , B. Attaf, Editor 2011, InTech: Morgantown, WV and Knoxville, TN.

Cheng, Y.S., et al., Applied Biochemistry and Biotechnology, 2010. **162**(6): p. 1768–1784.

Czaja, W., et al., Biomaterials, 2006. **27**: p. 145–151.

Das, K., et al., Cellulose, 2009. **16**: p. 783–793.

de Santa Maria, L.C., et al., Materials Letters, 2009. **63**(9–10): p. 797–799.

Dhakal, H.N., Z.Y. Zhang, and M.O.W. Richardson, Composites Science and Technology, 2007. **67**: p. 1674–1683.

Dimmel, R.D., , in 2008, McGraw-Hill Companies.

Dufrane, F.Y. . 2012; Available from: http://www.accessscience.com.

Dufresne, A., 2008, Oxford, UK: Elsivier.

Earle, J.M. and R.K. Seddon, Pure Applied Chemistry, 2000. **72**(7): p. 1391–1398.

Eichhorn, S.J., et al., Journal of material Science, 2010. **45**: p. 1–33.

Elanthikkal, S., et al., Carbohydrate Polymers, 2010. **80**: p. 852–859.

El-Sakhawy, M. and M.L. Hassan, Carbohydrate Polymers, 2007. **67**(1): p. 1–10.

Engel, P., L. Hein, and A. Spiess, Biotechnology for Biofuels, 2012. **5**(1): p. 77.

Eremeeva, T., Journal of Biochemical and Biophysical Methods, 2003. **56**(1–3): p. 253–264.

FitzPatrick, A.M., , in 2011, Queen's University: Kingston.

Fleming, K., D.G. Gray, and S. Matthews, Chemistry – A European Journal 2001. **7**(9): p. 1831–1835.

Foston, M. and A.J. Ragauskas, Biomass and Bioenergy, 2010. **34**(12): p. 1885–1895.

Frade, F.R., et al., Green Chemistry, 2009. **11**: p. 1660–1665.

Freire, C.S.R., et al., Journal of Colloid and Interface Science, 2006. **301**(1): p. 205–209.

Frone, N.A., M.D. Panaitescu, and D. Donescu, U.P.B.Sci.Bull, 2011. **73**(2).

George, J., S.S. Bhagawan, and S. Thomas, Composites Science and Technology, 1998. **58**: p. 1471–1485.

Goetz, L., et al., Carbohydrate Polymers, 2009. **75**: p. 85–89.

Goodger, E., 1976, London: Macmillan.

Grams, J., et al., International Journal of Mass Spectrometry, (2012).

Granström, M., 2009.

Habibi, Y., A.L. Lucia, and J.O. Rojas, Chemical Review, 2010. **110**: p. 3479–3500.

Habibi, Y., M. Mahrouz, and R.M. Vignon, Food Chemistry, 2009. **115**: p. 423–429.

Hadenström, M., et al., Molecular Plant, 2009. **2**(5): p. 933–942.

Hamelinck, N.C., V.G. Hooijdonk, and P.A. Faaij, Biomass and Bioenergy, 2005. **28**: p. 384–410.

Harun, Y.M., et al., Bioresource Technology, 2011. **102**: p. 5193–5199.

Hashaikeh, R. and R. Abushammala, Carbohydrate Polymers, 2011. **83**: p. 1088–1094.

Hassan, S.H.A., Y.S. Kim, and S.-E. Oh, Enzyme and Microbial Technology, 2012. **51**(5): p. 269–273.

Hendriks, A.T.W.M. and G. Zeeman, Bioresource Technology, 2009. **100**(1): p. 10–18.

Henriksson, M., et al., BioMacromolecules, 2008. **9**(6): p. 1579–1585.

Hernández-Pedro, N., et al., , in , L.C. Frewin, Editor 2012.

Hesse-Ertelt, S., et al., Magnetic Resonannce in Chemistry, 2008. **46**: p. 1030–1036.

Heydarzadeh, D.H., D.G. Najafpour, and A.A. Nazari-Moghaddam, World Applied Sciences Journal, 2009. **6**(4): p. 564–569.

Hoenich, N., Bioresources 2006. **1**(2): p. 270–280.

Hon, -.S.N.D. and N. Shiraishi, eds. 1991, Marcel Dekker, Inc.: New York, Basel.

Hubbe, A.M., et al., BioResources, 2008. **3**(3): p. 929–980.

Ibbet, R., et al., Bioresource Technology, 2011. **102**: p. 9272–9278.

Isarankura-Na-Ayudhya, C., et al., EXCLI Journal, 2007. **6**: p. 167–176

Jargalsaikhan, O. and N. Saraçoğlu, Pechia Stipitis Chemical Engineering Communications, 2008. **196**: p. 1–2, 93–103.

Jeffery, H.G., et al., 1978, London: John Wiley & Sons, Inc.

Jiang, M., et al., Industrial Crops and Products, 2011. **33**(3): p. 734–738.

Johnson, K.R., et al., Cellulose, 2009. **16**: p. 227–238.

Johnson, P.G. and D.A. French, Cellulose, 2009. **16**: p. 959–973.

Jonas, R. and L.F. Farah, Polymer Degradation and Stability, 1998. **59**(1–3): p. 101–106.

Kamel, S., eXPRESS Polymer Letters, 2007. **1**(9): p. 546–575.

Khalil, A.H.P.S., A.H. Bhat, and I. Yusra, Carbohydrate Polymers, 2012. **87**: p. 963–979.

Kim, J. and S. Yun, Macromolecules, 2006. **39**: p. 4202–4206.

Kontturi, E.J., 2005, University of Technology: Eindhoven.

Krässig, H., . Vol. 11. 1996, Amsterdam: Gordon and Breach Science.

Kroom-Batenburg, J.M.L., B. Bouma, and J. Kroon, Macromolecules, 1996. **29**(17): p. 5695–5699.

Kuhlmann, E., et al., Green Chemistry, 2007. **9**: p. 233–242.

Kumar, S., et al., Bioresource Technology, 2010. **101**: p. 1337–1347.

Kumar, S., 2006, Amritsar.

Kuntturi, E., T. Tammelin, and M. Österberg, Chemical Society Review, 2006. **35**(12): p. 1287–1304.

Kuroda, K., Y. Fukaya, and H. Ohno, The Electrochemical Society, 2012(53): p. 3708.

Lee, I., B.R. Evans, and J. Woodward, Ultramicroscopy, 2000. **82**: p. 213–221.

Leppänen, K., et al., Cellulose, 2009. **16**: p. 999–1015.

Li, M., et al., Ultrasonic Sonochemistry, 2012a. **19**: p. 243–249.

Li, R., L. Zhang, and M. Xu, Carbohydrate Polymers, 2012b. **87**: p. 95–100.

Li, S., et al., Carbohydrate Polymers, 2011. **86**: p. 441–447.

Liu, C., et al., Journal of Agricultural and Food Chemistry, 2006. **54**: p. 5742–5748.

Lu, P. and Y. Hsieh, Carbohydrate Polymers, 2010. **82**: p. 329–336.

Lu, P. and Y. Hsieh, Carbohydrate Polymers, 2012a. **87**: p. 2546–2553.

Lu, P. and Y. Hsieh, Carbohydrate Polymers, 2012b. **87**: p. 564–573.

Ludueña, L., A. Vázquez, and V. Alvarez, Carbohydrate Polymers, 2012. **87**: p. 411–421.

MacDonald, D.G., et al., Biotechnology and Bioengineering, 1983. **25**(8): p. 2067–2076.

Mansouri, S., et al., Industrial Crops and Products, 2012. **36**: p. 22–27.

Marques, G., J.C. del Río, and A. Gutiérrez, Bioresource Technology, 2010. **101**(1): p. 260–267.

Martin, K., , in , J. Clark and D. Macquarrie, Editors. 2002, Blackwellscience Ltd: Malden.

McKendry, P., Bioresource Technology, 2002. **83**(1): p. 37–46.

McMurry, J., ed. . 7 ed. 2004, Harris, D: London.

Methacanon, P., et al., Carbohydrate Polymers, 2010. **82**: p. 1090–1096.

Moràn, I.J., et al., . Cellulose, 2008. **15**: p. 149–159.

Mulinari, D.R., et al., BioResources, 2010. **5**(2): p. 661–671.

Nadanathangam, V. and P. Satyamurthy, , in 2011, IACSIP Press: Singapore.

Newman, R. and J. Hemmingston, Cellulose, 1994. **2**: p. 95–110.

Newman, R.H., Solid state nuclear magnetic resonance, 1999. **15**(1): p. 21–29.

Nishino, T., I. Matsuda, and K. Hirao, Macromolecules, 2004. **37**(20): p. 7683–7687.

Nishiyama, Y., P. Langan, and H. Chanzy, Journal of American Chemical Society, 2002. **124**: p. 9074–9082.

Obi, R.K., et al., Material Letters, 2012. **67**: p. 35–38.

Oke, I., Winter, 2010. **3**(2): p. 77–80.

Oksman, K., et al., Biomass and Bioenergy, 2011. **35**: p. 146–152.

Pakzad, A., et al., Journal of Materials Research, 2012. **1**(1): p. 1–9.

Panaitescu, M.D., et al., , in , B. Attaf, Editor. 2011, InTech: Romania. p. 642.

Pandey, K.J., et al., Composites:Part B, 2009. **40**: p. 676–680.

Pauly, M. and K. Keegstra, Current Opinion in Plant Biology, 2010. **13**: p. 305–312.

Peng, L.B., et al., The Canadian Journal of Chemical Engineering, 2011. **9999**: p. 1–12.

Petersson, L., P.A. Mathew, and K. Oksman, Journal of Applied Polymer Science, 2009. **112**: p. 2001–2009.

Piantanida, G., M. Bicchieri, and C. Coluzza, Polymer, 2005. **46**(26): p. 12313–12321.

Pinkert, A., et al., Green Chemistry, 2011. **13**: p. 3124–3136.

Qua, H.E., et al., Journal of Applied Polymer Science, 2009. **113**: p. 2238–2247.

Rai, R.S. and S. Subramanian, Progress in Crystal Growth and Characterization of Materials, 2009. **55**: p. 63–97.

Reddy, N. and Y. Yang, Polymer Engineering and Science, 2009. **49**(11): p. 2212–2217.

Rosa, F.M., et al., Carbohydrate Polymers, 2010. **81**: p. 83–92.

Saarinen, T., et al., BioResources, 2009. **4**(1): p. 94–110.

Saha, B., Journal of Industrial Microbiology & Biotechnology, 2003. **30**(5): p. 279–291.

Saha, P., et al., Carbohydrate Polymers, 2012. **87**: p. 1628–1636.

Samir, A.S.A.M., F. Alloin, and A. Dufresne, Biomacromolecules, 2005. **6**: p. 612–626.

Santos, N.C. and M.A.R.B. Castanho, Biophysical Chemistry, 2004. **107**(2): p. 133–149.

Saratale, G.D. and S.E. Oh, Biodegradation, 2011. **22**(5): p. 905–919.

Sari, E., et al., World Academy of Science, Engineering and Technology, 2011. **78**: p. 2011.

Sarkar, N., et al., Renewable Energy, 2012. **37**: p. 19–27.

Sarkar, P., E. Bosneaga, and M. Auer, Journal of Experimental Botany, 2009. **60**(13): p. 3615–3635.

Siqueira, G., J. Bras, and A. Dufresne, Biomacromolecules, 2009. **10**: p. 425–432.

Siqueira, G., J. Bras, and A. Dufresne, Polymers, 2010. **2**: p. 728–76.

Siqueira, G., et al., Cellulose, 2011. **18**: p. 57–65.

Siro´, I. and D. Plackett, Cellulose, 2010. **17**: p. 459–494.

Sjöström, E., 1993, New York, USA: Academic Press.

Skoog, A.D., et al., 2004, London: Harris, D.

Soykeabkaew, N., 2007, University of London: London.

Spoljaric, S., A. Genovese, and A.R. Shanks, Composites:Part A, 2009. **40**: p. 791–799.

Stenstad, P., et al., Cellulose, 2008. **15**: p. 35–45.

Strnad, S., O. Šauperl, and L. Fras-Zemljič, , in , M.M. Elnashar, Editor. 2010, Sciyo: India.

Sun, J.X., et al., Polymer Degradation and Stability, 2004. **84**: p. 331–339.

Sun, R., J.M. Lawther, and W.B. Banks, Industrial Crops and Products, 1995. **4**(2): p. 127–145.

Sun, Y.E. and J. Cheng, Bioresource Technology, 2002. **83**(1): p. 1–11.

Sundari, T.M. and A. Ramesh, Carbohydrate Polymers, 2012. **87**: p. 1701–1705.

Taylor, A.M., et al., Dendrochronologia, 2008. **26**(2): p. 125–131.

Teixeira, M.E., et al., Industrial Crops and Products, 2011. **33**: p. 63–66.

Terinte, N., R. Ibbet, and C.K. Schuster, , Lenzinger Berichte, 2011. **89**: p. 118–131.

Tharanathan, R.N., Trends in Food Science & Technology, 2003. **14**(3): p. 71–78.

Tonejc, A., Acta Chimica Slovenica, 1999. **46**(3): p. 435–461.

Turbak, F.A., et al., , 1978, International Telephone and Telegraph Corporation: US.

van der Werf, H.M.G., et al., Industrial Crops and Products, 1994. **2**(3): p. 219–227.

Vandamme, E.J., et al., Polymer Degradation and Stability, 1998. **59**(1–3): p. 93–99.

Vinardell, P.M., V. Ugartondo, and M. Mitjans, Industrial Crops and Products, 2008. **27**.

Vitz, J., et al., Carbohydrate Polymers, 2010. **82**: p. 1046–1053.

Wang, D., et al., Journal of Industrial and Engineering Chemistry, 2010. **16**: p. 152–156.

Wegner, H.T. and P.E. Jones, , in , L. Lucia and O. Rojas, Editors. 2009, Media.Wiley.com.

Xu, W., N. Reddy, and Y. Yang, Carbohydrate Polymers, 2009. **76**: p. 521–527.

Yang, B. and E.C. Wyman, Biofuels Bioproducts and Biorefining, 2008. **2**: p. 26–40.

Yang, B. and Y. Lu, Journal of Chemical Technology and Biotechnology, 2007. **82**: p. 6–10.

Yokoyama, T., J.F. Kadla, and H.m. Chang, Journal of Agricultural and Food Chemistry, 2002. **50**(5): p. 1040–1044.

Yun, S., et al., Journal of Applied Physics, 2008. **103**.

Zhang, B. and A. Shahbazi, Journal of Petroleum and Environmental Biotechnology, 2011.

Zhao, D., et al., Carbohydrate Polymers, 2012. **87**: p. 1490–1494.

Zhao, Q., et al., Cellulose, 2009. **16**: p. 217–226.

Zuluaga, R., et al., Carbohydrate Polymers, 2009. **76**: p. 51–59.

CHAPTER 4

STARCH-BASED BIODEGRADABLE COMPOSITES: SCOPE IN BIOMEDICAL AND AGRICULTURAL APPLICATIONS

N. JYOTHI*, P. PARVATHY CHANDRAN, SOUMYA B. NAIR, and G. SUJA

Division of Crop Utilization, Central Tuber Crops Research Institute, Sreekariyam, Thiruvananthapuram, Kerala, India

*Corresponding author: Tele: +91 471 298551; Fax: +91 471 2590063; E-mail: sreejyothi_in@yahoo.com

CONTENTS

ABSTRACT

Cassava starch-based blends and composites, suitable for biomedical and agricultural applications were developed and characterized. Starch-konjac glucomannan (KGM) blend films were prepared and their suitability as matrices for the controlled release of a model drug, theophylline was studied. Incorporation of KGM in cassava starch film could extend the drug release time to about 10 h. A superabsorbent polymer synthesized by graft copolymerization of poly(acrylamide) on to cassava starch, and subsequent saponification was evaluated for its potential for soil moisture retention, and the effect on soil properties and plant growth. The study showed that SAP amendment in the soil could reduce the irrigation frequency to once in 3–6 days without significantly affecting the plant growth. Superporous hydrogels (SPHs) with very fast swelling and superabsorbent properties were synthesized from cassava starch by solution polymerization in presence of an initiator pair, crosslinker, and a foaming agent. The polymers could reach equilibrium swelling in 8–9 min and the swelling was found to be highly sensitive to pH and ionic strength of the swelling medium. Because of the incorporation of starch, a safe, biodegradable and cheaper natural biopolymer, these materials are eco-friendly and cost-effective for exploitation in various applications.

4.1 INTRODUCTION

Starch is a natural biopolymer and was obtained from the roots, piths and seeds of the plants (Bhattacharya, 1995). It is a cheap and readily available raw material that is relatively easy to produce from sustainable sources. Starch is a precursor for a very large number of ingredients used in food, textile, paper, pharmaceutical, and adhesive industries. The presence of amorphous and crystalline regions and a large number of hydroxyl groups in starch offer better opportunities for starch modification to develop tailor-made products. Starch has vast scope for exploitation in the development of composite materials due to its biocompatibility with other polymer systems (Gomes and Reis, 2004; Murthy et al., 2006).

Recent studies have shown that natural polysaccharides such as starch, chitosan, cellulose, and their derivatives are safer, cheaper, and environmentally friendly substrates for incorporation in the development of a number of polymer materials, which are used in the biomedical field, food industry and many other industries, where, presently petroleum-based synthetic polymers are used in abundance. Due to the harmful effects such as toxicity, non-biodegradability and the disposal problems associated with such materials, natural polysaccharides have been investigated as substitutes for synthetic polymers (Berger et al., 2005; Bhuniya et al., 2003; Chen et al., 2004a; Coviello et al., 2005; Reis et al., 2008).

Owing to their outstanding merits, natural polysaccharides have been used in controlled drug delivery for many years (Panyam and Labhasetwar, 2000). These are highly stable, safe, non-toxic, hydrophilic and biodegradable, and can form non-covalent bonds with biological tissues forming bio adhesion. These bio adhesive polysaccharides can prolong the residence time and therefore, increase the absorption of loaded drugs (Lee et al., 2000; Zonghua et al., 2008). The biodegradable polymers, including starch degrade *in vivo* as a result of natural biological processes, eliminating the need to remove the drug-delivery system after release of the active agent has been completed (Kotwala et al., 2006). Starch is one of the most interesting polymers used in this field because of an attractive combination of availability, low price, and good performance. It is biodegradable and is naturally metabolized by the human body.

Superabsorbent polymers (SAP) are three-dimensional polymeric networks that have the ability to swell in water and retain a significant amount of water within their structure. The majority of the superabsorbents manufactured are synthetic polymers (essentially acrylics) due to their superior price to efficiency balance. But, due to their non-biodegradability and toxicity, presently there is a growing interest on natural polymer-based SAPs. SAPs can be effectively used in agricultural application due to their capacity to absorb water and expand under pressure, thus not only providing plants with water but also helping to aerate the soil (Johnson, 1984). Uz et al. (2008) have studied the relations between soil texture and water-holding capacity as a response of increasing polyacrylamide (PAM) concentration. The SAP hydrogels reduce irrigation frequency and compaction tendency, stop erosion and water runoff, and increase the soil aeration and microbial activity (El-Rehim et al.., 2004; Zohuriaan-Mehr and Kabiri, 2008). They act as "miniature water reservoirs"' in soil, releasing water into the soil and maintaining the moisture balance. They have the capacity to release water when the moisture equilibrium of the soil changes or depending on the requirement of plant roots. The structure of most SAPs resembles that of humus in soil (Huttermann et al., 2009). Although both, SAPs and humus, have hydrophilic and cation binding functional groups, SAPs possess higher density of these groups and do not possess aromatic moieties. Chen et al. (2004b) have earlier reported the synthesis of SAPs by grafting acrylic acid and acrylamide on to starch, which exhibited good water retention at high temperature and thus can be used in agricultural application, especially in arid and desert regions.

Superporous hydrogels (SPHs) are materials developed to demonstrate fast-swelling and superabsorbent properties (Omidian et al., 2005). These were initially proposed as gastric retention devices; however, SPHs may be tailor-made for other applications in pharmaceutical and biomedical industries (Omidian et al., 2007). The high swelling pressure of SPHs can potentially be used to trigger an alarm system upon the incursion of water and hence, can be utilized as smart hydrogels or sensors for various applications.

Tuber crops, especially cassava (tapioca) are rich sources of starch. Cassava starch is produced mainly in the tropical countries, and in India, it is a major industrial starch (Jyothi et al., 2010). Compared to cereal starches, cassava starch is relatively purer and possesses bland taste and flavor due to the absence of proteins, lipids and other such compounds as in cassava roots. With an objective to explore new areas of utilization of cassava starch, in this paper, we report some biomaterials developed from cassava starch, which can be exploited in biomedical and agricultural applications.

4.2 MATERIALS AND METHODS

Cassava starch was extracted from the freshly harvested tubers according to a reported procedure (Sajeev et al., 2003). Konjac glucomannan was obtained from a commercial manufacturer (Green food, Yunnana Fuyuan Guanghua Konjac Exploitation Co., LTD, China). Acrylamide (AM), acrylic acid, ceric ammonium nitrate (CAN), potassium persulphate, N,N'-methylene bisacrylamide (MBA) and N,N,N',N'-tetramethylethylenediamine (TEMED) were purchased from Sigma Aldrich Chemicals (St. Louis, USA). All the other chemicals used were of analytical grade.

4.2.1 CASSAVA STARCH-KONJAC GLUCOMANNAN BLEND FILMS FOR SUSTAINED DRUG RELEASE

Konjac glucomannan (KGM) is a natural hetero polysaccharide isolated from the tubers of *Amorphophallus konjac* plant. Glucomannan is a water soluble polysaccharide consisting of D-glucose and D-mannose in a molar ratio of 1:1.6 joined by β-(1-4)-linkages (Dong-bao et al., 2002). KGM has received much attention in the field of controlled drug delivery and tissue coating due to its biodegradability, good biocompatibility, gel-forming ability, and excellent film-forming capacity (Zhang et al., 2005).

Cassava starch-KGM blend films were prepared according to a reported method (Chen et al., 2008) with slight modifications (Nair and Jyothi, 2011). Gelatinized starch paste was mixed with KGM, glycerol and the model drug, Theophyllin (400 mg dispersed in 50 mL of water) and casted on Teflon sheets. Starch and KGM were used in four different concentrations *viz.*, 1.0, 1.5, 2.0 and 3.0 g. The drug loaded films were stored at 43% relative humidity for 2 weeks in a desiccator at room temperature, before being tested. A weighed quantity of drug loaded film was placed in 100 mL of dissolution medium (phosphate buffer, pH – 7.4) maintained at 37 °C. The samples were kept under constant stirring at 100 rpm. The amount of drugs released into the medium was determined by withdrawing 10mL of the dissolution fluid at predetermined time intervals. The volume withdrawn was replaced with an equal volume of the release medium to maintain sink conditions. The absorption of withdrawn solutions was determined at 275 nm using a UV-visible Spectrophotometer after filtration and suitable dilution (Zhang et al., 2008). To find out the mechanism of drug release, the initial 60% drug release data was fitted to Korsmeyer -Peppas model (Avachat and Kotwal, 2007).

4.2.2 STARCH-BASED SAP FOR SOIL MOISTURE RETENTION

A superabsorbent polymer (SAP) was synthesized by a two-step process which involved graft copolymerization of acrylamide (AM) onto cassava starch in presence of ceric ammonium nitrate (CAN) as the free-radical initiator and the subsequent saponification of the grafted starch with sodium hydroxide solution (Parvathy and Jyothi, 2012). The factors used were: concentration of AM = 20g/10g starch (dry wt. basis), concentration of CAN = 4.93g/L, duration of reaction = 120 min and temperature = 45 °C. After completion of the reaction, the grafted starch was washed several times with water followed by methanol and dried at 55 °C. The starch graft copolymer (5g) was then mixed with 125 mL of 5% sodium hydroxide solution. The mixture was stirred gently at 400 rpm at room temperature (30 ± 2 °C) for 60 min on a magnetic stirrer. The pH of the solution was then adjusted to 7.0 using glacial acetic acid and precipitated in methanol. The precipitate was washed with distilled water, dried at 55–60 °C and ground.

An experiment was conducted to determine the effect of starch-based superabsorbent polymer on soil moisture retention. The soil used in the experiment was laterite sandy clay loam soil collected from 20–25 cm depth from CTCRI farm. The soil was dried, ground and passed through 500 μm sieve. In the first experiments to study the effect of optimum SAP concentration and release pattern of moisture, 200 g portion of the soil mixture was taken in 300 cm^3 pots. Powdered superabsorbent polymer was mixed at the rate of 0.1, 0.2, 0.3 and 0.5% with the soil. Triplicate pots of each soil mixture were saturated with tap water by placing the pots in a tray of water for 24 h. Excess water in each pot was drained out by raising the pots gravimetrically. The pots were placed under laboratory conditions at 28 ± 2 °C. The experiment was done for a period of 1 month and weights of the pots were recorded at an interval of 5 days to determine the percentage of moisture retained.

Another experiment was carried out with chilly plants in the pots with different watering intervals. A weighed quantity of finely powdered SAP (0.5% based on soil weight) was mixed with soil and taken in pots. Then chilly seeds were sown, watered and the pots were kept in field conditions. The pots watered daily were taken as control and those watered once in 5 days, once in 6 days, once in 9 days and once in 12 days were taken as treatments A, B C and D respectively. The moisture content of the soil was determined at regular intervals by drying a weighed quantity of the soil from replicated pots at 105 °C overnight and weighing. The plant shoot length and number of leaves were also determined at regular intervals.

4.2.3 CASSAVA STARCH-BASED POROUS SAP

Cassava starch-poly (acrylic acid-*co*-acrylamide) superabsorbent polymer with fast-swelling property was synthesized by the acid-induced decomposition of the bicarbonate compound to give a homogeneous pore structure. An initiator pair (ammonium persulphate and N,N,N′,N′-tetramethylethylenediamine), cross-linker (N,N′-methylene bisacrylaide), and a foaming agent (sodium bicarbonate) were used in the synthesis of the polymer by solution polymerization. The sensitivity of the hydrogel to pH and ionic strength was studied by allowing the dry polymer (xerogel) to swell in aqueous solutions of pH ranging from 3.0 to 11.0 and sodium chloride with different ionic strength.

4.3 RESULTS AND DISCUSSION

4.3.1 CASSAVA STARCH-KONJAC GLUCOMANNAN BLEND FILMS FOR SUSTAINED DRUG RELEASE

Theophylline, a xanthine drug, which is used in the therapy for respiratory diseases (Yu et al., 1996), was used for this study. Theophylline (Thp) is widely used as a model drug in various controlled-release studies (Antal et al., 1997; Coviello et al., 2003). Rapid release and short elimination half-life of theophylline encourage the drug to be formulated in sustained-release dosage form (Selim et al., 2002). In the present study, the film containing 1.5 g each of ST and KGM showed the optimum sustained-release profile at pH 7.4. Figure 4.1 shows the *in vitro* drug release pattern from the drug incorporated neat starch and ST-KGM blend films at 37 °C. The drug release after 10 min was 47.5% and 19.2% from the neat starch and the blended films, respectively. Thus, the initial burst release was significantly lower in the KGM blended starch films. After 6 h, the drug release from ST film was almost completed; however, only 61% of the drug was released from the ST-KGM blend film. After 10 h, the release was 92% from the ST-KGM film. The results indicated that drug release could be controlled and extended with incorporation of KGM in the starch film. The release of drug from the blend films apparently followed Higuchi's kinetic model, and the mechanism corresponded to anomalous transport with non-Fickian kinetics corresponding to coupled diffusion/polymer relaxation (Nair and Jyothi, 2013).

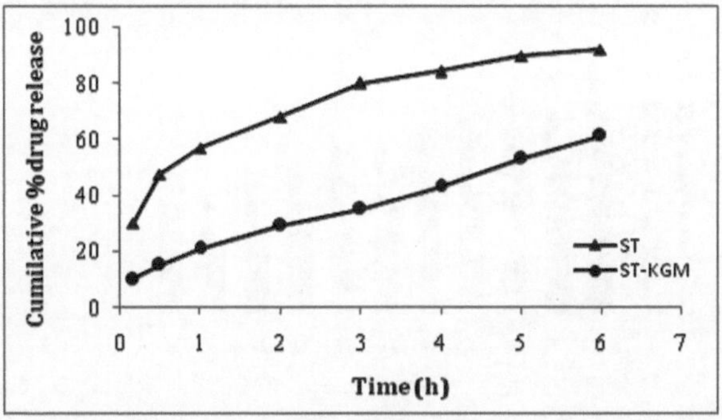

FIGURE 4.1 Drug release profile from neat starch and starch-KGM blend films.

4.3.2 EFFECT OF STARCH-BASED SAP ON SOIL MOISTURE RETENTION

Unlike superabsorbent polymers used in hygienic applications, which must possess fast rate of fluid absorption and ability to retain it under high load; the agricultural SAPs should have the ability to absorb water and to release it gradually as per the plant specific requirements. It was found that the moisture retention is significantly higher in the hydrogel amended soil, and it was higher in the soil samples treated with higher concentrations of the SAP (Figure 4.2 (a)). The soil moisture content over a 1-month period, in the control and SAP (0.5%) treated soils, is presented in Figure 4.2 (b). The amount of water available in the soil as well as the period of its availability are also important and were determined by the rate of evaporation from the soil. The evaporation of water from untreated soil was quick compared with that of the soil amended with the hydrogel.

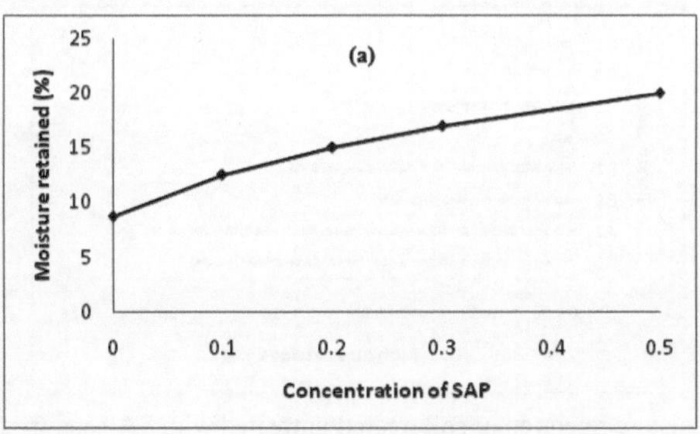

(Continued)

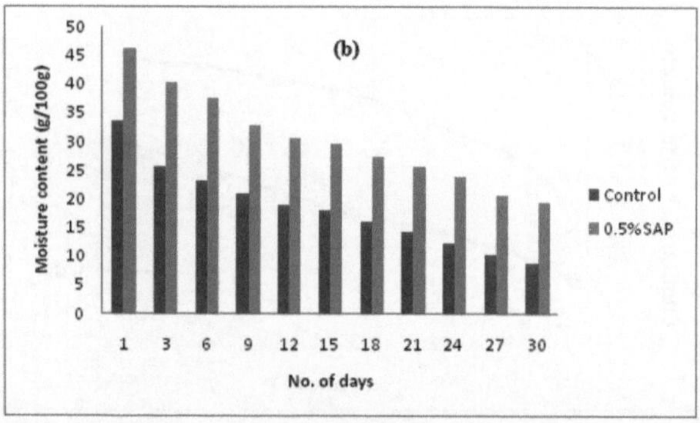

FIGURE 4.2 Soil moisture retention (a) after 30 days of watering in the control soil and treated soil amended with different concentrations of starch-based SAP and (b) in the control and treated soils during different time intervals.

The pot study with chilly plants in field conditions showed that the soil amended with the SAP retained higher amounts of moisture compared to the treatments without SAP (Figure 4.3). The average moisture content of the soil treated with SAP and watered once in 3 days was about 21%, whereas, that of the soil containing SAP and watered once in 6 days was about 16–17%. Moisture contents of the soil in treatments with SAP at watering intervals of 9 and 12 days were comparatively lower and found insufficient for the optimum growth of the plants. Figure 4.4 shows the effect of SAP on plant growth in various treatments. It was found that the plants grown in soil treated with SAP and watered once in 3 days showed higher plant growth compared to the corresponding treatments without SAP as well as other treatments with SAP (Figure 4.5). The plants in the treatments D and C dried up after the first 30 days and 50 days respectively.

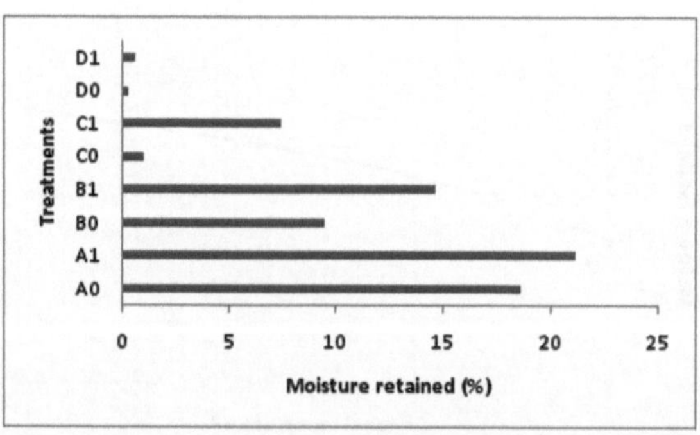

FIGURE 4.3 Moisture retention of soil treated/untreated with SAP and irrigated at different intervals. (A1, B1, C1 and D1: soil samples treated with SAP and at watering intervals of 3,

6, 9 and 12 days respectively; A0, B0, C0 and D0: soil samples without SAP and at watering intervals of 3, 6, 9 and 12 days respectively).

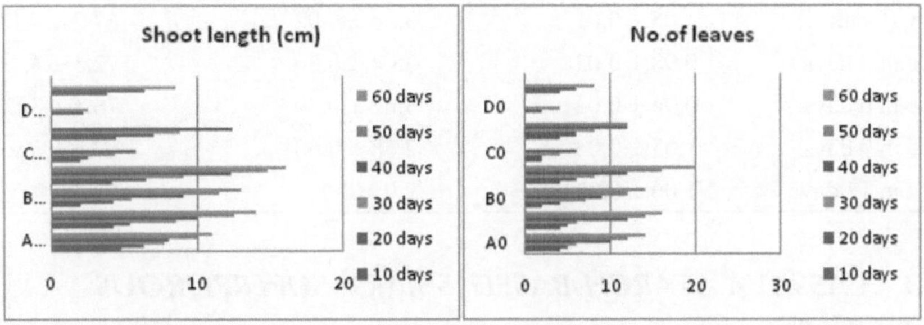

FIGURE 4.4 Effect of SAP on shoot length and number of leaves of plants (average temperature: 28 ± 2 °C).

FIGURE 4.5 Chilly plants in (a) SAP treatment with watering once in 6 days, (b) treatment with watering once in 3 days, and (c) control without SAP, with daily watering.

The addition of hydrogel in the soil resulted in a favorable alteration in the soil physical characteristics, such as porosity and bulk density (Table 4.1). The increased dimension in the swollen form of the hydrogel resulted in an increased porosity of the soil and increased availability of oxygen to the roots (Fannery and Busscher, 1982). The expansion and contraction of SAP in the soil during water absorption and evaporation helps to improve the air content in the soil (Buchholz and Graham, 1997). The study showed that the use of the starch-based SAP hydrogel amendment is useful for increased plant establishment, especially, in drought-prone areas.

TABLE 4.1 Physical properties of the control and superabsorbent polymer amended soils

Watering frequency	Bulk density ($Mg\ m^{-3}$)	Particle density ($Mg\ m^{-3}$)	Porosity (%)
Daily (Control)	1.05 ± 0.01	3.36 ± 0.02	67.2 ± 4.1
Once in 3 days	0.98 ± 0.02	3.04 ± 0.36	72.5 ± 4.3
Once in 6 days	0.94 ± 0.04	3.62 ± 0.30	76.1 ± 3.2
Once in 9 days	1.03 ± 0.03	2.68 ± 0.60	67.4 ± 3.9
Once in 12 days	1.06 ± 0.03	2.0 ± 0.5	52.4 ± 4.3

4.3.3 CASSAVA STARCH-BASED SMART SUPERPOROUS HYDROGELS

The porous superabsorbent polymers synthesized from cassava starch exhibited very fast water absorption properties and spongy texture (Figure 4.6). The water absorption capacity was found to be about 200 g/g and maximum absorbency was reached in 8–9 min. The fast swelling of SPHs in aqueous solution is due to the absorption of water by capillary forces through interconnected pores. The fast-swelling property is a desirable characteristic in the development of personal-care products such as baby diapers, feminine hygiene products, athletic garments, and so forth.

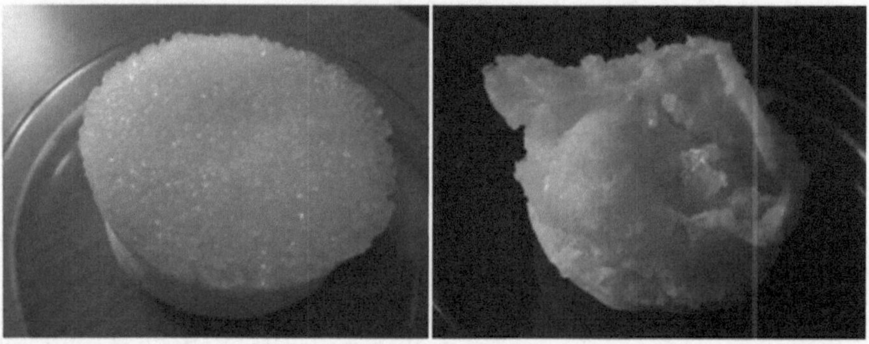

FIGURE 4.6 Cassava starch-based superporous hydrogels.

The present study showed that the swelling property of the polymers was highly sensitive to pH and presence of cations in the swelling medium (Figure 4.7) and can be regarded as smart hydrogels, which swell or shrink in response to the external stimuli such as temperature, pH, ionic strength, pressure, light, electric field and so forth (Gemeinhart et al., 2000). The absorbency of the hydrogel measured at different pH ranging from 3.0 to 11.0 showed that maximum absorbency was obtained at pH 7.0. At acidic and alkaline pH conditions there was a significant decrease in the swelling of these hydrogels. In the presence of sodium chloride in the swelling medium, there was a considerable reduction in the swelling capacity of these hydrogels (Figure 4.6). The increase in concentration of the cation in the medium was also found to have a significant lowering effect on the absorption properties.

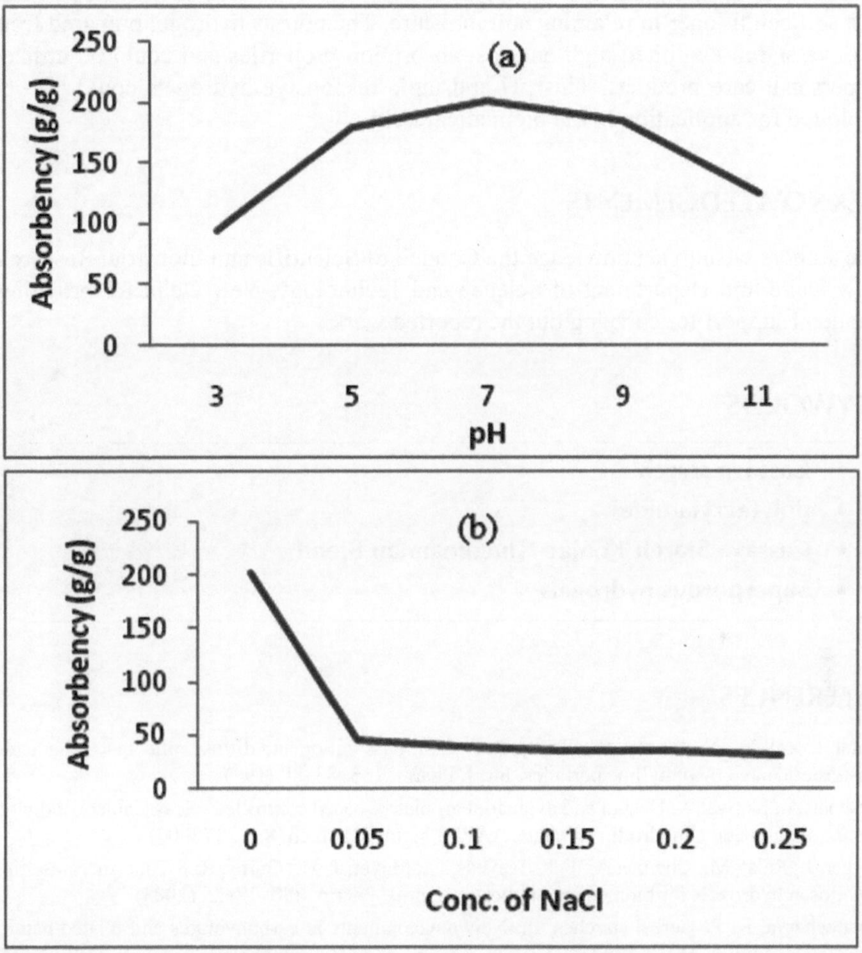

FIGURE 4.7 Sensitivity of superporous hydrogel to (a) different pH conditions and (b) ionic media.

4.4 CONCLUSIONS

Cassava starch-based blend films, and superabsorbent polymers were developed and their potential in specific applications was studied. Starch-konjac glucomannan blend film was found to be effective as a sustained-release matrix for the drug, theophyllin. Since both the ingredients are biodegradable, natural and edible, these blend films can be safely used for biomedical applications. Incorporation of starch in poly (acrylamide) by graft-copolymerization and subsequent saponification was used to develop a superabsorbent polymer, which was found to have been potential as a soil conditioner in retaining soil moisture. The porous hydrogel prepared from cassava starch exhibited high and fast-absorption properties and could be utilized in personal-care products. This pH and ionic responsive hydrogels could also be exploited for application in the biomedical field.

ACKNOWLEDGEMENTS

The authors wish to acknowledge the Council of Scientific and Industrial Research, New Delhi and Department of Science and Technology, New Delhi for providing financial support for carrying out the reported work.

KEYWORDS

- **Cassava starch**
- **poly(acrylamide)**
- **Cassava Starch-Konjac Glucomannan Blend**
- **Superporous hydrogels**

REFERENCES

Antal, I.; Zelk'o, R.; Roczey, N.; Plachy, J.; Racz, I. Dissolution and diffuse reflectance characteristics of coated theophylline particles. Int. J. Pharm. **155**: 83–89 (1997).

Avachat, A.; Kotwal, V. Design and evaluation of matrix-based controlled release tablets of diclofenac sodium and chondroitin sulphate. AAPS Pharm. Sci. Tech. **8**: E88 (2007).

Berger, J.; Reist, M.; Chenite, A.; Felt-Baeyens, O.; Mayer, J. M.; Gurny, R. Pseudo-thermosetting chitosan hydrogels for biomedical application. Int. J. Pharm. **288**: 17–25 (2005).

Bhattacharya, K. R. Cereal starches, In: New developments in carbohydrates and related natural products; Mulky, M. J.; Pandey, A.; Eds.; Oxford and IBH Publishing Co.: New Delhi, 1995, 39–54.

Bhuniya, S. P.; Rahman, S.; Satyanand, A. J.; Gharia, M. M.; Dave, A. M. Novel route to synthesis of allyl starch and biodegradable hydrogel by copolymerizing allyl-modified starch with methacrylic acid and acrylamide. J. Polym. Sci. Part A: Poly. Chem. **41(11)**: 1650–1658 (2003).

Buchholz, F. L.; Graham, A. T. Modern superabsorbent polymer technology, John Wiley & Sons Inc., New York, (1997).

Chen, J.; Liu, C.; Chen, Y.; Chen, Y.; Chang, P. R. Structural characterization and properties of Starch/Konjac Glucomannan blend films. Carbohydr. Polym. **74**: 946–952 (2008).

Chen, S. C.; Wu, Y. C.; Mi, F. L.; Lin, Y. H.; Yu, L. C.; Sung, H. W. A novel pH-sensitive hydrogel composed of N,O-carboxymethyl chitosan and alginate cross-linked by genipin for protein drug delivery. J. Control Release **96**: 285–300 (2004a).

Chen, S. L.; Zommorodi, M.; Fritz, E.; Wang, S.; Huttermann, A. Hydrogel modified uptake of salt ions and calcium in populus euphratica under saline conditions. Trees Struct. Funct. **18**: 175 (2004b).

Coviello, T.; Grassi, M.; Palleschi, A.; Bocchinfuso, G.; Coluzzi, G.; Banishoeib, F. A new scleroglucan/borax hydrogel: Swelling and drug release studies. Int. J. Pharm. **289**: 97–107 (2005).

Coviello,T.; Grassi, M.; Lapasin, R.; Marino,A.; Alhaique, F., Scleroglucan/borax: Characterization of a novel hydrogel system suitable for drug delivery. Biomaterials **24**: 2789–2798 (2003).

Dong-bao, H.; Xiao-ming, S.; Yi, S.; Dong-feng, Z. Gelation and swelling behavior of oxidized konjac glucomannan/chitosan hydrogel. Wuhan University J. of Nat. Sci. **7(4)**: 481–485 (2002).

El-Rehim, H. A.; Hegazy, E. S.; El-Mohdy, H. L. Radiation synthesis of hydrogels to enhance sandy soils water retention and increase plant performance. J. Appl. Polym. Sci. **93:** 1360–1371 (2004).

Fannery, R. L.; Busscher, W. J. Use of a synthetic polymer in potting soils to improve water holding capacity. Commn. Soil. Sci. Plant Anal. **13:** 103–111 (1982).

Gemeinhart, R. A.; Chen, J.; Park, H.; Park, K. pH-sensitivity of fast responsive superporous hydrogels. J. Biomater Sci. Polym. Edn. **11(12)**: 1371–1380 (2000).

Gomes, M. E.; Reis, R. L. Biodegradable polymers and composites in biomedical applications: from catgut to tissue engineering, Part 1 Available systems and their properties. Int. Mater. Rev. **49(5)**: 261–273 (2004).

Huttermann, A.; Orikiriza, L. J. B.; Agaba, H. Application of superabsorbent polymers for improving the ecological chemistry of degraded or polluted lands. Clean-Soil Air Water **37:** 517–526 (2009).

Jyothi, A. N.; Sajeev, M. S.; Sreekumar, J. Hydrothermal modifications of tropical tuber starches. 1. Effect of heat-moisture treatment on the physicochemical, rheological and gelatinization characteristics. Starch/Stärke **62**: 28–40 (2010).

Kotwala, D.; Raval, A.; Choubey, A.; Engineer, C.; Kotadia, H. Paclitaxel drug delivery from cardiovascular stent. Trends Biomater Artif. Organs **19:** 88–92 (2006).

Lee, J. W.; Park, J. H., Robinson, J. R. Bioadhesive-based dosage forms: the next generation. J. Pharm. Sci. **89**: 850–866 (2000).

Murthy, P. S. K.; Mohan, Y. M.; Sreeramulu, J.; Raju, K. M. Semi-IPNs of starch and poly(acrylamide-*co*-sodium methacrylate): Preparation, swelling and diffusion characteristics evaluation. React. Funct. Polym. **66:** 1482–1493 (2006).

Nair, S. B.; Jyothi, A. N. Cassava starch-konjac glucomannan biodegradable blend films: *In vitro* study as a matrix for controlled drug delivery. Starch/Stärke (In Press). Article first published online: 1 Nov 2012, doi: 10.1002/star.201200070 (2012).

Nair, S. B.; Jyothi, A. N.; Sajeev, M. S.; Mishra, R. Rheological, mechanical and moisture sorption characteristics of cassava starch-konjac glucomannan blend films. Starch/Starke **63**: 728–739 (2011).

Omidian, H.; Park, K.; Rocca, J. G. Recent developments in superporous hydrogels. J. Pharm. Pharmacol. **59**: 317–327 (2007).

Omidian, H.; Rocca, J. G.; Park, K. Advances in superporous hydrogels. J. Control. Release **102:** 3–12 (2005).

Panyam, J.; Labhasetwar, V. Biodegradable nanoparticles for drug and gene delivery to cells and tissue. Adv. Drug. Deliver Rev. **55**: 329–347 (2000).

Parvathy, P. C.; Jyothi, A. N. Synthesis, characterization and swelling behaviour of superabsorbent polymers from cassava starch-graft-poly(acrylamide). Starch/Stärke 64(3): 207–218 (2012).

Reis, A. V.; Guilherme, M. R.; Moia, T. A.; Mattoso, L. H. C.; Muniz, E. C.; Tambourgi, E. B. Route to synthesis of ally starch and biodegradable hydrogel by copolymerizing allyl-modified starch with methacrylic acid and acrylamide. J. Polym. Sci. Part A: Polym. Chem. **41**: 1650–1658 (2003).

Reis, A. V.; Guilherme, M. R.; Moia, T. A.; Mattoso, L. H. C.; Muniz, E. C.; Tambourgi, E. B. Synthesis and characterization of a starch-modified hydrogel as potential carrier for drug delivery system. J. Polym. Sci. Part A: Polym. Chem. **46**: 2567–2574 (2008).

Sajeev, M.S.; Moorthy, S.N.; Kailappan, R.; Rani, V.S. Gelatinization characteristics of cassava starch settled in the presence of different chemicals. Starch/ Stärke **55:** 213–221 (2003).

Selim, R. M. D.; Quadir, M. A.; Haider, S. S. Development of theophylline sustained release dosage form based on Kollidon SR. Pak. J. Pharm. Sci. **15:** 63–70 (2002).

Uz, B.Y.; Ersahin, S.; Demiray, E.; Ertas. A. Analyzing the soil texture effect on promoting water holding capacity by polyacrylamide, In: International Meeting on Soil Fertility Land Management and Agroclimatology. Turkey. 209–215 (2008).

Yu, Z.; Schwartz, J. B.; Sugita, E.T. Theophylline controlled release formulations: *In vivo–in vitro* correlations. Biopharm. Drug Dispos. **17**: 259–272 (1996).

Zhang, Y. Q.; Xie, B. J.; Gan, X. Advance in the applications of konjac glucomannan and its derivatives. Carbohydr. Polym. **60**: 27–31 (2005).

Zhang,W. F.; Chen,X. G.; Li, P. W.; He, Q. Z.; Zhou, H. Y. Preparation and characterization of theophylline loaded chitosan/b-cyclodextrin microspheres. J. Mater Sci – Mater. Med. **19**:305–310 (2008).

Zohuriaan-Mehr, M. J.; Kabiri, K. Superabsorbent polymer materials: A review. Iran. Polym. J. **17(6):** 451–477 (2008).

Zonghua, L.; Yanpeng, J.; Yifei, W.; Changren, Z.; Ziyong, Z. Polysaccharides-based nanoparticles as drug delivery systems. Adv. Drug Deliver Rev. **60:** 1650–1662 (2008).

CHAPTER 5

SEAWEED POLYSACCHARIDES-BASED NEW HYDROGEL MATERIALS: A GREEN APPROACH

JAI PRAKASH CHAUDHARY and RAMAVATAR MEENA*

Scale-Up and Process Engineering Unit, CSIR-Central Salt and Marine Chemical Research Institute, G.B.Marg, Bhavnagar-364002, Gujarat, India

*Corresponding author: Tel: +91 278 2567760; Fax: +91 278 2567562; E-mail: rmeena@csmcri.org

CONTENTS

ABSTRACT

Seaweed polysaccharides-based hydrogels constitute a cluster of potential materials, used in several biomedical areas, and are still developing for new promising applications. The seaweed polysaccharides namely agar, carrageenan, agarose, and alginate are the attractive materials for food, textile, pharmaceutical, and biomedical applications. These are obtainable only from the seaweed resource, except for alginates which can also be produced from bacteria. Agar and carrageenans are obtained from the red seaweeds (Rhodophyta), while the alginates are isolated from brown seaweeds (Phaeophyta). Recently, several new hydrogel materials with new functionality, pH-sensitive hydrogel networks, and thermally stable hydrogel network consisting of seaweed polysaccharides and their blends/composites were synthesized in our laboratory. Several techniques, such as grafting, blending/composites, and cross linking were developed for the preparation of new hydrogel materials. These hydrogels were investigated using dynamic oscillatory measurements (Rheology), thermal (DSC, TGA), syneresis measurement in gel samples, moisture content, morphological (SEM), XRD, swelling measurements, bulk density, true density, pore volume, and porosity on the dried samples. These hydrogel materials exhibited considerable thermal stability, pH-responsiveness in aqueous media, and showing super absorbency in different pH solutions. All these new properties predispose these modified products to potential applications in various domains, including ingestible and non-ingestible ones. Hydrogel materials with balanced hydrophobicity/hydrophilicity, can offer desirable release rates, swelling, and dissolution profiles.

5.1 INTRODUCTION

Biopolymers are most abundant in the higher orders of land plants and in seaweeds, and they play diverse roles in the all leaving organisms. They constitute approximately 75% of the dry weight. Polysaccharides serve as structural material contributing largely to the mechanical strength of the tissues in plants. They play an additional role in the seaweeds by preventing them from desiccation as they get exposed to the sun in intertidal regions during low tide conditions. Biopolymers, including seaweed polysaccharides/galactans-based hydrogels are in increasing demand in the biomedical and pharmaceutical areas due to their biodegradability and biocompatibility (Ratner and Hoffman, 1976; Ratner, 1981; Meena et al., 2010; Prajapati et al., 2013). The words gels and hydrogels are usually used interchangeably by group of scientists to describe polymeric cross-linked network structures. Hydrogels are defined as a considerably weaken cross-linked system, and are classified mainly into weak or strong gels depending on their flow behavior (Ferry, 1980). Phillips and Williams have mentioned that edible gels are used widely in the food sectors and mostly refer to gelling polysaccharides (i.e. hydrocolloids) (Phillips and Williams, 2000). Hydrogels are three-dimensional network structures obtained from a class of synthetic and/or natural polymers, which can absorb and retain a significant amount of water (Rosiak and Yoshii, 1999; Rodrigues et al., 2014). The hydrogel structure is formed by the hydrophilic groups or domains present in a polysaccharide network upon the hydration in water or in an aqueous medium (Morris et al., 1980; Iain, 1989).

Agar, agarose, alginate and carrageenan (Figure 5.1a–c) are used in numerous applications such as confectionery, bakery, dairy products, pastry, sauces and dressings, meat products, spreads, beverages, thickeners and to stabilize food systems (Meena et al., 2010). Seaweed polysaccharides namely agar, agarose, and carrageenan were extracted from the red seaweeds namely *Gelidiella acerosa, Gracilaria dura, Kappaphycus alvarezii* and alginate were extracted from the brown seaweed *sargassum whitii* occurring in Indian waters (Meena et al., 2010); oat spelt and beech wood xylans were extracted as described (Meena et al., 2011). The basic disaccharide repeating units of agar/agarose are (1→3) linked β-D-galactose and (1→4) linked α-L-3,6-anhydrogalactose, while *kappa*-carrageenan consists of alternating α-1,3-linked D-galactopyranose and β-1,4-linked 3,6-anhydro-D-galactopyranose. Alginates are linear anionic polysaccharides of (1, 4) linked α-guluronic (G units) and β-D-mannuronic acid (M units) residues (Meena et al., 2010).

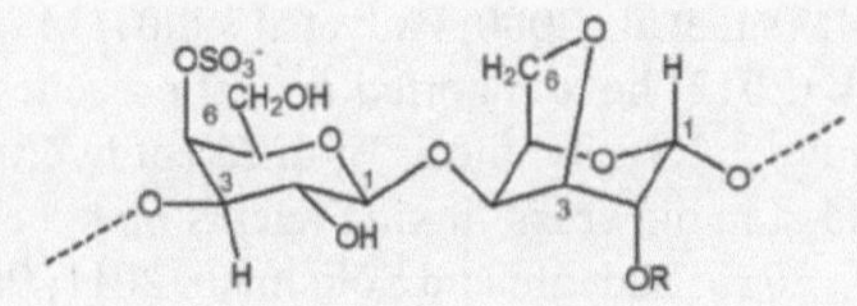

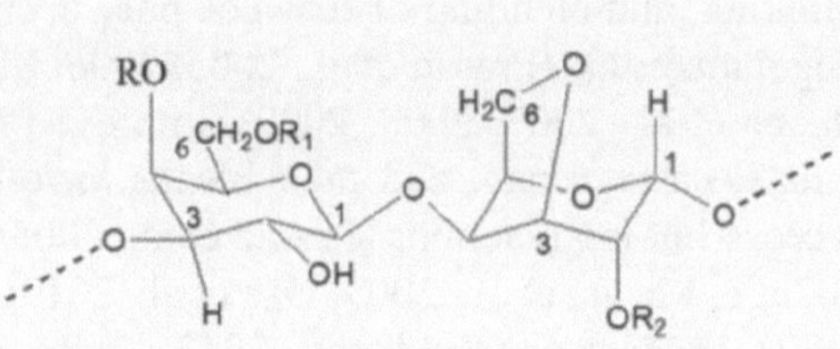

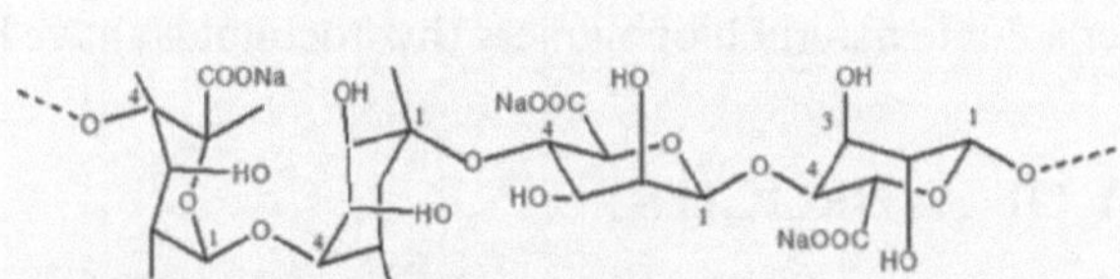

FIGURE 5.1 (a–c) Structure of *kappa*-carrageenan (a), agar/agarose (R = H or SO_3^-, R_1 = H or Me, R_2 = H or Me) (b), and alginate (c), used for hydrogel preparation.

The conjugates of alginate with cysteine and pyrrole have been reported in the literatures (Bernkop-Schnurch et al., 2001; Ionescu et al., 2005). Alginate-based modified hydrogel materials are usually made by cross-linking reaction with suitable cationic crosslinkers such as calcium chloride, barium chloride, strontium or poly-L-lysine (Martinsen, 1989; Draget et al., 1997). Numerous reports are offered in the literature on chemical cross-linking method to produce functional hydrogel materials with improved properties, using polysaccharide. The cross-linking reaction of konjac glucomannan with organic borates has been reported (Gao et al., 2008a, b). Glutaraldehyde, a chemical crosslinker, has been used for cross-linking of blends of collagen and algal sulphated polysaccharides (Figueiró et al., 2006). Although utilization of the chemical crosslinkers has been successful in yielding promising polysaccharide-based materials the chemical nature of the crosslinkers may not be good for targeted pharmaceutical applications.

Genipin is a water soluble naturally occurring cross-linking agent and widely used in herbal medicine. In contrast, it has also now emerged as an effective cross-linking agent for biopolymers (Akao et al., 1994; Mwale et al., 2005; Nickerson et al., 2006). Genipin participates in both short and extensive range covalent cross linking of ε-amino groups in amine-containing materials and amino acids or proteins (Touyama et al., 1994; Tsai et al., 2000; Paik et al., 2001; Mi et al., 2002; Butler et al., 2003). Uliniuc et al. (2012) have reported new approaches for the synthesis of hydrogel with or without catalyst. Synthesis of different hydrogels using γ-rays, graft copolymerization and suitable cross-linking agents have been reported in the literature (Pourjavadi et al., 2010; Sadeghi and Soleimani, 2011; Pasqui et al., 2012).

Recently, we have developed different functional modification methods, for example, grafting, cross linking, and blending of seaweed polysaccharides to prepare improved quality hydrogel materials (Prasad et al., 2005, 2006; Meena et al., 2006; Prasad et al., 2006; Meena et al., 2007, 2008, 2009). Improved hydrogel systems based on agarose, alginate, carrageenan, and their blends have been synthesized following grafting and cross-linking reactions (Prasad et al., 2005a, b; Prasad et al., 2006; Meena et al., 2007a, b; Meena et al., 2009; Oza et al., 2010; Oza, Chhatbar et al., 2011; Meena et al., 2011; Meena and Siddhanta, 2012). In this article, the preparation of new hydrogel materials from hydrophilic polymers of natural origin, with emphasis on water soluble natural biopolymers (hydrocolloids) have been reported.

5.2 HISTORY OF HYDROGELS

A variety of hydrogels have been used in numerous biomedical applications, for example, as material for contact lenses in the field of ophthalmology and as absorbable sutures in surgery, as well as in several new areas of medical practice to cure such disease as osteoporosis, diabetes mellitus, heart diseases, and neoplasms. Professors Lim and Wichterle of Prague, Czech Republic, synthesized the first hydrogel material based on synthetic poly-2-hydroxyethyl methacrylate in 1955. The synthetic poly-2-hydroxyethyl methacrylate hydrogel material was used in contact lens production (Wichterle and Lim, 1960; Nierzwicki and Prins, 1975; Hoffman, 2002; Swarbrick, 2006). The stability of hydrogel materials in varying pH solutions and temperature was the main advantage of such biomaterials. The preparation of calcium alginate microcapsules for cell engineering has been reported by Lim and Sun (1980s). Yannas' group has reported the modification of synthetic hydrogels with some natural substances, for example, collagen and shark cartilage to obtain the new hydrogel materials for newer applications (Iwona and Helena, 2010). At present, hydrogel materials continue to interest scientists. The safe and non-toxic biomaterials are produced by using the advance technology from the new materials. Hydrogel materials might be used in the advanced applications, for example, tissue engineering and regeneration, post-operative adhesion formation, drug-delivery systems, and in biosensors and cell transplantation (Sawhney and Pathak, 1994; Peppas and Bures, 2000; Miyata et al., 2002; Nguyen and West, 2002; Chang et al., 2010).

5.3 WHAT ARE HYDROGELS?

Biopolymers and biopolymers-based hydrogel materials are in increasing demand in the medical and pharmaceutical areas due to their biodegradability and biocompatibility. According to the medical and pharmaceutical encyclopedias, there is still no perfect definition of the term hydrogel. Hydrogel material's polymer hydrogels can be of either synthetic or natural origin, homopolymers or copolymers (Langer and Peppas, 2003; Iwona and Helena, 2010). Hydrogels are three-dimensional hydrophilic cross-linked networks, which are able to absorb and preserve water or biological solutions without dissolution several times over their weight (Qiu and Park, 2001; Truong and West, 2002).

Hoffman (2002) has reported that the hydrogel is a permanent or chemical gel stabilized by covalently cross-linked networks (Hoffman, 2002), and these hydrogels may be prepared by cross linking of water-soluble polymers. Cross-linked hydrogels are able to absorb a large amount of water without the polymer dissolving and showed similar characteristics to those of soft tissue (Truong and West, 2002; Langer and Peppas, 2003). The swelling or absorbing property of stimuli sensitive hydrogel materials depend on the surrounding environment and respond to the presence of stimuli (Qiu and Park, 2001).

There are two common classes of hydrogels—physical gels or pseudogels or reversible gels, which are linked by hydrogen bonds, electrostatic forces, hydrophobic interactions or chain entanglements (such gels are non-permanent, and usually they can be converted to polymer solutions by heating) and chemical or true or permanent hydrogels in which chains are linked covalently (Figure 5.2a). The interest of society and community towards environmental protection issues sensitized some producers to development of biodegradable super absorbents/hydrogels. The plausible gel formation mechanism of seaweed polysaccharides has also been described in Figure 5.2b–d (Morris et al., 1980; Iain, 1989).

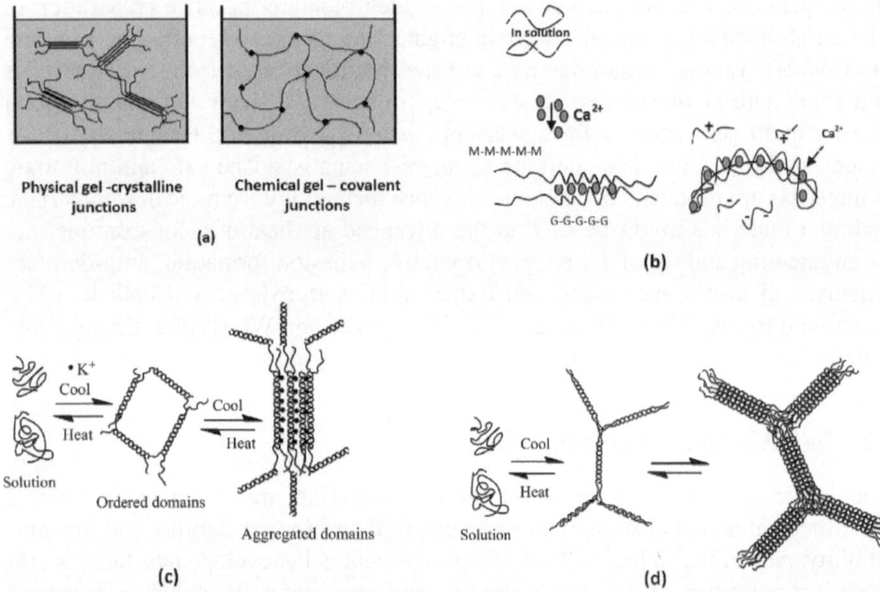

FIGURE 5.2 (a–d) Types of gels (a), Mechanism of gel formation through the interaction between calcium ions and sodium alginate (b), Carrageenan gel formation—Domain model (c), and Agar/agarose gel formation in water through hydrogen bonding (d).

5.4 GELLING SEAWEED POLYSACCHARIDES

Agar, agarose, alginate, and carrageenan are commercially important seaweed polysaccharides due to their excellent gel forming ability. Agar and agarose formed gel in water without adding any gelling agent. Carrageenan formed the strongest gel in the presence of KCl, while alginate formed the strongest gel in presence of $CaCl_2$. The phycocolloids/galactans (e.g. agar, agarose, alginate and *kappa*-carrageenan; Figures 5.1a–c) were extracted from the different seaweeds, namely *Gelidiella acerosa, Gracilaria dura, Sargassum spp.* and *Kappaphycus alvarezii* occurring in Indian waters. Agar, agarose, alginate, and *kappa*-carrageeenan were prepared using methods reported by us earlier (Ghosh et al., 2006; Meena, 2007; Prasad et al., 2007; Chhatbar et al., 2009). Genipin was purchased from M/s Challenge Bioproducts Co. Ltd., Taiwan. Iso-propanol (Laboratory Reagent grade) was procured from Ranbaxy Chemicals Ltd., Mohali (Punjab), India. Sodium carbonate, hydrochloric acid, and *o*-tolidine (AR grade) were purchased from SD Fine Chemicals Ltd, Mumbai (India). Potassium persulphate, methyl methacrylate, isopropanol, toluene used were of analytical grade and were purchased from Sigma-Aldrich, Mumbai and Ranbaxy Chemicals, Mumbai respectively.

5.5 PREPARATION OF NEW HYDROGEL MATERIALS

5.5.1 CROSS-LINKED HYDROGEL MATERIALS

Cross-linked hydrogels of seaweed polysaccharide/s (SP/s), namely agar, agarose, carrageenan, and their blend with naturally-occurring cross-linker genipin were prepared under ambient conditions, using water as a solvent as described in our previous works (Meena et al., 2007a, b; Meena et al., 2009). In brief, SP or blend dissolved in boiling water. To the solutions of the SP or blend were added different aliquots of the genipin stock solution and were kept at room temperature for cross linking. The reaction mixture was poured in isopropanol (1:2 w/w) for 24 h. The cross-linked hydrogel product was then isolated and was dried. In this article, based on these backgrounds and view points, the preparation of cross-linked hydrogels based on carrageenan (Figure 5.1a), agar (Figure 5.1b) and a blend (Agar/kC) within naturally-occurring in cross-linker genipin is described.

FIGURE 5.3 (a–d) SEM images (a, b) and optical micrographs (c, d) of non-cross-linked and cross-linked Agar/kC blend.

The cross-linked Agar/kC blend hydrogels were blue in appearance and revealed the cross-linking reaction between genipin and blend (Figure 5.3a–d) (Meena et al., 2009). The cross-linked Agar/kC blend hydrogel with 0.8 wt% genipin showed greatest swelling ratio (%), 8600 ± 32.56, 9380 ± 22.15 and 7000 ± 23.87 in acidic, water and alkaline media, respectively (Meena et al., 2009). An important observation is the remarkable stability of the Genprin-fixed-galactans hydrogels, in acidic (pH 1.2) and alkali (pH 12.5), while in such solutions, the parent galactans get readily depolymerized and dispersed. The cross-linked blend exhibited greater thermal stability and low gel thinning (Figures 5.4 and 5.5). Cross-linked hydrogel materials might be utilized in specific food applications as well as in other applications, which demand pH-resistance, including a controlled release of molecules.

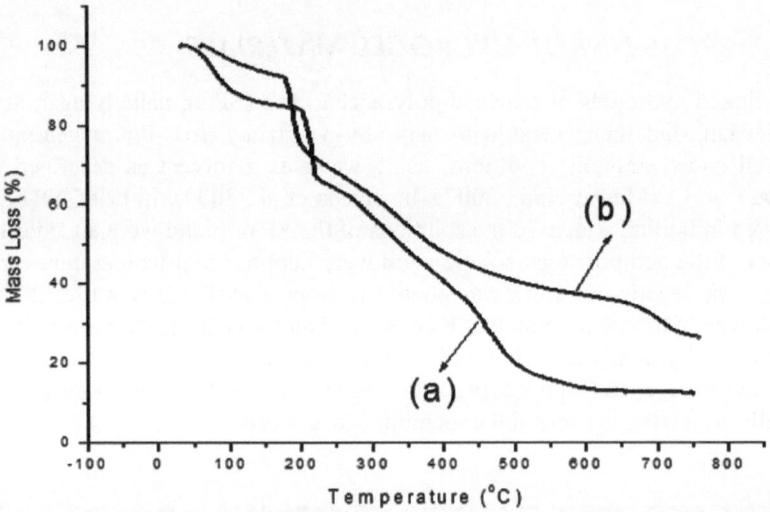

FIGURE 5.4 TGA graphs (a) non-modified Agar/kC blend and (b) cross-linked Agar/kC blend.

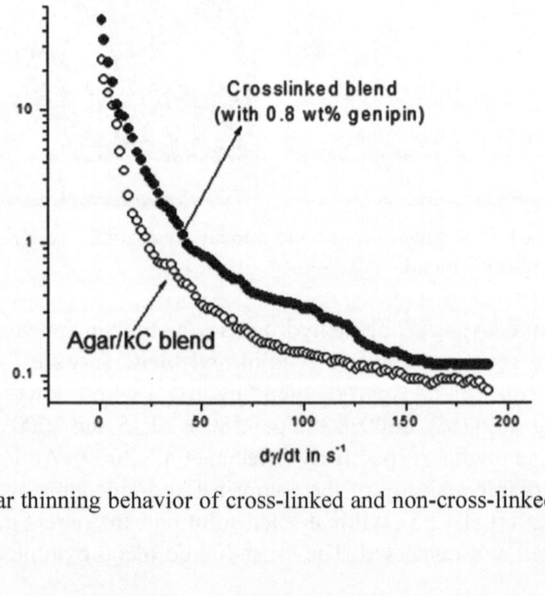

FIGURE 5.5 Shear thinning behavior of cross-linked and non-cross-linked Agar/kC blend (at 45 °C).

5.5.2 FLUORESCENT HYDROGEL MATERIALS

Under functional modification of seaweed polysaccharides, a simple microwave-induced method was developed for synthesizing water-soluble fluorescent derivatives of water insoluble alginic acid (ALG (Figure 5.6) with four different diamines, namely hydrazine (HY), ethylenediamine (EDA), 1,6-hexanediamine (HDA), and 1,4-cyclohexanediamine (CHDA), followed by a cross-linking reaction with a genipin, (see Figure 5.6 for chemical structures) (Chhatbar et al., 2011). The ethylenediamine derivative of alginic acid (ALG-EDA) exhibited superior fluorescent activity without cross linking, which upon cross linking was enhanced threefold. ALG-HY, ALG-HDA, and ALG-CHDA, were not fluorescent, but their respective cross-linked products exhibited good fluorescent activity (Figures 5.7 and 5.8). These fluorescent polysaccharide products are of potential utility in the domain of sensor applications. Mechanism for preparation of fluorescent derivatives has been given in Figure 5.9.

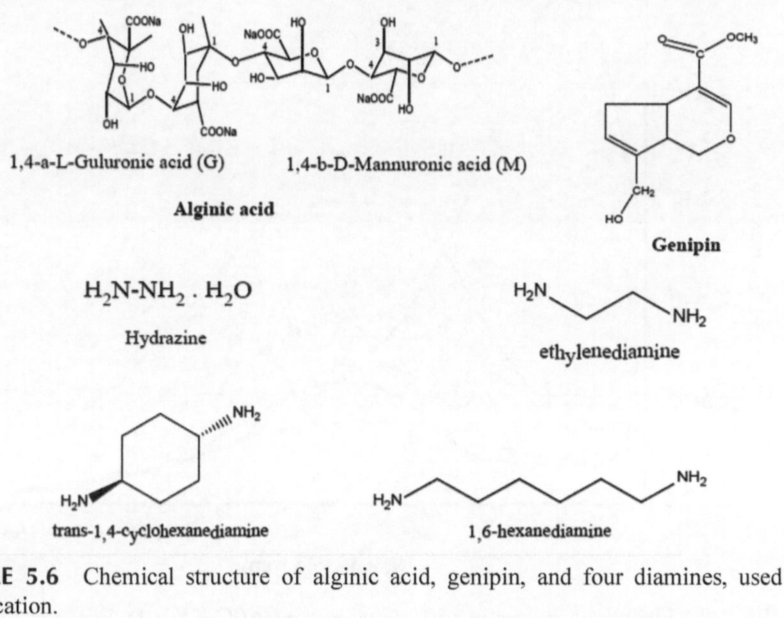

FIGURE 5.6 Chemical structure of alginic acid, genipin, and four diamines, used for modification.

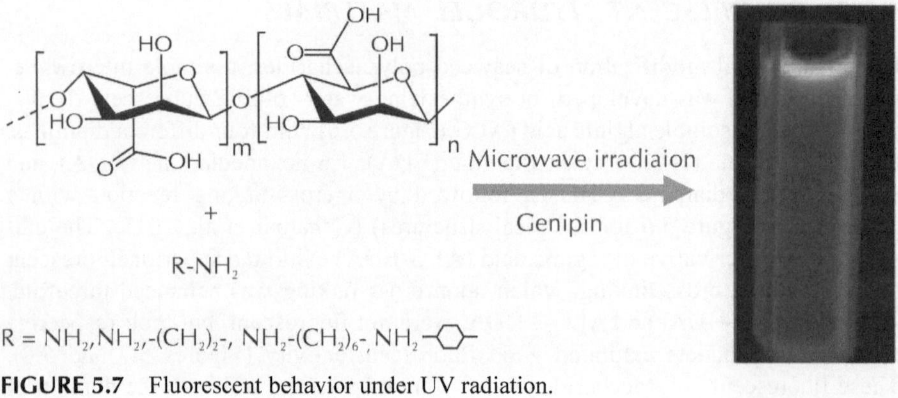

FIGURE 5.7 Fluorescent behavior under UV radiation.

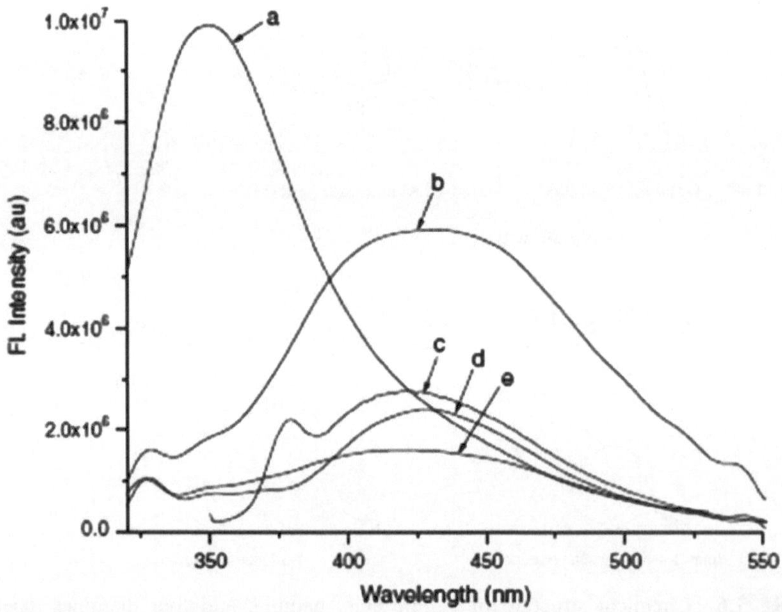

FIGURE 5.8 Fluorescence spectra of (I) cross-linked ALG-HY, (II) cross-linked ALG-EDA, (III) ALG-EDA, (IV) cross-linked ALG-HAD and (V) cross-linked ALG-CHDA at 4 × 10^{-4} M concentration.

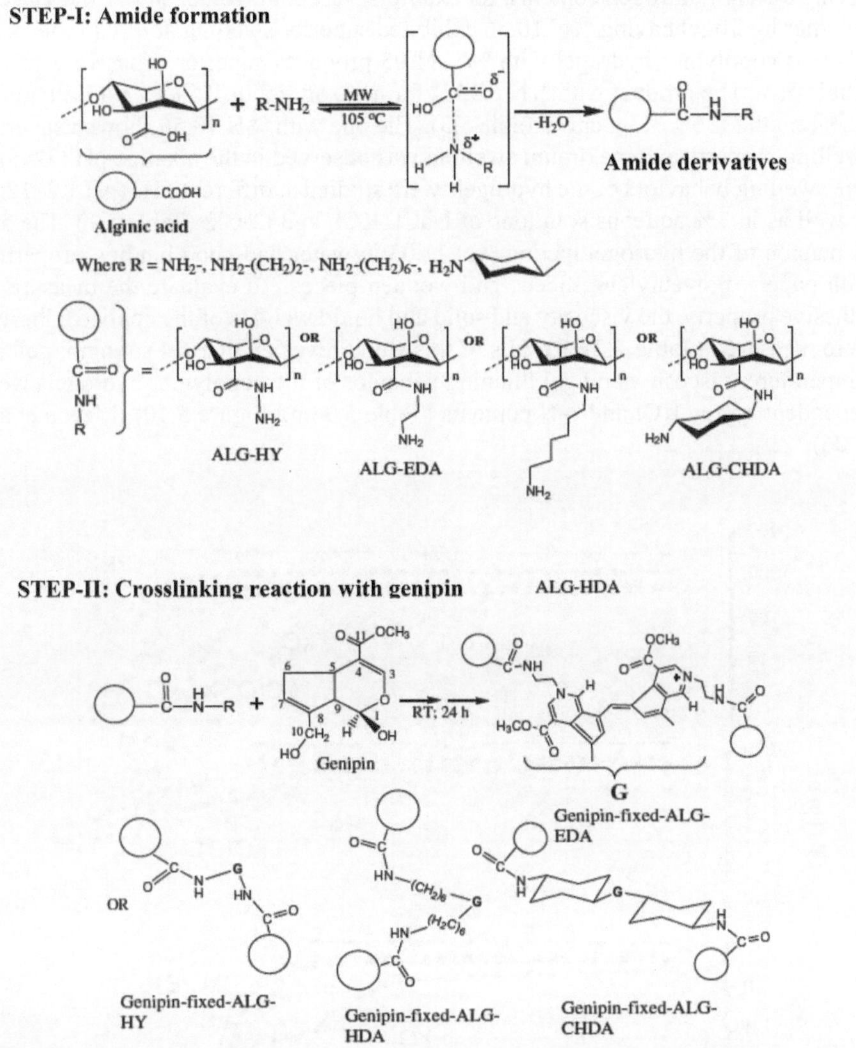

FIGURE 5.9 Mechanism for the formation of fluorescent derivatives and cross-linked products.

5.5.3 COPOLYMER HYDROGEL MATERIALS

The copolymer hydrogel of κ-carrageenan (kC) and acrylamide (AAm), has been synthesized in an aqueous medium in the presence of the water-soluble initiator KPS, under microwave irradiation conditions. The reaction conditions were optimized by varying the concentration of AAm and KPS to obtain copolymer hydrogels having different nitrogen contents, for example, %N 6.35, 10.56, and 11.05. The copolymer hydrogel having %N 10.56 exhibited superior swelling ability (Table 5.1), whereas copolymer hydrogel with %N 11.05 produces superior adhesive capacity (Table 5.2). The product with %N 6.35 is formed soft gel in 1% KCl exhibiting low gel strength (135 ± 3.3 g cm^{-2}; Table 5.3). The one with %N 10.56 showed superior swelling property and maximum swelling was observed in the alkaline pH (22 g/g). The swelling behaviors of the hydrogels were studied at different pHs (pH 1.2–12.5) as well as in 1% aqueous solutions of NaCl, KCl, and CaCl2 (Table 5.1). The 5% dispersion of the hydrogel having %N 11.05 in water had good binding properties with papers, polyethylene sheets, and wooden pieces. To evaluate the measure of adhesive property, the viscosity and solid and liquid weights of the applied adhesive were measured (Table 5.2). Results of this study revealed that gel strength, gelling temperature, viscosity, and gel thinning behavior of the copolymer hydrogels were dependent on the KCl and %N contents (Table 5.3 and Figure 5.10) (Meena et al., 2006).

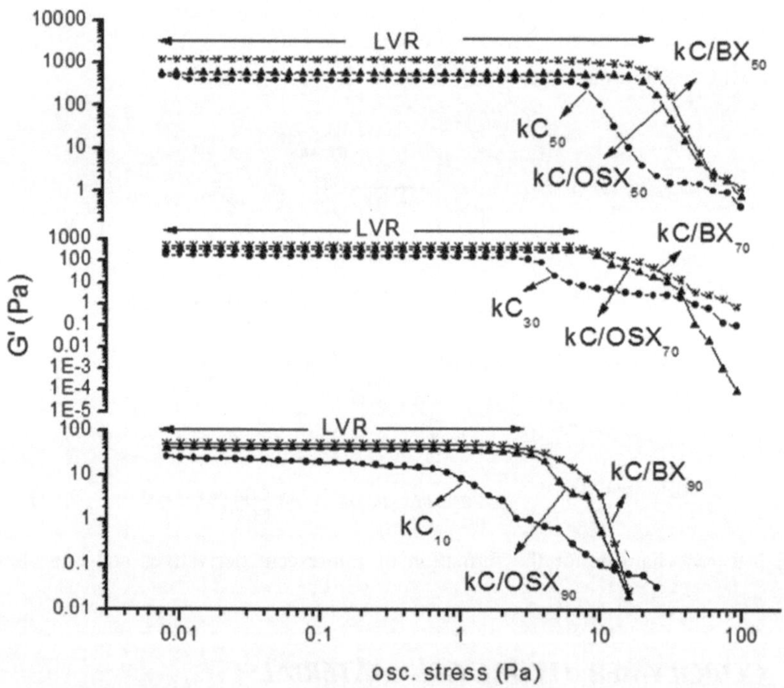

FIGURE 5.10 Shear thinning behavior of the copolymers with %N 6.35 and %N 11.05 in absence and presence of KCl.

TABLE 5.1 Swelling behavior of modified hydrogel materials

Nature of solution	Swelling (g/g)			
	kC (%N 0.21)	kC-*graft*-PAAm copolymer hydrogels with		
		%N 6.35	%N 10.56	%N 11.05
Water	N.S.	8.5 ± 0.05	14.5 ± 0.07	12.7 ± 0.06
Alkaline (12.5)	N.S.	12.6 ± 0.06	22.0 ± 0.06	15.4 ± 0.07
Acidic (1.2)	N.S.	8.0 ± 0.05	14.0 ± 0.05	11.0 ± 0.06
1% NaCl	N.S.	8.0 ± 0.07	12.6 ± 0.07	10.3 ± 0.05
1% KCl	N.S.	9.0 ± 0.06	13.5 ± 0.06	12.8 ± 0.07
1% $CaCl_2$	N.S.	8.5 ± 0.07	12.5 ± 0.06	10.8 ± 0.06

N.S. = Control kC does not swell in all solutions

TABLE 5.2 Liquid and solid weights of modified carrageenan hydrogel (adhesive) materials (with %N 11.05), using Fevicol as a reference adhesive

Specimens/materials	Liquid and solid weight of adhesives (in pound per thousand sq. feet of surface area)			
	kC-*graft*-PAAm (%N 11.05)		Fevicol	
	Liquid wt.	Solid wt.	Liquid wt.	Solid wt.
Paper	6.55 ± 0.05	1.82 ± 0.05	7.22 ± 0.06	2.56 ± 0.06
Plastic sheet	4.01 ± 0.04	2.1 ± 0.05	4.2 ± 0.05	1.6 ± 0.05
Wooden piece	16.9 ± 0.05	4.7 ± 0.06	18.6 ± 0.05	4.8 ± 0.05

TABLE 5.3 Physical parameters of modified carrageenan hydrogel materials

Sample name[a]	Apparent viscosity (cP, 80 °C)	Gel strength (g/cm^2)	Gelling temperature (°C)
kC (in 5% KCl)	1100 ± 7.4	>1400	71 ± 0.54
kC-*graft*-PAAm[b] gel (in 1% KCl)	65 ± 0.75	135 ± 3.3	28 ± 0.45
kC-*graft*-PAAm[c] gel (in 1% KCl)	69 ± 0.75	120 ± 5.4	22 ± 0.54
kC-*graft*-PAAm[d] gel (in 1% KCl)	74 ± 0.54	<100 ± 5.4	19 ± 0.54

[a]Measured in 5% sol/gel; [b] %N = 6.35; [c] %N = 10.56; [d] %N = 11.05

5.5.4 COMPOSITE HYDROGEL MATERIALS

Composite hydrogels of seaweed polysaccharide *viz. kappa*-carrageeenan (kC) with wood polysaccharides, for example, oat spelt (OSX) and beech (BX) xylan, were prepared. Swelling capacity was increased drastically, after mixing with xylans (Meena et al., 2011). Gelling and melting temperatures of kC gels were increased in kC/$BX_{50–90}$, while decreased in kC/$OSX_{50–90}$ hydrogels (Meena et al., 2011). The rheology results indicated that kC_{10} (01%) solution behaves like gel materials when mixed with 90% (w/w) OSX or BX (Meena et al., 2011). This study presented novel results on the interactions between kC and xylan. Mixed kC/OSX or kC/BX gels were stronger and elastic than the parent kC gels (Figure 5.11). The SEM results revealed that OSX might be used for introducing porosity into the parent kC gels (Figure 5.12a–e). Interaction among kC and OSX is weaker than that between kC and BX. Results of this study suggest that we can achieve a desirable gel structure or texture by mixing seaweed polysaccharide, for example, kC with OSX or BX in adequate amounts (Meena et al., 2011). This study opened the new research area of these polymers with emerging new applications, obtainable from a renewable bioresource.

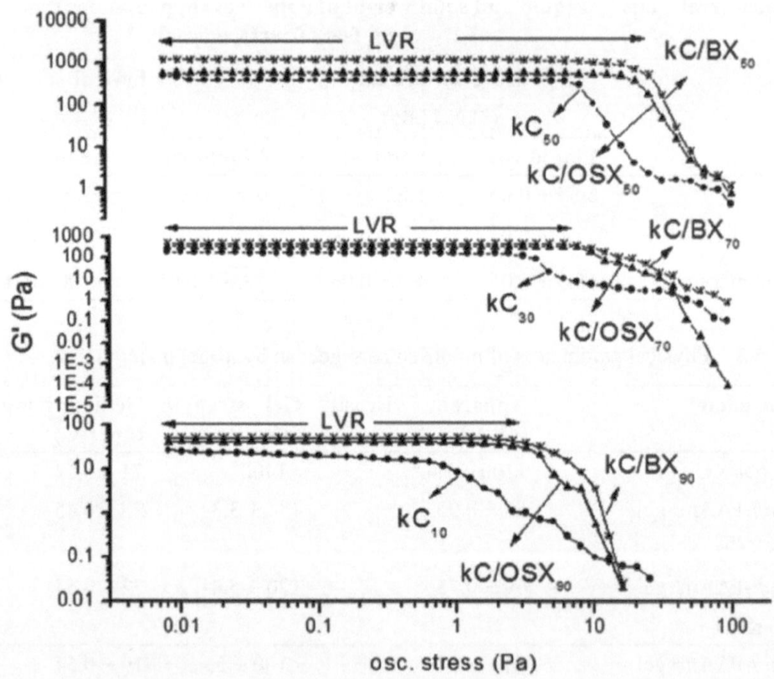

FIGURE 5.11 Stress dependence of storage modulus [50, 70, 90 represented composite hydrogels with 50, 70 and 90 wt% (w. r. to dry kC) OSX or BX, and kC_{10}, kC_{30}, and kC_{50} represented 0.1, 0.3 and 0.5% kC].

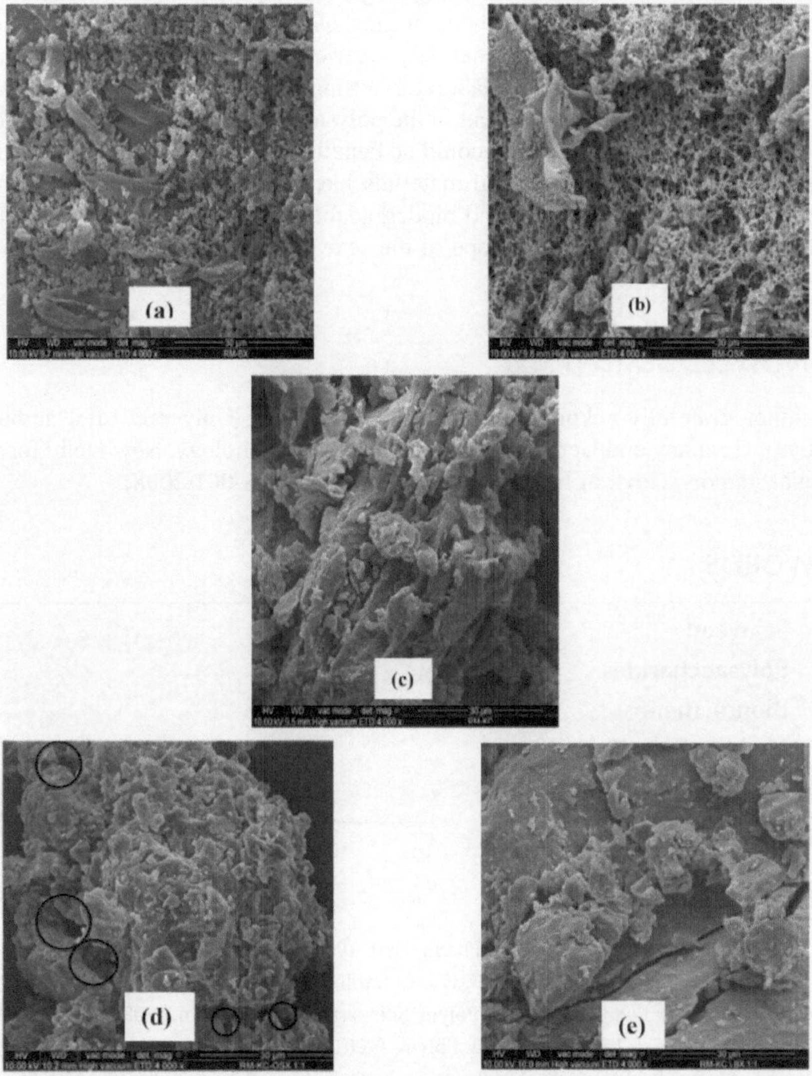

FIGURE 5.12 (a–e) SEM images of (a) BX, (b) OSX, (c) kC, (d) kC/BX$_{50}$, and (e) kC/OSX$_{50.}$ kC/OSXorBX$_{50}$ represented hydrogels with 50 wt% (w. r. to dry kC) OSX or BX.

5.6 CONCLUSIONS

Seaweeds polysaccharides used in this article are gelling in nature, and have been used for a wide variety of applications in food, pharma, biotech, and personal-care industries. This group of biopolymers is emerging as important substitutes for other polymers of plant, animal, and synthetic origins. Under a value addition program of Indian seaweeds and their polysaccharides, superior grades of agar, agarose, and κ-carrageenans were prepared in our laboratory, using new methods. It was found that different monosaccharide was present in the polysaccharides in various proportions and combinations, which were and could be beneficially modified or derivatized in a targeted fashion to prepare useful materials like gelling from non-gelling ones, absorbents from non-abosorbent and biodegradable films to suit many new applications. This study expands the scope of the seaweeds producing gelling polysaccharide.

ACKNOWLEDGEMENT

The author gratefully acknowledges Prof. Bodo Saake, University of Hamburg, Hamburg, Germany and Department of Science and Technology, New Delhi for the financial support (Govt. of India, Sanction No. SR/FT/CS-001/2008).

KEYWORDS

- **Seaweed**
- **Polysaccharides**
- **Biopolymers**
- **Hydrogels**
- **Functional modification**

REFERENCES

Akao, T.; Kobashi, K.; Aburada, M. Biol. Pharm. Bull. 1994, 17, 1573–1576.

Bernkop-Schnurch, A.; Kast, C. E.; Richter, F. J. Control. Release. 2001, 71, 277–285.

Butler, M. F.; Ng, Y.-F.; Pudney, P. D. A. J. Polym. Sci. Part A: Polym. Chem. 2003, 41, 3941–3953.

Chang, C.; Duan, B.; Cai, J.; Zhang, L. Eur. Polym. J. 2010, 46, 92–100.

Chhatbar, C.; Meena, R.; Prasad, K.; Siddhanta, A. K. Carbohydr. Polym. 2009, 76, 650–656.

Chhatbar, M. U.; Meena, R.; Prasad, K.; Chejara, D. R.; Siddhanta, A. K. Carbohydr. Res. 2011, 346, 527–533.

Draget, K. I.; Skjak-braek, G.; Smidsrød, O. Int. J. Biol. Macromol. 1997, 21, 47–55.

Ferry, J. D. Viscoelstic properties of polymers, John Wilet & sons, New York,1980, pp. 486–544.

Figueiró, S. D.; Macêdo, A. A. M.; Melo, M. R. S.; Freitas, A. L. P.; Moreira, R. A.; de Oliveira, R. S.; Góes, J. C.; Sombra, A. S. B. Biophy. Chem. 2006, 120, 154–159.

Gao, S.; Guo, J.; Wu, L.; Wang, S. Carbohydr. Polym. 2008a, 73, 498–505.

Gao, S.; Guo, J.; Nishinari, K. Carbohydr. Polym. 2008b, 72, 315–325.

Ghosh, P. K.; Siddhanta, A. K.; Prasad, K.; Meena, R.; Bhattacharya, A. US Patent No. 2006, 7,067,568.

Hoffman, A. Adv. Drug Deliv. Rev. 2002, 43, 3–12.

Iain, C. M. D. Pure &App/. Chern. 1989, 61, 1315–1322.

Ionescu, R. E.; Abu-Rabeah, K.; Cosnier, S.; Marks, R. S. Electrochem. Comm. 2005, 7, 1277–1282.

Iwona, G.; Helena, J. Chemistry & Chem. Technol. 2010, 4, 297–304.

Langer, R.; Peppas, N. Bioeng. Food & Nat. Prod. 2003, 49, 2990–3006.

Martinsen, A.; Skjak-braek, G.; Smidsrød, O. Biotechnol. Bioeng. 1989, 33, 79–89.

Meena, R.; Chhatbar, M.; Prasad, K.; Siddhanta, A. K. Polym. Int. 2008, 57, 329–336.

Meena, R., Chhatbar, M. U.; Prasad, K.; Siddhanta, A.K. Carbohydr. Polym. 2011a, 83, 1402–1407.

Meena, R.; Lehnen, R.; Schmitt, U.; Saake, B. Carbohydr Polym. 2011b, 85, 529–540.

Meena, R.; Prasad, K.; Mehta, G.; Siddhanta, A. K. J. Appl. Polym. Sci. 2006a, 102, 5144–5153.

Meena, R.; Prasad, K.; Siddhanta, A. K. Food Hydrocoll. 2006b, 20, 1206–1215.

Meena, R.; Prasad, K.; Siddhanta, A. K. Food Hydrocoll. 2009, 23, 497–509.

Meena, R.; Prasad, K.; Siddhanta, A. K. Int. J. Biol. Macromol. 2007a, 41, 94–101.

Meena, R.; Prasad, K.; Siddhanta, A. K. J. Appl. Polym. Sci. 2007b, 107, 290–296.

Meena, R.; Prasad, K.; Siddhanta, A.K. Int. J. Biol. Macromol. 2007c, 41, 94–101.

Meena, R.; Siddhanta, A. K.; Prasad, K.; Ramavat, B. K.; Eswaran; K.; Thiruppathi, S.; Ganesan, M.; A. Mantri, V.; Subba Rao, P. V. *Carbohydr. Polym.* 2007d, 69, 179–188.

Meena, R.; Siddhanta, A. K.; Prasad, K.; Chhatbar, M. U. Modification and characterization of some Indian seaweed polysaccharides. *In:* Proceeding: 11th European Workshop on Lignocellulosics and Pulp, August 16–19, 2010, Hamburg, Germany, 2010, pages 513–516.

Mi, F-L.; Sung, H-W.; Shyu S-S. Carbohydr. Polym. 2002, 48, 61–72.

Miyata, T.; Uragami, T.; Nakamae, K. Adv. Drug Deliv. Rev. 2002, 54, 79–98.

Morris, E. R.; Rees D. A.; Robinson G. R. J. Mol. Biol. 1980, 138, 349.

Mwale, F.; Iordanova, M.; Demers, C. N.; Steffen, T.; Roughley, P.; Antoniou, J. Tissue Eng. 2005, 11, 130–140.

Nguyen, K.; West, J. Biomaterials. 2002, 23, 4307–4314.

Nickerson, M. T.; Paulson, A. T.; Wagar, E.; Farnworth, R.; Hodge, S. M.; Rousseau, D. Food Hydrocoll. 2006, 29, 1072–1079.

Nierzwicki, W.; Prins, W. J. Appl. Polym. Sci. 1975, 19, 1885–1892.

Oza, M. D.; Meena, R.; Prasad, K.; Paul, P.; Siddhanta A. K. Carbohydr. Polym. 2010, 81, 878–884.

Oza, M. D.; Meena, R.; Siddhanta, A. K. Carbohydr. Polym. 2012, 87, 1971–1979.

Paik, Y.-S.; Lee, C.-M.; Cho, M.-H.; Hahn, T.-R. J. Agric. Food Chem. 2001, 49, 430–432.

Pasqui, D.; Cagna M. D.; Barbucci, R. Polymers. 2012, 4, 1517–1534.

Peppas, N.; Bures, P. Eur. J. Pharm. Biopharm. 2000, 50, 27–46.

Phillips, G. O.; Williams, P. A. Handbook of hydrocolloids. In *Starch*, (Eds., P. Murphy), Woodhead Publishing limited, Cambridge, England, 2000.

Prasad, K.; Trivedi, K.; Meena, R.; Siddhanta A. K. Polym. J. 2005a, 37, 1–7.

Prasad, K.; Siddhanta, A. K.; Rakshit, A. K.; Bhattacharya, A.; Ghosh, P. K. Int. J. Biol. Macromol. 2005b, 35, 135–144.

Prasad, K.; Meena, R.; Siddhanta, A. K. J. Appl. Polym. Sci. 2006a, 101, 161–166.

Prasad, K.; Mehta, G.; Meena, R.; Siddhanta, A. K. J. Appl. Polym. Sci. 2006b, 102, 3654–3663.

Prasad, K.; Siddhnata, A. K.; Ganesan, M.; Ramavat, B. K.; Jha, B.; Ghosh, P. K. Bioresource Technol. 2007, 98, 1907–1915.

Pourjavadi, A.; Soleyman, R.; Bardajee, G. R.; Seidi, F. Transactions C: Chem and Chemical Eng. 2010, 17, 15–23.

Qiu, Y.; Park, K. Adv. Drug Deliv. Rev. 2001, 53, 321–339.

Ratner, B. D. Biocompatibility of clinical implant materials; Williams, D.F.; Ed.; CRC Press: Boca Raton, FL, 1981, Chapter 7.

Ratner, B. D.; Hoffman, A. S. Hydrogels for medical and related applications; Andrade, J. D.; Ed.; American Chemical Society: Washington, DC, 1976; Vol. 31, pp 1–36.

Rodrigues, F. H.; Spagnol, C.; Pereira, A. G.; Martins, A. F.; Fajardo, A. R.; Rubira, A. F.; Muniz, E. C. J. Appl. Polym. Sci. 2014, 131(2).

Rosiak, J. M.; Yoshii, F. Nucl. Instrum. Meth. Phy. Res. B. 1999, 151, 56–64.

Sadeghi, M.; Soleimani, F. Int. J. Chem. Eng. Appl. 2011, 2, 304–306.

Sawhney, A.; Pathak, C. J. Biomed. Mat. Res. 1994, 28, 831–838.

Swarbrick, J. (Ed): Encyclopedia of Pharmaceutical Technology, 3rd edn. Informa Healthcare, New York 2006.

Touyama, R.; Takeda, Y.; Inoue, K.; Kawamura, I.; Yatsuzuka, M.; Ikumoto, T.; Shingu, T.; Yokoi, T.; Inouye, H. Chem. Pharm. Bull. 1994, 42, 668–673.

Tsai, C. C.; Huang, R. N.; Sung, H. W.; Liang, H. C. J. Biomed. Mater. Res. 2000, 52, 58–65.

Uliniuc, A.; Popa, M.; Hamaide, T.; Dobromir, M. Cellulose Chem. Technol. 2012, 46, 1–11.

Wichterle, O.; Lim, D. Nature 1960, 185, 117–118.

CHAPTER 6

EDIBLE NANOCOMPOSITE FILMS BASED ON HYDROXYPROPYL METHYL CELLULOSE REINFORCED WITH BACTERIAL CELLULOSE NANOCRYSTALS

JOHNSY GEORGE[1*], S. N. SABAPATHY[1], and SIDDARAMAIAH[2]

[1]Food Engineering and Packaging, Defence Food Research Laboratory, Siddarthanagar, Mysore 570 011, India; [2]Sri Jayachamarajendra College of Engineering, Mysore 570 006, India; *E-mail: g.johnsy@gmail.com

[2]Department of Polymer Science & Technology, Sri Jayachamarajendra College of Engineering, Mysore-570 006, Karnataka, India

CONTENTS

ABSTRACT

Edible food packaging films are gaining a lot of interest as an effort to reduce packaging waste created by non-degradable conventional plastics. Edible films are those films used for coating or wrapping various foods to extend their shelf life and may be eaten together with the food. Hydroxypropyl methylcellulose (HPMC) is widely used as an edible film; however, it is having moderate strength and high moisture sensitivity leading to relatively high permeability. The addition of cellulose nanocrystals as a reinforcing agent in HPMC is a viable method to improve these properties. Cellulose nanocrystals obtained from food-grade bacterial cellulose fibers were incorporated into HPMC at various concentrations and the resultant nanocomposite films were characterized for its properties. Atomic Force Microscopic (AFM) images confirmed the formation of rod like cellulose nanocrystals. The addition of highly crystalline cellulose nanocrystals also helped to reduce the moisture affinity of HPMC, which is desirable for edible packaging applications. The thermal properties of these nanocomposite edible films were also evaluated. Results of this study demonstrated that bacterial cellulose nanocrystals are capable of fabricating high-performance edible polymer nanocomposite films.

6.1 INTRODUCTION

Several biopolymeric materials were used as edible food packaging films, in order to maintain the quality of fresh produces like fruits and vegetables. These films will act as moisture barriers that prevent loss of moisture in fresh produces to a certain extent and thereby extend their shelf life (Kamper and Fennema, 1984). Among such edible films, cellulose and its derivatives are having great potential as they are biodegradable, non toxic, and easily available at low cost. HPMC is one such material that is widely used in food industry for various coating and packaging applications due to their excellent film forming ability (Tharanathan, 2003). HPMC is flexible, transparent, resistant to oils and fats, and also having sufficient gas barrier properties, which is beneficial for extending the shelf life of food products. HPMC has a unique property of undergoing thermal gelation at higher temperatures (Chen, 2007; Espinel et al., 2014) and it imparts almost negligible flavor to food (BeMiller and Whistler, 1996). However, the main limitation of HPMC is its poor resistance to water vapor and comparatively fewer mechanical properties (Krochta and Mulder-Johnston, 1997). One of the possible methods to overcome this challenge is the fabrication of nanocomposites using cellulose nanocrystals. Cellulose nanocrystals are stiff, rod-shaped nanoparticles, which are highly crystalline in nature. Cellulose nanocrystals synthesized from a bacterium *Gluconacetobacter xylinus* using acid hydrolysis can be used for edible applications (George and Siddaramaiah, 2012). These bacterial cellulose nanocrystals (BCNC) are capable of reinforcing polymeric materials due to their superior aspect ratio (Rusli et al., 2011). The present research investigation reports the fabrication of edible nanocomposites using HPMC and BCNC and their possible application as edible food packaging films. The effect of BCNC on the physico-mechanical, moisture sorption and thermal properties of these nanocomposites were also evaluated.

6.2 EXPERIMENTAL SECTION

HPMC powder with hydroxy propyl content of 8.2% was procured from M/s Loba Chemie, Mumbai, India. Bacterial cellulose (BC) was produced from the bacterium, *Gluconacetobacter xylinus* as previously reported (George et al., 2005). Bacterial cellulose fibrils were hydrolyzed using HCl (2.5 N) at boiling conditions for 3 h to obtain cellulosic nanocrystals. The obtained nanocrystals at different concentrations (1, 2, 3 and 4 wt%) were added to 10% HPMC solution. The nanocomposite was solution casted and dried at 37 °C for 24 h. AFM images were obtained using a Solver PRO-M scanning probe microscope from NT-MDT, Ireland. Mechanical properties of nanocomposite films were evaluated in tensile mode using Lloyd Instruments (Model: LRX Plus, UK), UTM at a strain rate of 100 mm/min. Moisture sorption analysis was carried out using a moisture sorption analyzer (Q 5000 SA, TA Instruments, USA) under controlled conditions of temperature and humidity. Differential Scanning Calorimetry (DSC) measurements were studied using DSC 2010, TA instruments, USA, while thermogravimetric analysis (TGA) studies were carried out using TGA Q500, TA Instruments, USA.

6.3 RESULTS AND DISCUSSIONS

Cellulose fibers obtained from bacteria was subjected to acid hydrolysis, which resulted in a preferential digestion of amorphous fractions leaving behind highly crystalline cellulose nanocrystals. AFM imaging was used to characterize these cellulose nanocrystals. Figure 6.1 shows the AFM image of cellulose nanocrystals, which revealed that acid hydrolysis resulted in the formation of nanometer sized, rod-shaped nanocrystals, which were further used for reinforcing HPMC matrix. These rod-shaped nanocrystals were found to have an average length of 100–300 nm and width of 10–30 nm. Edible HPMC films reinforced with BCNC at various concentrations were characterized for investigating their properties.

FIGURE 6.1 AFM images of bacterial cellulose nanocrystals obtained by HCl hydrolysis.

Figure 6.2 shows the mechanical behavior of pure HPMC and their BCNC nanocomposites. The addition of nanocrystals resulted in increasing the tensile strength and modulus of HPMC, while it reduced the percentage of elongation. The tensile strength of HPMC increased from 54.2 MPa to 73.8 MPa by the addition of 4 wt% of BCNC, while percentage elongation reduced from 54.2% to 36.8%. The results suggested that the nanocrystals were having a reinforcing effect on HPMC, and the effective load transfer between polymer chains and nanocrystals resulted in increasing the strength of the nanocomposites, which is in agreement with our previous reports (George et al., 2011). Figure 6.3 shows the moisture sorption plot of HPMC and its nanocomposites. From the plot, it is evident that the addition of cellulose nanocrystals resulted in decreasing the moisture sorption at higher relative humidities of 70% and above. The sorption isotherms of HPMC and its BCNC nanocomposites are shown in Figure 6.4. The curves show an increase in equilibrium moisture content with increasing relative humidity, at a constant temperature of 25 °C. At low and intermediate humidity ranges, moisture content increases linearly, whereas at higher humidity ranges, moisture content increases rapidly. The addition of nanocrystals reduced the equilibrium moisture content of HPMC at RH above 70%, possibly reducing the water binding capacity of HPMC. This could be due to the fact that BCNC is having better interaction with the hydrophilic sites of HPMC, and thereby reduces the interaction of the hydrophilic polymer with water. Moreover, the absorbed water, due to its interference with intermolecular hydrogen bonding of carbohydrates, induces relaxation of polymer chains. Cellulose nanocrystals having better interaction with the polymer chains tend to prevent the dislocation of polymer chains and subsequently reduce the adsorption of more water. Similar results were reported for HPMC reinforced with nanofibrils of plant cellulose (Bilbao-Sainz et al., 2011).

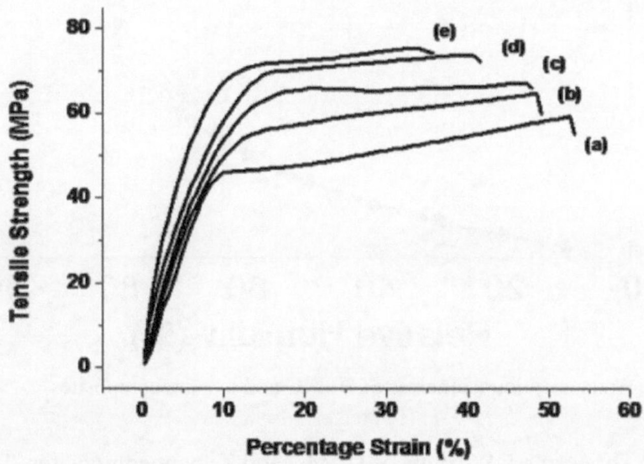

FIGURE 6.2 Stress strain curves of HPMC and its nanocomposites.

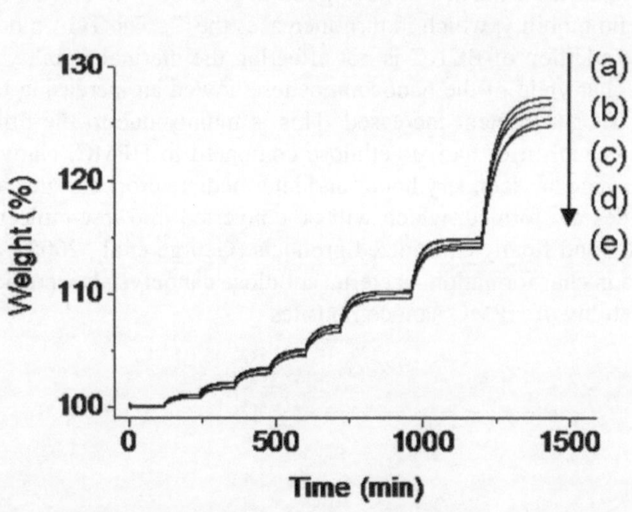

FIGURE 6.3 Weight percentage changes of HPMC and its nanocomposites associated with a stepwise increase in relative humidity.

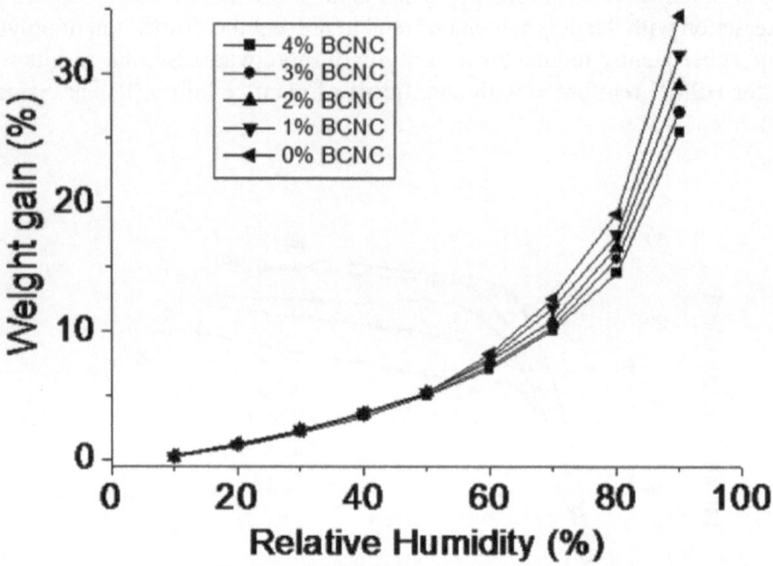

FIGURE 6.4 Water sorption isotherms of HPMC and its nanocomposites.

Figure 6.5 shows the DSC plots of HPMC and its nanocomposites. The addition of cellulose nanocrystals is found to increase the glass transition temperature. Cellulose nanocrystals are expected to have good interaction with HPMC and thereby restrict the chain mobility, which in turn increases the T_g. The TGA plots (Fig. 6.6) reveal that the addition of BCNC is not affecting the thermal stability of HPMC. However, the char yield of the nanocomposites showed an increasing trend as the percentage of BCNC content increased. This is mainly due to the differences in degradation mechanism of native cellulose compared to HPMC. Native cellulose undergoes cleavage of secondary bonds and intermediate products such as anhydromonosaccharides are formed, which will be converted into low-molecular-weight polysaccharides and finally carbonized products (George et al., 2008). Apart from the differences in char formation, bacterial cellulose nanocrystals were not affecting the thermal stability of HPMC nanocomposites.

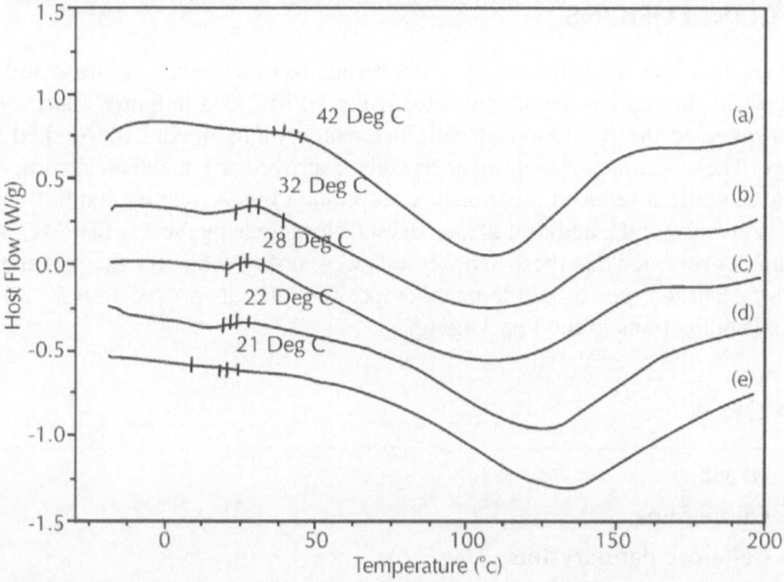

FIGURE 6.5 DSC plots of HPMC (a) and its BCNC nanocomposites (b) 1%, (c) 2%, (d) 3% & (e) 4%.

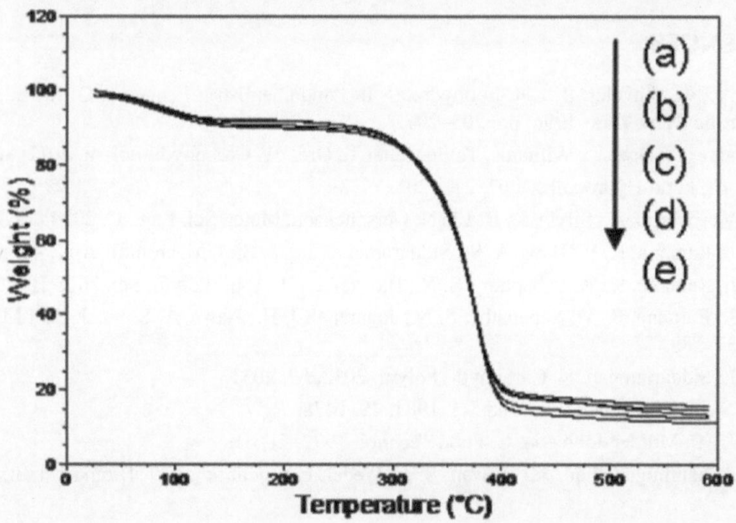

FIGURE 6.6 TGA plots of HPMC (a) and its BCNC nanocomposites (b) 1%, (c) 2%, (d) 3% & (e) 4%.

6.4 CONCLUSIONS

This work involves the synthesis of nanocrystals from bacterial cellulose and converting them into edible nanocomposites using HPMC as a polymer matrix. AFM images revealed the formation of cellulose nanocrystals having rod-shaped morphology. The addition of these nanocrystals improved the mechanical properties of HPMC, while it reduced the moisture sorption at higher relative humidity. DSC studies showed that the addition of nanocrystals is increasing the T_g of HPMC, while TGA studies revealed that these nanocrystals were not affecting the thermal stability of HPMC. These types of edible nanocomposite materials proved to have a lot of potential applications in food packaging.

KEYWORDS

- **HPMC**
- **Edible films**
- **Cellulose nanocrystals**
- **Sorption**
- **Bacterial cellulose**

REFERENCES

BeMiller, J. N.; Whistler, R. L. Carbohydrates. In Food chemistry; Fennema, O. R., Ed.; Marcel Dekker Inc: New York, 1996; pp. 205–207.

Bilbao-Sainz, C.; Bras, J.; Williams, T.; Senechal T., Orts, W. Carbohydr. Polym. 2011, 86, 1549.

Chen, H. H.; Food Hydrocoll., 2007, 21, 1201.

Espinel, V.; Ricardo, A.; Silvia, K. F.; Lía N. Gerschenson. Mater. Sci. Eng.: C, 2014, 36, 108–117.

George, J.; Ramana, K. V.; Bawa A. S.; Siddaramaiah. Int. J. Biol. Macromol. 2011, 48, 50.

George, J.; Ramana, K. V.; Sabapathy, S. N.; Bawa, A. S. J. Appl. Polym. Sci. 2008 108, 1845.

George, J.; Ramana, K. V.; Sabapathy, S. N.; Jagannath J. H.; Bawa, A. S. Int. J. Biol Macromol. 2005, 37, 189.

George, J.; Siddaramaiah, K. Carbohydr. Polym. 2012, 87, 2031.

Kamper, S. L.; Fennema, O. J. Food Sci. 1984, 49, 1478.

Krochta J. M.; Mulder-Johnston, C. Food Technol. 1997, 51, 61.

Rusli, R.; Shanmuganathan, K.; Rowan, S. J.; Weder, C.; Eichhorn, S. J. Biomacromol. 2011, 2, 1363.

Tharanathan, R. N. Trends Food Sci. Technol. 2003, 14, 71.

CHAPTER 7

SYNTHESIS AND CHARACTERIZATION OF PH SENSITIVE POLY(VINYL ACETATE-*CO*-METHACRYLIC ACID) MICROSPHERES FOR CONTROL RELEASE OF ACEBUTOLOL HYDROCHLORIDE

V. RAJINEE KANT[1*], S. RAVINDRA[1], K. CHOWDOJIRAO[2], S. G. ADOOR[3], F. ANTOINE MULABA-BAFUBIANDI[1]

[1]School of Mining, Metallurgy and Chemical Engineering, Doornfontein Campus, University of Johannesburg, South Africa

[2]Department of Polymer Science & Technology, SKU, Anantapur, India

[3]Department of Chemistry GSS College, Belgaum, India;
*E-mail: rajineekanth@gmail.com

CONTENTS

ABSTRACT

Novel poly (Vinyl acetate-co-Methacrylic acid) poly (VAc-co-MAA) copolymeric microspheres have been prepared by free radical emulsion polymerization using varying amounts of methacrylic acid, vinyl acetate (VAc) and N, N'-methylene bis acrylamide (NNMBA). These microspheres are crosslinked with N, N'-methylenebisacrylamide. Acebutolol hydrochloride an antihypertensive drug was loaded into these microspheres during in situ polymerization. The microspheres have been characterized by differential scanning calorimetry (DSC) and x-ray diffractometry (X-RD) to understand about the drug dispersion in microspheres and noticed that the drug dispersed in molecular level in microspheres. Scanning electron microscopy (SEM) was used to assess the surface morphology of particles prepared. In vitro release of Acebutolol hydrochloride has been studied in terms of copolymer composition, amount of cross-linking agent and amount of Acebutolol hydrochloride in the microspheres. Microspheres with different copolymer compositions have been prepared with yield up to ranging 80–85%. In vitro dissolution experiments performed in phosphate buffer pH 7.4 medium indicated the controlled release of Acebutolol hydrochloride from the microspheres up to 14 h.

7.1 INTRODUCTION

The unique and novel properties of environmentally sensitive polymers have begun to find several applications in new chemical and biological processes. Because these cross-linked polymers exhibit dramatic swelling and collapsing behavior, they have been proposed as devices to facilitate pH-specific membrane separations, promote solvent recovery, improve purification and recovery of pharmaceutical products from solution, serve as temperature-dependent thermal insulation systems, and act as carriers for controlled drug delivery devices. Much research has been done on polymer carriers capable of releasing an active agent at a constant rate over a long period of time (zero-order release). Swelling controlled release systems have been utilized for responsive release when certain environmental conditions, such as temperature, pH, ionic strength, or glucose level, are changed. Advantages of synthetic polymers, to derive new properties is by graft copolymeric microspheres for the oral delivery of drugs over the conventional dosage forms, have been reported (Babu et al., 2010; Döring et al., 2013; Elzoghby et al., 2011; Lam and Gambari, 2014; Ravindra et al., 2011).

Methacrylic acid polymers exhibit pH-sensitivity, swelling to high degrees in basic solutions, and collapsing in acidic solutions. This behavior is due to electrostatic repulsions between the carboxylic acids and ions present in the buffered solution. At high pH values, the pendant carboxylate side chains are repelled by the anions in the solution and expand to minimize charge concentration. Polymers of methacrylic acid have been noted (Klier and Scranton, 1990; Seno et al., 1991;

Weiss et al., 1991) to display sharp sensitivity to external pH, with capability of controlling solute permeation due to corresponding changes in swelling behavior. Osada et al. (1991) developed a chemo mechanical valve from poly-(ethylene glycol)-modified PMAA, which was capable of controlling solute penetration with pulses in electric current. Because the heat transfer properties of such swollen and collapsed gels are different, they can act as insulation materials (Chahroudi, 1990). Finally, such environmentally sensitive hydrogels have been shown to display unique on/off diffusional behavior which responds to changes in physiological conditions (Dong and Hoffman, 1991; Gutowska et al., 1992; Kim et al., 1994; Mukae et al., 1990; Yoshida et al., 1992a; Yoshida et al., 1992b).

Vinyl acetate (VAc) copolymer is a heat processable, flexible, and inexpensive material (Miyazaki et al., 1982; Sang-Chul et al., 2002). The safety and biocompatibility of EVA copolymer are reflected in its use as a biomaterial for artificial hearts and as an antithrombogenic material. The usefulness of EVA copolymer as a drug delivery system for pilocarpine, progesterone, hydrocortisone (Johnson, 1980), fluoride ion (Gennaro et al., 1976), 5-fluorouracil (Miyazaki et al., 1984), isosorbidedinitrat (Ocak and Agabeyoglu, 1999), nicardipine (Morimoto et al., 1988), and macromolecules, such as proteins, has been described. Acebutolol hydrochloride is an antihypertensive drug widely used in pharmaceuticals. Because of their innumerable applications these two polymers under study are used to prepare copolymeric microspheres and loaded with antihypertensive drug and studied for drug delivery application and the results are presented here.

7.2 MATERIAL AND METHODS

Sodium lauryl sulfate, Potassium per sulfate, Vinyl acetate, Methacrylic acid, and N, N'-methylene bis acrylamide—all chemicals purchased by S. D. Fine chemicals, Mumbai, India. Acebutolol hydrochloride was given as gift sample by Waksman Saleman Pharmaceuticals, Anantapur, India.

7.2.1 SPREPARATION OF POLY (VAC-CO-MMA) COPOLYMERIC MICROSPHERES FOR CONTROLLED RELEASE OF ACEBUTOLAL-HCL

Sodium lauryl sulfate (1 g) and sodium hydrogen phosphate (100 mg) were dissolved in 80 mL of water taken in a three-necked round-bottom flask equipped with a mechanical stirrer, a condenser, and a gas inlet to maintain the inert nitrogen atmosphere. The flask was immersed in an oil bath with a thermostatic control to maintain the desired temperature accurate to ±1 °C. The solution was stirred at 800 rpm speed until it became clear and 100 mg of potassium per sulfate was added. Required amount of monomers, cross-linking agent, NNMBA and Acebutolol hydrochloride were dissolved separately in 20 mL of water. This mixture was added to the reaction mixture drop wise using a dropping funnel and the reaction was continued for 8 h at 70 °C to obtain the maximum yield. The reaction mixture was taken out after 8 h and added to 1% calcium chloride solution drop wise to break the emulsion. Particles were then isolated by centrifuging the product at the rotor speed of 12,000 rpm, washed with water and dried under vacuum at 40 °C for 24 h. The blank microspheres without drug incorporation were prepared by the above method.

The development of copolymeric microspheres of poly (Vinyl Acetate-Co-Mehtacrylic acid) for in vitro release of Acebutolol hydrochloride is an antihypertensive drug. These copolymeric microspheres were crosslinked with *N,N*-methylene bis acrylamide have been prepared by the free radical emulsion polymerization using varying amounts of Vinyl Acetate (VAc), Mehtacrylic acid (MAc) and *N,N*-methylene bis acrylamide (NNMBA), which, for controlled release of Acebutolol hydrochloride were loaded into these microspheres during in situ polymerization.

7.2.2 DIFFERENTIAL SCANNING CALORIMETRY STUDIES

Differential scanning calorimetry studied DSC curves of the plain NFD, MC-g-AAm/G microspheres, and NFD-loaded MC-AAm/G microspheres were recorded using Rheometric Scientific differential scanning calorimeter (Model-DSC SP, UK). The analysis was performed by heating the samples at the rate of 10 °C/min under inert atmosphere.

7.2.3 X-RAY DIFFRACTION (X-RD) STUDIES

X-ray diffraction (X-RD) patterns of the plain 5-fluorouracil, placebo AAm-*co*-MMA core-shell microspheres and 5-fluorouracil-loaded core-shell microspheres were recorded using Rigaku Geigerflex diffractometer equipped with Ni filtered Cu K radiation (λ = 1.5418A°). Dried core-shell microspheres of uniform size were mounted on a sample holder and the patterns were recorded in the angle range of 2–65° at the speed of 5 °C/min.

7.2.4 SCANNING ELECTRON MICROSCOPIC STUDIES

SEM images of the microspheres were recorded using a QUANTA scanning electron microscope, IISC, Bangalore, at the required magnification. A working distance of 33.5 mm was maintained and the acceleration voltage used was 10 kV with the secondary electron image as a detector.

7.2.5 PARTICLE SIZE ANALYSIS

Particle size of the microspheres was measured using a particle size analyzer (Mastersizer, 2000, Malvern Instruments, UK). About 500 mg of the microspheres were transferred to the dry sample holder and stirred vigorously to avoid the agglomeration of particles during the measurements. For measurement of sizes of different formulations/batches, the sample holder was cleaned by vacuum. The particle size was also measured using an optical microscopy.

7.3 ESTIMATION OF DRUG LOADING AND ENCAPSULATION EFFICIENCY

Specific amount of dry microspheres were vigorously stirred in a beaker containing 10 mL of ethanol to extract the drug from the copolymeric microspheres. A 10 mL of phosphate buffer pH 7.4 containing 0.02% Tween-80 solution was added to the above solution and ethanol was evaporated with gentle heating and continuous shaking. The aqueous solution was then filtered and assayed using a UV spectrophotometer (Lab India, Mumbai, India) at the fixed λ_{max} value of 270 nm. The results of percentage drug loading and encapsulation efficiency were calculated using equations (1) and (2).

$$\%\ \text{Drug loading} = \left(\frac{\text{Weight of drug in microspheres}}{\text{Weight of microspheres}}\right)\times 100 \tag{7.1}$$

$$\%\ \text{Encapsulation efficiency} = \left(\frac{\text{Actual loading}}{\text{Theoretical loading}}\right)\times 100 \tag{7.2}$$

7.3.1 IN VITRO RELEASE

In vitro release studies have been carried out by dissolution experiments using the tablet dissolution tester (Lab India, Mumbai, India) equipped with eight baskets. Dissolution rates were measured at 37 °C under 100 rpm speed. Drug release from the microspheres was studied in an intestinal (7.4 pH phosphate buffer) fluid. At regular intervals of time, sample aliquots were withdrawn and analyzed by a UV spectrophotometer (Lab India, Mumbai, India) at the fixed λ_{max} value of 270 nm.

7.3.2 CONVERSION OF COPOLYMER

The yield of the copolymeric microspheres was determined gravimetrically. After copolymerization, the latex solution was added to 1% calcium chloride solution and centrifuged to isolate the particles from the mixture. The copolymeric microspheres were washed several times successively with water and methanol solvents to remove the remaining monomer and initiator, and then dried in a vacuum oven at 50 °C till attainment of constant weight. The conversion of monomers was calculated by the equation (3)

$$\text{Conversion} = (W/M) \times 100 \tag{7.3}$$

Where W is weight of the dry copolymer obtained from the latex sample and M is weight of the monomers taken. The yield of copolymeric microspheres varied between 80% and 85% for various formulations prepared in this study.

7.4 RESULTS AND DISCUSSION

7.4.1 DIFFERENTIAL SCANNING CALORIMETRY (DSC) STUDIES

DSC thermograms of plain Acebutolol hydrochloride (A), plain VAc-co-MAA microspheres (B) and drug load VAc-co-MAA microspheres (C) were recorded using Rheometric Scientific differential scanning calorimeter (Model-DSC SP, UK) and are shown in Figure 7.1(A, B & C). The analysis was performed by heating the samples at the rate of 10 °C/min under inert atmosphere. It is noticed from the Figure 7.1(A) that the drug melts at 152.8 cal, but this is not observed in Figure 7.1.C. This suggests that Acebutolol hydrochloride is molecularly dispersed in the polymer matrix.

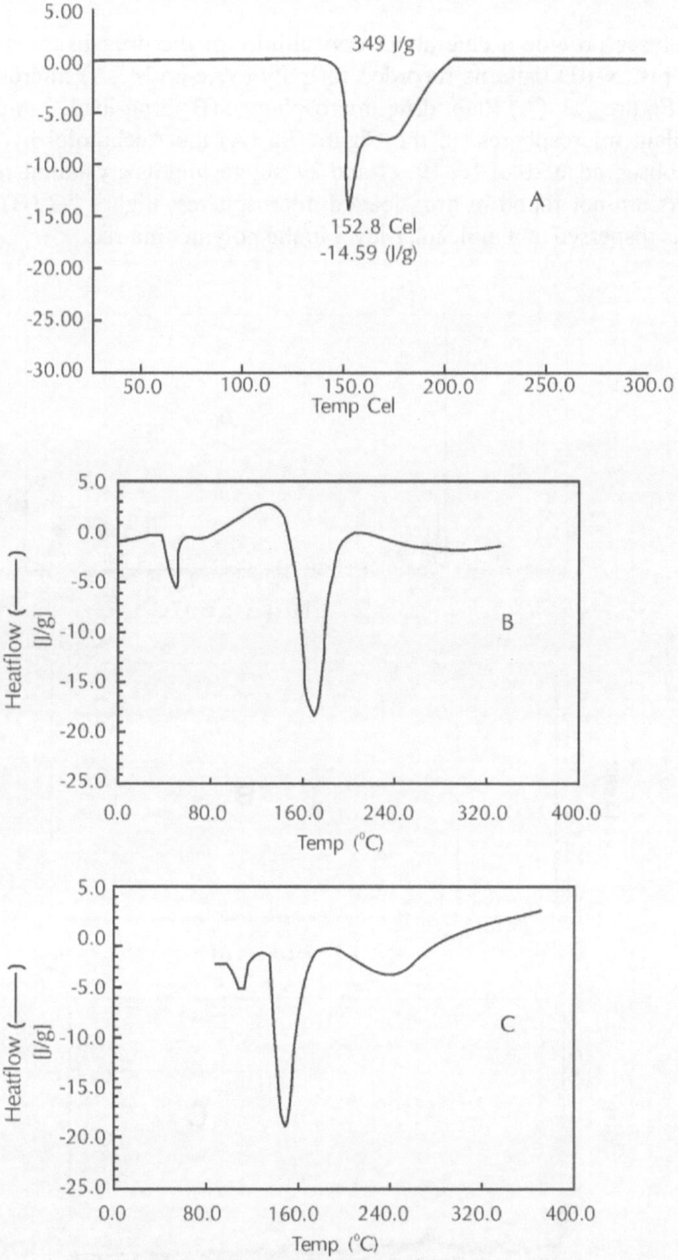

FIGURE 7.1 DSC thermograms of plain Acebutolol hydrochloride (A), plain Poly (VAc-co-MAA) Microspheres (B), drug-loaded Poly (VAc-co- MAA) microspheres(C).

7.4.2 X-RAY DIFFRACTION (X-RD) STUDIES

X-RD analyses provide a clue about crystallinity of the drug in the cross-linked microspheres. X-RD patterns recorded for poly (VAc-co-MAA) microspheres are shown in Figure 7.2. (A) Plain drug microspheres, (B) drug-loaded microspheres, and (C) plain microspheres. In the Figure 7.2 (A) the Acebutolol hydrochloride peaks are observed at 2θ of 16, 19, 21 and 29, suggesting its crystalline nature. But, these peaks are not found in drug-loaded microspheres, Figure 7.2 (B) indicating that drug is dispersed at a molecular level in the polymer matrix.

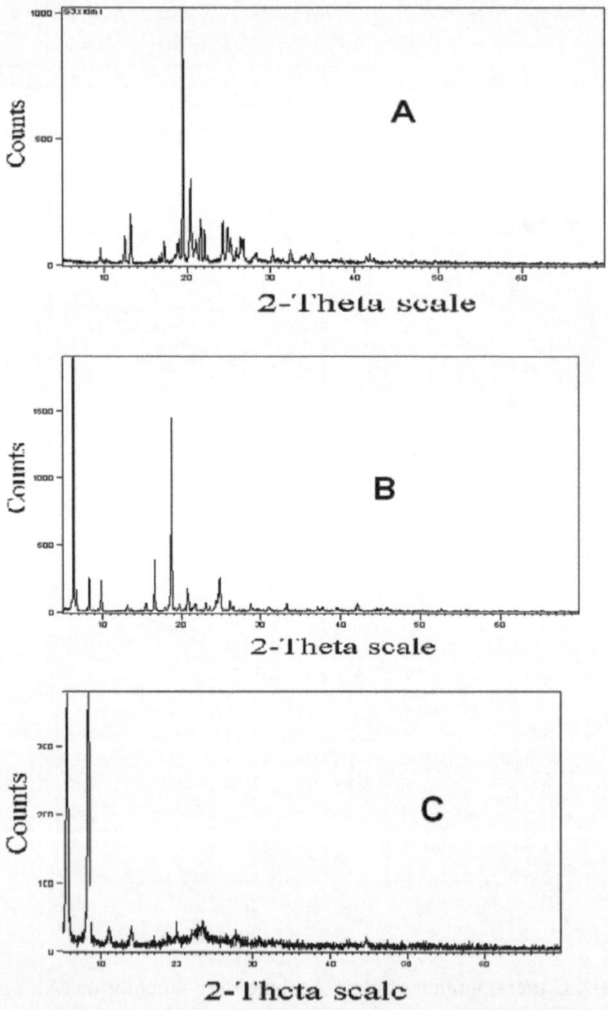

FIGURE 7.2 (A) plain drug (B) Drug-loaded microspheres and (C) pristine microspheres.

7.4.3 SCANNING ELECTRON MICROSCOPIC (SEM) STUDIES

SEM images of the microspheres were recorded using a (QUANTA, IISC, Bangalore) at the required magnification. Working distance of 33.5 mm was maintained and the acceleration voltage used was 10 kV with the secondary electron image (SEI) as a detector. Figure 7.3 shows the SEM micrograph of Acebutolol hydrochloride-loaded poly (VAc-co-MAc) microspheres and they are spherical in shape with rough surface.

FIGURE 7.3 Scanning electron micrograph of Acebutolol hydrochloride-loaded poly (VAc-co-MAc) microspheres.

7.4.4 PARTICLE SIZE ANALYSIS

Mean particle diameter and size distributions have been analyzed using a particle size analyzer (Mastersizer, 2000, Malvern Instruments, UK). Results of mean particle diameter of the microspheres produced by taking three different amounts of cross-linking agent (VAc-co-Mac-2, VAc-co-Mac-4, and VAc-co-Mac-5) are included in Table 7.1. These results suggest that, as the extent of crosslinking increases the volume mean diameter decreases. On a population basis, particle size distribution is unimodel. Microspheres used in preparing drug-loaded formulations were selected from a uniform size distribution range as displayed in Figure 7. 4. A narrow size distribution of microspheres was observed with particle size 80–400 μm, but majorities of particles are in the size range between 150 and 220 μm.

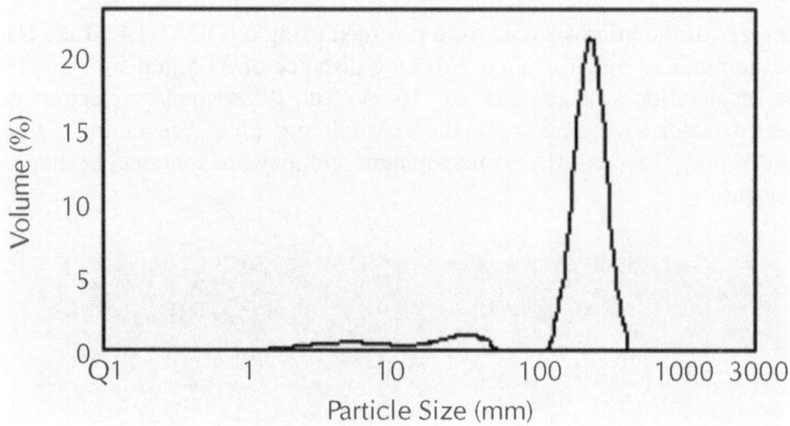

FIGURE 7.4 Particle size distribution curve for microspheres of poly (VAc-co-MAc).

7.5 IN-VITRO DRUG RELEASE

7.5.1 ENCAPSULATION EFFICIENCY

Results of percentage of encapsulation efficiencies are given in Table 7.1. The percentage of encapsulation efficiency varied depending upon the initial loading of the drug. In general, for formulations VAc-MAc-1, VAc-MAc-2 and VAc-MAc-3, the percentage of encapsulation efficiency increased systematically with increasing drug content of the matrices. From the Table 7.1 it is also noticed that as the amount of NNMBA increasing in the matrix (VAc-co-MAc-2, VAc-co-MAc-4, and VAc-co-MAc-5), the percentage of encapsulation efficiency decreased. It is observed that with increasing amount of cross-linking agent NNMBA in the matrix, the percentage of encapsulation decreased due to the lesser free volume available in the matrix; this could be because at higher cross-linking density, microspheres would become more rigid. A similar report was made by Krishna Rao et.al (Krishna Rao et.al., 2006). The highest percentage of encapsulation efficiency of 79 was observed for VAc-co-MAc -3 containing 15% of Acebutolol hydrochloride with a higher content of VAc-co-MAc in the copolymer matrix and its size was also more, that is, 31μm.

TABLE 7.1 Results of percentage of encapsulation efficiency and mean particle diameter of Poly (VAc-co-MAc) microspheres with different amounts of cross-linking agent, monomer concentration and Acebutolol hydrochloride

Sample code	% (VAc)	% MAc	% NNM-BA	Acebuto-lol hydro-chloride	% Encap-sulation efficiency ± SD	Mean particle diameter (mm) ± SD
VAc-co-MAc -1	10	90	1	5	71 ± 1	29 ± 6
VAc-co-MAc -2	10	90	1	10	77 ± 2	31 ± 8
VAc-co-MAc -3	10	90	1	15	79 ± 2	34 ± 6
VAc-co-MAc -4	10	90	2	10	76 ± 9	28 ± 4
VAc-co-MAc -5	10	90	3	10	72 ± 8	16 ± 2
VAc-co-MAc -6	20	80	1	10	72 ± 6	30 ± 4
VAc-co-MAc -7	30	70	1	10	69 ± 5	24 ± 1

7.5.2 DRUG RELEASE KINETICS

In the area of pharmaceutics, it has been the usual practice to understand the release kinetics of a drug through a polymeric matrix using the empirical relationship proposed by Ritger and Peppas (Ritger and Peppas, 1987). Following this practice, in the present study, we have analyzed the cumulative release data using the following equation (4).

$$\left(\frac{M_t}{M_\infty}\right) = kt^n \tag{7.4}$$

Here, the ratio, M_t/M_∞ represents the fractional drug release at time, t; k is a constant that is a characteristic of the drug-polymer system, and n is an empirical parameter characterizing the release mechanism. Using the least-squares procedure, we have estimated the values of n and k for all the nine formulations developed at 95% confidence limit; these data are given in Table 7.2. If the value of $n = 0.5$, then the drug diffuses and releases out of the microsphere matrix following a Fickian diffusion. If $n > 0.5$, anomalous or non-Fickian transport occurs, for $n = 1$, non-Fickian or more commonly called Case II release kinetics is operative. The values of n ranging between 0.5 and 1 indicate the anomalous type transport (Harogoppa et al., 1991).

The values of *k* and *n* have shown a dependence on the extent of cross-linking, percentage of drug loading and MAc content of the matrix. Values of *n* for microspheres prepared by varying the amount of VAc-co-MAc in the microspheres of 90, 80 and 70% by keeping Acebutolol hydrochloride (10%) and NNMBA constant, ranged from 0.51 to 0.70 leading to a shift of transport from Fickian diffusion to non-Fickian type. This indicated the shift from erosion type release to a swelling-controlled, non-Fickian mechanism. This could possibly be due to a reduction in the regions of low micro viscosity and closure of micro cavities in the swollen state. Similar findings have been observed elsewhere, wherein the effect of different polymer ratios on dissolution kinetics was studied. On the other hand, the values of *k* are quite smaller for the drug-loaded microspheres, suggesting their lesser interactions compared to microspheres containing varying amount of MAc.

TABLE 7.2 Release kinetics parameters of microspheres with different amounts of cross-linking agent, monomer concentration and Acebutolol hydrochloride at 37 °C

Formulation code	*K*	*n*	Correlation coefficient '*r*'
VAc-co-MAc -1	0.0076	0.7793	0.9762
VAc-co-MAc -2	0.0230	0.5164	0.9878
VAc-co-Mac -3	0.0263	0.5105	0.9895
VAc-co-MAc -4	0.0208	0.5743	0.9764
VAc-co-MAc -5	0.0108	0.6518	0.9687
VAc-co-MAc -6	0.0137	0.6340	0.9637
VAc-co-MAc -7	0.0108	0.7014	0.9619

7.5.3 EFFECT OF CROSS-LINKING AGENT

The percentage of cumulative release versus time curves for varying amounts of NNMBA are displayed in Figure 7.5. The percentage of cumulative release is quite fast and large at lower amount of NNMBA, whereas the release is quite slower at higher amount of NNMBA. The cumulative release is somewhat smaller when lower amount of NNMBA was used probably because at higher concentration of NNMBA, the polymeric chains would become rigid due to the contraction of microvoids, thus decreasing the percentage of cumulative release of Acebutolol hydrochloride through the polymeric matrices.

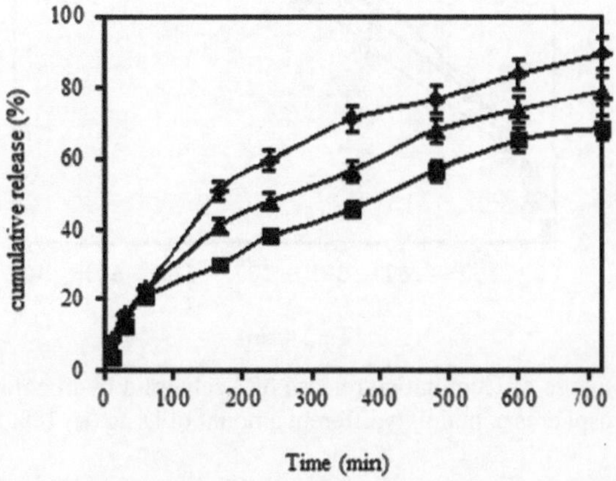

FIGURE 7.5 Percentage of Cumulative release of Acebutolol hydrochloride through Poly (VAc-co-MAc) microspheres containing different amount of cross-linking agent (■) 3%, (▲) 2% and (¨) 1% at 37 °C.

7.5.4 EFFECT OF DRUG CONCENTRATION

Figure 7.6 shows the release profiles of poly (VAc-*co*-MAc) microspheres that are loaded with different amounts of Acebutolol hydrochloride. It is noticed that initially, during the first hour, the release is quite fast in all the formulations, but later it slowed down. Release data suggests that those formulations containing a highest amount of drug (i.e., 15 wt.%) displayed the higher release rates than those containing smaller amounts of Acebutolol hydrochloride (i.e., 10 and 5 wt.%). A prolonged and slow release was observed for formulation containing a lower amount of Acebutolol hydrochloride (i.e., 5 wt.%) at 37 °C, this is due to the large free volume spaces available in the matrix through which, a lesser number of Acebutolol hydrochloride molecules would transport. It is noticed that for all the Acebutolol hydrochloride-loaded formulations, the almost complete release of Acebutolol hydrochloride was observed after 720 min.

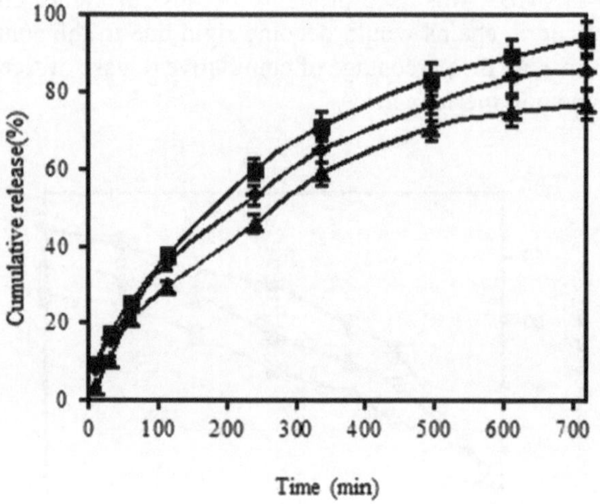

FIGURE 7.6 Percentage of Cumulative release of Acebutolol hydrochloride through poly (VAc-co-MAc) microspheres containing different amount of Drug (■) 15%, (") 10%, and (▲) 5% at 37 °C.

7.5.5 EFFECT OF PH

To investigate the effect of pH and ionic strength of the external medium on the swelling of microspheres, we have measured the percentage cumulative release in both pH 1.2 and 7.5 media cumulative release. Data presented in Figures 7.7 and 7.8 indicates that by increasing the pH from 1.2 to 7.5, a considerable increase in the cumulative release is observed for all microspheres. With increase in the pH of the buffer solution from 1.2 to 5, the swelling capacity of microspheres increases and hence, cumulative release increases. At lower pH the swelling capacity of the microspheres decrease because the carboxylic groups in methacrylic acid is not that much ionized and acquired negative charges; the columbic repulsive force among carboxyl anion also caused the intra molecular hydrophobic interactions. Hence cumulative release decreased and also prolonged. These copolymeric microspheres show excellent pH sensitivity 1.2 to 7.5 and it would be used for carrier drug systems for human bodies because of the acidic degree in stomach (Chen et al., 2005).

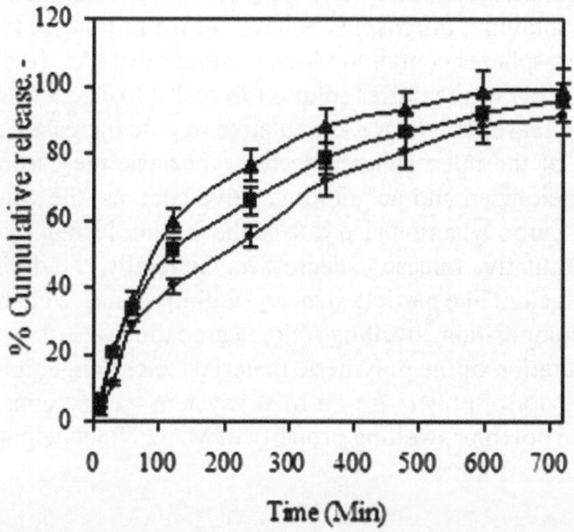

FIGURE 7.7 Percentage of Cumulative release of Acebutolol hydrochloride through Poly (VAc-co-MAc) microspheres with pH of 7.5 at 37 °C, (■) 15% (♦) 10% and (▲) 5%.

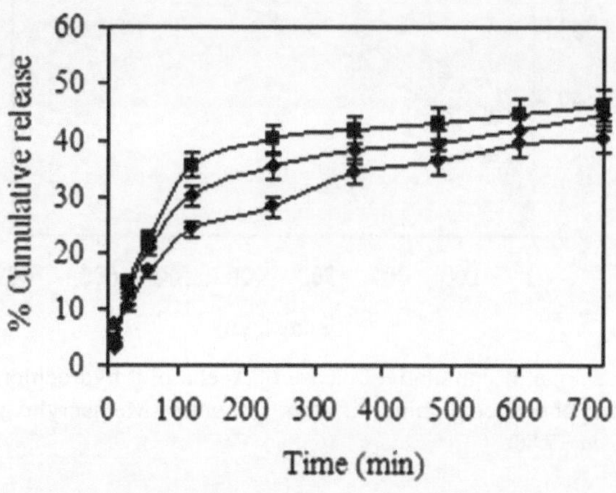

FIGURE 7.8 Percentage of cumulative release of Acebutolol hydrochloride through poly (VAc-co-MAc) microspheres of pH 1.2, at 37 ˚C, (■) 15%, (♦) 10% and (▲) 5%.

7.5.6 EFFECT OF METHACRYLIC ACID

Figure 7.9 shows the in vitro release data of Acebutolol hydrochloride from poly (VAc-co-MAc) particles performed with different ratio of MAc in the polymeric microspheres. The data shows that higher amount of MAc containing microspheres have more encapsulation efficiency and also the release studies show that higher amount of MAc containing microspheres have shown prolonged release characteristics than the microspheres containing lower amount of MAc. In all microspheres, with increase in the pH of the buffer solution from 1.2 to 7.5, the swelling capacity of microspheres increases and hence, cumulative release increases. At lower pH the swelling capacity of the microspheres decrease because the carboxylic groups in methacrylic acid deionized and acquired negative charges; the columbic repulsive force among these carboxyl anion also caused the intramolecular hydrophobic interactions. Hence cumulative release is decreased. Generally, the drug release pattern depends on many factors like particle size, crystallinity, surface character, molecular weight, polymer composition, swelling ratio, degradation rate, drug binding affinity and the rate of hydration of the polymeric materials, etc. (Ritger and Peppas, 1987) In the release behavior of poly (VAc-co-MAc) system we can consider the binding affinity of drug and polymer swelling property of MAc. Ninety eight percent of drug release was observed within 12 h.

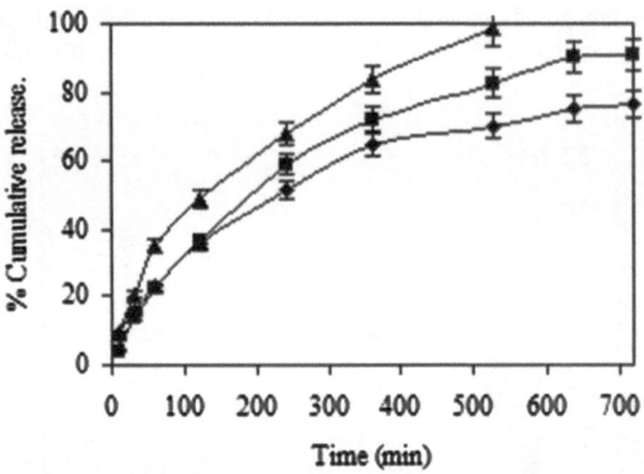

FIGURE 7.9 Percentage of cumulative release of Acebutolol hydrochloride through Poly (VAc-co-MAc) microspheres containing different amount of Methacrylic acid, (■) 90%, (¨) 80%, and (▲) 70%, at 37 °C.

7.6 CONCLUSIONS

Poly (Vinyl Acetate-co-Methacrylic acid) copolymeric microspheres crosslinked with N, N'-methylene bisacrylamide have been prepared by free radical emulsion polymerization. The microspheres have been characterized by differential scanning calorimetry (DSC) and x-ray diffractometry (X-RD) to understand about the drug dispersion in microspheres. Scanning electron microscopy (SEM) was used to assess the surface morphology of particles prepared, and observe the spherical nature with rough surface. Microspheres with different copolymer compositions have been prepared in yields ranging 80–85%. The in vitro drug release indicated that particle size and release kinetics shows the Fickian diffusion to non-Fickian type composition, depending on amount of cross-linking agent used and amount of Acebutolol hydrochloride present in the microspheres. The microspheres could be retained in the gastric environment for more than 14 h, which might help to improve the bioavailability of Acebutolol hydrochloride.

KEYWORDS

- **Drug delivery**
- **Copolymer**
- **pH Sensitive**
- **Microspheres**

REFERENCES

Babu, V. R.; Rajinee Kanth, V.; Jadhav M.; Mukund, Tejraj M. Aminabhavi. J Appl. Polym. Sci. 115: 3542–3549, 2010.

Chahroudi, D., Proceedings *of the 33rd Annual Technical Conference of the Society of Vacuum Coaters*, Society of Vacuum Coaters: Washington, DC, 1990; 212.

Chen, S.; Liu, M.; Jin, S.; Chen, Y. J. Appl. Polym. Sci. 2005, 98, 1720.

Dong, L.-C.; Hoffman, A. S. J. Controll. Rel.. 1991, 15, 141.

Döring, A., Birnbaum, W.; Kuckling, D. Chem. Soc. Rev., 2013. 42, 7391.

Elzoghby, A. O.; El-Fotoh, W. S. A.; Elgindy, N. A. J. Controll. Rel.: Official J. Controll. Rel. Soc.. 2011, 153.

Gennaro, B. D.; Solomon, O.; Kopec, L.; Korostoff, E.; Ackerman, J. L. Controlled release polymeric formulations, in: Paul, D. R.; Harris F. W. (Eds.), American Chemical, Washington, DC, 1976, 135.

Gutowska, A., Bae, Y. H.; Feijen, J.; Kim, S. W. J. Controll. Rel..1992, 22, 95.

Harogoppa, S.B.; Aminabhavi, T.M. Macromolecules. 1991, 24, 2598.

Johnson, J.C. (Ed.), Sustained Release Medications Noyes Data Co, Park Ridge, 1980, 96.

Kim, Y.-H.; Bae, Y. H.; Kim, S. W. J. Controll. Rel..1994, 28, 143.

Klier, J.; Scranton, A. B.; Peppas, N. A. Macromolecules. 1990, 23, 4944.

Krishna, R. K. S. V.; Vijaya, K. N.; Subha, M. C. S.; Sairam, M.; Aminabhavi, T. M. Cabohydr. Polym.. 2006, 66, 344.

Lam, P. L.; Gambari, R. J. Controll. Rel.: Official J.Controll. Rel. Soc.. 2014, 178, 25.

Miyazaki S.; Ishii, K.; Sugibayashi, K.; Morimoto, Y.; Takada, M. Chem. Pharm. Bull. 1982, 30, 3770.

Miyazaki S.; Takeuchi S.; Takada K.; Chem. Pharm. Bull. 1984, 32, 1633.

Morimoto Y.; Seki, T.; Sugibayashi, K.; Juni, K.; Miyazaki, S. Chem. Pharm. Bull. 1988, 36, 2633.

Mukae, K.; Bae, Y. H.; Okano, T.; Kim, S. W. Polym. J. 1990, 22, 250.

Ocak, F.; Agabeyoglu I.; Int. J. Pharm. 1999, 180, 177.

Osada, Y.; Gong J. P.; Sawahata K.; Macromol J. Sci. Chem. A 1991, 28, 1189.

Ravindra, S.; Varaprasad, K. N.; Reddy N.; Vimala, K.; Raju, K. M. J. Polym. Environ. 2011, 19, 413–418.

Ritger, P. L.; Peppas, N.A. J. Control. Rel. 1987, 5, 37.

Shin, S. C.; Lee, H. J. Eur. J. Pharm. Biopharm. 2002, 54, 201.

Seno M.; Len, M. L.; Iwamoto, K. Colloid. Polym. Sci. 1991, 269, 873.

Weiss, A.; Adler, M.; Grodzinskv, K. P.; Yarmush, A. J. J. Membr. Sci. 1991, 58, 153.

Yoshida, R.; Sakai, K.; Okano, T.; Sakurai, Y. Ind. Eng. Chem. Res. 1992a, 31, 23.

Yoshida, R.; Sakai, K.; Okano, T.; Sakurai, Y. J. Biomater. Sci. Polym.Ed. 1992b, 3, 243.

CHAPTER 8

CARBON NANOPARTICLE-BASED FLEXIBLE LLDPE COMPOSITES FOR ENHANCED ELECTROMAGNETIC SHIELDING EFFECTIVENESS

BYRON S. VILLACORTA and AMOD A. OGALE*

Department of Chemical Engineering and Center for Advanced Engineering Fibers and Films, Clemson University SC, 29634-0909; *E-mail: ogale@clemson.edu

CONTENTS

ABSTRACT

This chapter investigated the electromagnetic shielding effectiveness (EM SE) and electrical properties of heat-treated CNFs dispersed in a flexible linear low-density polyethylene (semicrystalline) matrix, and explored the effect of two other carbon-based modifiers on the EM SE of composites prepared by multiple melt-mixing routes with LLDPE for potential use in ductile/flexible EMC applications. Attention is also directed to the electrical and mechanical properties of such composites in relation with their electromagnetic shielding performance. The composite with 20 wt% of filler content displayed an increasing dependence with the in-plane electrical conductivity and permittivity, consistent with the electromagnetic shielding theory.

8.1 INTRODUCTION

Electromagnetic compatibility (EMC) has become a significant challenge for electronic products to protect electronic circuits from electrostatic discharge (ESD) and radiated electromagnetic interference (EMI) (Paul, 1992; Ramo et al., 1984). Because of their superior properties, conductive polymer composites can be used to shield electronic devices from receiving or emitting external electromagnetic signals (Al-Saleh and Sundararaj, 2009; Al-Saleh et al., 2011; Dani and Ogale, 1996; Gelves et al., 2006). Polymeric matrix composites are light-weight, corrosion-resistant and flexible materials, and often possess better processability, as compared with metallic counterparts. Electrical modifiers have been incorporated into polymer matrices to form resistive,inductive, and capacitive (RLC) electrical networks (Al-Saleh and Sundararaj, 2009; Al-Saleh et al., 2011; Gelves et al., 2006). Different types of conductive agents have been added for this purpose, such as conductive compounds, intrinsic conductive polymers, as well as metal- and carbon-based fillers including carbon black and fibers (Al-Saleh et al., 2011; King et al., 2006). There is a special interest in developing conductive polymer nanocomposites from carbon nanoforms due to their nano-scale and superior transport properties (Al-Saleh and Sundararaj, 2011; Al-Saleh et al., 2011; Keith et al., 2006; King et al., 2010). This enhances the entropic contributions of the material such as the electrical conductivity, and polarization loss, (i.e., the DC electric loss and AC dielectric loss, respectively) as well as the polarization (dielectric storage), which is an enthalpic term mainly arising from the interfacial polarization between the electrical modifiers and the insulating matrix (Nanni and Valentini, 2011). Electrical conductivity of composites is known to increase as the modifier concentration increases, and at the percolation threshold, the conductivity undergoes a power-law increment (Dani and Ogale, 1997; Wang and Ogale, 1993). Although electrical conductivity of the composite increases, its processability, flexibility, strength, and ductility typically decrease with modifier concentration (King et al., 2006, 2009; Kum et al., 2006; Xiao et al., 2007). Therefore, an optimal formulation that provides ease of processing as well as good electrical and mechanical properties is desirable.

Carbon nanoforms such as carbon nanofibers (CNFs) and nanotubes (CNTs) may offer a size advantage to polymeric nanocomposites due to their very small diameter (<0.5 μm), high aspect ratio, and excellent transport properties when compared with traditional carbon fibers (Dani and Ogale, 1996, 1997; Janda et al., 2005; Lee et al., 2010; Ling et al., 2009; Lozano et al., 2004; Nanni and Valentini, 2011; Villacorta and Ogale, 2011; Yang et al., 2005). Studies with carbon nanoparticles have shown that, to maximize their composite properties, the nanoparticles not only have to form an electrical network, but also be adequately "homogenized" in the matrix (Kasaliwal et al., 2011; Pegel et al., 2011). Such combination is particularly difficult to achieve, because the nanoparticles themselves tend to form agglomerates due to the intermolecular forces and their large specific surface area (Bhattacharya et al., 2010; Kasaliwal et al., 2011). Moreover, to assure uniform composite properties it is not only a matter of dispersing the nanoparticles, but at the same time obtaining a good distribution and electrical interconnection among the nanoparticles is essential. This means that due to the nature of the physical phenomenon of electrical percolation, certain level of clustering is also necessary (Kasaliwal et al., 2011; Yang et al., 2005). Thus, the electrical properties of a polymer composite depend on three factors: (a) the intrinsic transport properties of the electrical modifiers and nature of the polymeric matrix, (b) the concentration of such modifiers, and (c) the level of network formation impressed by the processing conditions.

Previous studies on nanocomposites made with highly conductive nanoparticles and amorphous polymers have been reported (Jimenez and Jana, 2007; Mathur et al., 2008); such nanocomposites possessed a strain-to-failure of less than 5%. In a recent study (Villacorta et al., 2012), we have investigated the EM SE and electrical properties of heat-treated CNFs dispersed in a flexible linear low-density polyethylene (semicrystalline) matrix. This chapter explores the effect of two other carbon-based modifiers on the EM SE of composites prepared by multiple melt-mixing routes with LLDPE for potential use in ductile/flexible EMC applications. Attention is also directed to the electrical and mechanical properties of such composites in relation with their electromagnetic shielding performance.

8.2 *EXPERIMENTAL*

8.2.1 *MATERIALS*

The matrix polymer used throughout this study was Poly(ethylene-co-1-octene), (Dowlex™ 2045), a film grade flexible, linear, low-density polyethylene (LLDPE). It has a density of 0.920 g/cm^3, DSC melting point of 122 °C, and melt flow index of 1 g/10 min. Carbon nanofibers, Pyrograf® III PR-19 (Applied Science

Inc.), and multi-walled carbon nanotubes (MWNT) from CheapTubes Inc., were used as nanoparticles. PR-19 CNFs are made by chemical vapor deposition (CVD) from natural gas as precursor by using a Fe-sulfide catalyst at about 900 °C, and have a specific surface area of 15–25 m^2/g (Lee et al., 2007, 2010). MWNT are also produced by CVD from methane with a Ni-Fe catalyst and have about 60 m^2/g of specific surface area. With the purpose of comparing the performance of nanocomposites with traditional carbon fiber composites, mesophase pitch-based carbon fibers (CF) [Cytec Thornel® P-55 length: 1–3 mm, diameter ~10 μm) were also used. For electrical composite properties, one of the most important factors is the intrinsic electrical conductivity of the carbon particles. This, in turn, is a strong function of its graphitic crystallinity, which can be increased by heat-treating the nanoparticles to temperatures exceeding 2000 °C as demonstrated in our previous studies (Villacorta et al., 2012; Lee et al., 2007, 2008). Therefore, ultra-high thermal treatment (HT) at 2500 °C was carried out for the as-received nanoparticles in a Thermal Technology HP50-7010 furnace in helium atmosphere prior to compounding (Lee et al., 2008).

8.2 PROCESSING

A Rheomix 600 batch mixer (BM) was used to melt-mix LLDPE with 20 wt% of all three types of particles at 190 °C and 20 rpm for 2 min. In addition, continuous twin-screw extrusion in counter-rotating mode at the same conditions listed above (i.e., 190 °C and 20 rpm) was used to compare different processing techniques, that is, batch versus continuous. For this purpose, a DSM Xplore conical, twin-screw micro-compounder (DSM) with a residence time of about 2 min (at 20 rpm in counter-rotation mode) was used. The composites were processed by thermal compaction at 190 °C into circular sheets about 2.5 mm thick and 133 mm diameter utilizing a Carver laboratory press. With regards to the processability of the above nanocomposites, rheological studies on the same formulation of PR-19 HT CNF/LLDPE processed by continuous twin-screw extrusion have been carried out (Villacorta et al., 2012). These studies indicate that the presence of nanoparticles at 20 wt% level did not increase the viscosity significantly. Additionally, about 15 g of each type of the heat-treated nanoparticles were ultra-sonicated (Fisher Scientific FS20H) in 200 mL of a mixture of hexanes for 45 min, in a deionized water bath at 20 °C. Hexane was preferred by the virtue of its hydrophobicity, since PR-19 CNF and MWNT are also hydrophobic nanoparticles. After sonication, the suspension was dried (Thermolyne Oven Series 9000) at 75 °C for about 18 h. Ultra-sonicated nanoparticle-based nanocomposites at 20 wt% were also produced by batch mixing at the same processing conditions listed above.

8.2.3 CHARACTERIZATION

The initial state of agglomeration of the nanoparticles themselves was assessed by scanning electron microscopy (Hitachi S-4800, SEM). Each nanoparticle type was inspected before and after ultra-sonication. About 20 different locations were observed for each nanoparticle type. In addition, a dimensional analysis of the nanoparticles was carried out by analyzing their high-magnification SEM micrographs using computational image analysis (ImagePro®). For both nanoparticle types, a set of about 150 representative imaged nanoparticles were measured. It has been observed in the literature that optical microscopy (OM) enables the assessment of microscopic dispersion of nanocomposites (Pegel et al., 2011). However, this can be difficult to at such high levels of solids content (i.e., 20 wt% or about 10 vol%) due to the darkness of the composite in thick samples. Therefore, for each type of nanocomposite, thin films of 7–12 μm thickness were specially processed at temperatures exceeding 250 °C, and nine different microscopic locations were investigated by OM in transmission mode (BX60 Olympus Optical Microscope). For each location, two different magnifications were imaged.

The DC in-plane volume electrical conductivity, σ (S/m), of the nanocomposites was measured using a Keithley 6517B High Resistance Meter (current range: 1 pA–20 mA) connected to a Keithley 8002A Resistivity Test Fixture modified with external electrodes (ASTM D257). The measurements were performed with help of the Keithley 6524 software by which a DC voltage of ±1V was applied across the composite samples. The relative humidity and temperature of the experimental area were monitored using a 6517-RH humidity probe and a 6517-TP thermocouple, respectively. The conductivity was obtained from resistance measurements of die-cut specimens that were 12.5 mm wide, 2.5 mm thick and with lengths of about 20 mm. Silver paint was applied on the surfaces at each end of the samples and their in-plane resistance was measured. From the sample geometry and the resistance measurements, the conductivity was computed from four replicates ($n = 4$). The measurements were performed on a 6-mm thick Teflon® sheet to insulate the area for proper isolated measurements.

The bulk electrical resistivity (BER) of the particles themselves was measured using a Keithley 196 System digital multi-meter (DMM) while compressing the particles at 50 MPa in an insulating fixture. The bulk electrical resistivity of the CNFs was computed from the measured resistance and sample geometry while being pressurized (Singjai et al., 2007). Raman spectroscopy was conducted on the particles to analyze the disordered (D) and graphitic (G) bands observed in carbon materials at 1315 cm^{-1} and 1580 cm^{-1}, respectively. A Renishaw micro-Raman spectroscope equipped with a 785 nm wavelength diode laser was used to determine the ID/IG Raman ratio (Lee et al., 2007). The complex electrical permittivity (real ε' and imaginary ε'') of the nanocomposites, in their sheet form (2.5 mm thick), was measured utilizing an Agilent 4291B RF Impedance/Material Analyzer and an Agilent 16453A Dielectric Material Test Fixture. Prior to the measurements, the analyzer was calibrated utilizing an Agilent calibration kit (short (0 Ω), open (0 S), load (50 Ω)). Short/open/load fixture compensation was also applied to increase the accuracy of the measurements. The analysis frequency range was 30 MHz to 1.5 GHz.

The static decay time was measured using an Electro-Tech Systems, Inc. 406D Static-Decay Meter that complies with the Federal Test Method 101D, Method 4046 and Military Standard Mil-B-81705C. This standard requires that 99% of the initial induced charge be dissipated is less than 2 sec for qualifying material per Mil-B-81705C. The Static-Decay Meter was calibrated by the ESD Testing Laboratory of Electro-Tech Systems, Inc. (Glenside PA) prior to the measurements. An Electro-Metrics EM-2107A coaxial transmission line test fixture was used to apply a far-field plane electromagnetic wave to the nanocomposite specimens (ASTM D4935). The electromagnetic shielding effectiveness (EM SE) was determined from the measured incident power P_0, and the transmitted power, P_T (Janda et al., 2005):

$$EM\,SE = -10\,log\frac{PT}{P0} \tag{8.1}$$

The calibrated unit has a dynamic range greater than 80 dB. The accuracy of the calibration was checked by measuring a set of Mylar®-gold composite standard specimens. The EM-2107A test fixture was connected through coaxial cables to an Agilent Technologies N5230A PNA Series Network Analyzer that was calibrated with 85033D 3.5 mm calibration kit. For measuring EM SE, each circular nanocomposite specimen was aligned between the test fixtures and measured at frequencies from 30 MHz to 1.5 GHz. Two true replicates were tested. The sample diameter was 133 mm, whereas the thickness was 2.5 mm. To determine the shielding components for the nanocomposites, the two main scattering parameters, S_{11} and S_{21}, that define transmittance and reflectance were measured (Park et al., 2010):

$$T = \left|\frac{P_T}{P_0}\right| = |S_{21}|^2 \tag{8.2}$$

$$R = \left|\frac{P_R}{P_0}\right| = |S_{11}|^2 \tag{8.3}$$

Then, the total shielding effectiveness (EM SE) and reflective shielding effectiveness (EM SER) were obtained as:

$$EM\,SE = -10\,log\,T \tag{8.4}$$

$$EM\,SE_R = -10\,log\,(1-R) \tag{8.5}$$

The absorptive shielding effectiveness (EM SEA) was computed from their difference as:

$$EM\,SE_A = EM\,SE - EM\,SE_R \tag{8.6}$$

Tensile tests were conducted using the ASTM D638 Type V technique at room temperature on dog bone-shaped specimens of 25 mm gauge-length and 3 mm width. The specimens were 1 mm thick and were die-cut into the ASTM dog bone shape. An ATS Universal 900 tensile tester at a cross-head speed of 25 mm/min was used to test six replicates per nanocomposite type.

8.3 RESULTS AND DISCUSSION

8.3.1 MORPHOLOGY

Figure 8.1 displays representative high-magnification SEM micrographs and their histograms for the diameter and length distributions of both heat-treated nanoparticles measured by computational image analysis (ImagePro®). PR-19 HT CNFs have a diameter of 119 ± 8 nm and a length of 10 ± 2 μm, whereas MWNTs HT have a diameter of 42 ± 3 nm and length of 6 ± 1 μm (i.e., average ±95% confidence intervals), for an aspect ratio (L/D) of about 85 and 145, respectively.

Figure 8.2 contains representative SEM micrographs that compare the nanoparticles before and after ultra-sonication. Similar agglomeration levels and morphology were seen post-sonication, with the presence of agglomerates as large as about 50 μm. Thus, from SEM analysis, no clear evidence was found about reduction in the level of agglomeration after ultra-sonication. Possibly, a re-agglomeration process might have taken place during the evaporation of hexane during drying, along with the mechanical interlocking and entanglements that prevented ultra-sonication from accomplishing thorough deagglomeration. Representative optical micrographs of the nanocomposites are presented in Figure 8.3. For both kinds of nanoparticles, clusters as big as 50 μm could be identified, which are similar to the size of the nanoparticle agglomerates. For PR-19 HT and PR-19 HT-Sonicated nanocomposites similar level of clustering and distribution were observed. For MWNT HT nanocomposites, the morphology appears a little denser (darker) when compared to PR-19 HT nanocomposites, but similar levels of cluster-size and distribution are seen for both unsonicated and sonicated cases. When scanning electron microscopy was used to analyze the cryo-fractured cross-section of the nanocomposites, clusters could not be clearly identified after inspecting about 20 different locations (Figure 8.4). Generally, the nanoparticles appeared distributed all over the nanocomposites. Thus, dispersion/distribution assessment must be carried out at different magnification levels, which is consistent with similar observations made in other literature studies (Pegel et al., 2011).

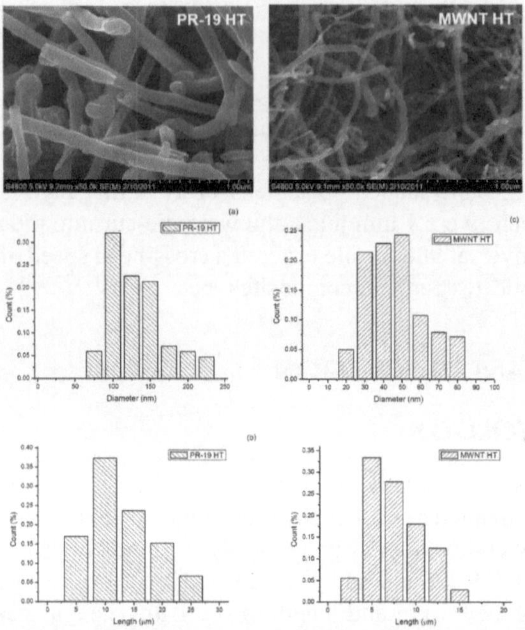

FIGURE 8.1 Representative scanning electron micrographs used for dimensional analysis of nanoparticles and their histograms for: (a) diameter and (b) length distributions of PR-19 HT and (c) diameter and (d) length distributions of MWNT HT.

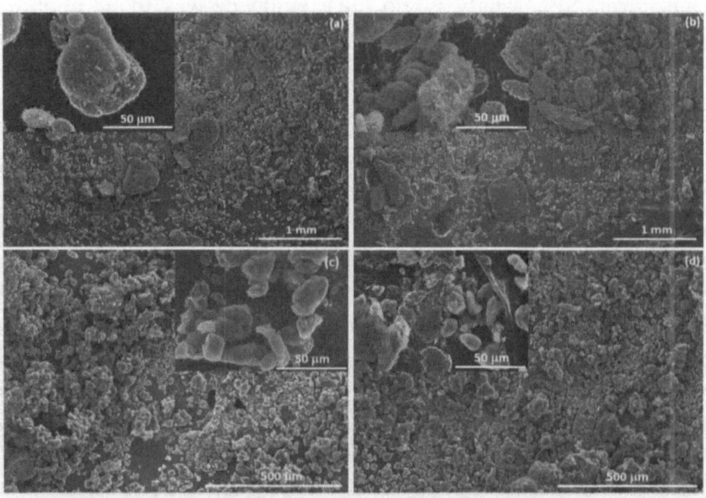

FIGURE 8.2 Representative scanning electron micrographs (SEM) of nanoparticles: (a) PR-19 HT, (b) PR-19 HT-Sonicated, (c) MWNT HT and (d) MWNT HT-Sonicated. Insets display the microstructure at higher magnification.

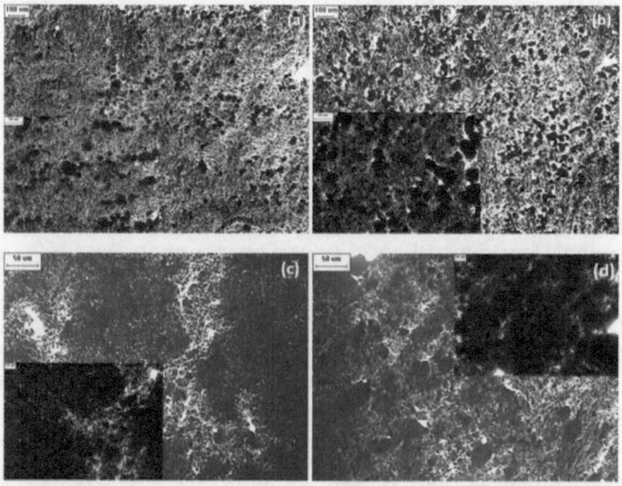

FIGURE 8.3 Representative optical micrographs (OM) of batch-mixed nanocomposite films (7–12 μm thick) at 20 wt%: (a) PR-19 HT, (b) PR-19 HT-Sonicated (c) MWNT HT and (d) MWNT HT-Sonicated. Insets display the microstructure at higher magnification.

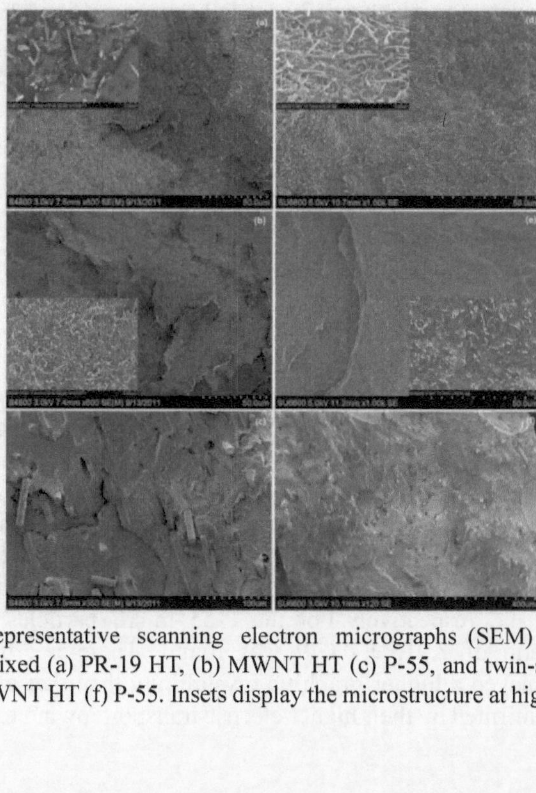

FIGURE 8.4 Representative scanning electron micrographs (SEM) of composites at 20 wt%: batch-mixed (a) PR-19 HT, (b) MWNT HT (c) P-55, and twin-screw extruded (d) PR-19 HT, (e) MWNT HT (f) P-55. Insets display the microstructure at higher magnification.

8.1.2 ELECTRICAL PROPERTIES

The DC electrical conductivity σ of the 20 wt% (11 vol%) nanocomposites was measured at 33.5 ± 5.6, 20.4 ± 3.3, and 5.0 ± 1.8 S/m (n = 4 in all cases) for the MWNT HT, PR-19 HT and P-55 batch-mixed composites, respectively. The composites were electrically percolated systems when compared with a conductivity of 7.0 ± 1.1 × 10^{-15} S/m for pure LLDPE, which is 15 orders of magnitude lower than that of the composites. For polyolefins nanocomposites, percolation thresholds ranging from 5 to 15 wt% have been reported for CNFs and CNTs in literature studies (Al-Saleh and Sundararaj, 2009; Lee et al., 2008; Morcon and Simon, 2011). However, even though at the percolation threshold (e.g., 10^{-5} to 10^{-2} S/m) the conductivity of the composite is much higher than that of the pure resin (e.g., 10^{-17} to 10^{-15} S/m), it is typically not high enough for shielding purposes. Therefore, for attaining shielding effectiveness, it is necessary to increase loading levels beyond the percolation threshold, but not so high such that the processability and mechanical properties can be still retained. Because the objective of the present work is to develop a nanomaterial with an optimal combination of shielding performance, flexibility, ductility and processability, the selection of 20 wt% particle content (~11 vol%) used in this study was based on our previous work (Villacorta et al., 2012). The conductivities of batch-mixed nanocomposites made of ultra-sonicated nanoparticles were 32.1 ± 2.2 S/m (n = 4) and 19.6 ± 2.8 S/m (n = 4), respectively for MWNT HT and PR-19 HT nanocomposites, which are not significantly different from the conductivities of their unsonicated counterparts. This indicates that the ultra-sonication method used did not provide a significant improvement in the electrical conductivity of the nanocomposites. This outcome is consistent with the fact that the morphology of the composites obtained from unsonicated and sonicated particles was not different.

The conductivity of MWNT HT, PR-19 HT and P-55 twin-screw extruded composites was 16.7 ± 1.3 S/m (n = 4), 13.3 ± 1.4 S/m (n = 4) and 0.8 ± 0.3 S/m (n = 4) respectively. The continuously twin-screw extruded composites displayed lower conductivities than those of the batch-mixed composites, confirming that different processing techniques can lead to different composite electrical properties. In our studies, twin-screw extruder applied a higher level of shear than batch mixing, which generally reduced the electrical network and so the electrical conductivity of the processed composites (Kasaliwal et al., 2011; Villmow et al., 2011).

The bulk electrical resistivity (BER) of the nanoparticles was 0.031 ± 0.010 Ω·cm for PR-19 HT and 0.028 ± 0.015 Ω·cm for MWNT HT. In contrast, short P-55 carbon fibers showed a higher BER of 0.116 ± 0.036 Ω·cm. Similarly, an I_D/I_G Raman ratio of only 0.961 ± 0.044 and 0.782 ± 0.105 were found for PR-19 HT and MWNT HT, respectively. For the P-55 micro- particles, a significantly higher Raman ratio of 2.915 ± 0.079 was found. Therefore, the heat- treated nanoparticles displayed a higher graphitic crystallinity than that of the micro-particles, which is confirmed by their higher electric transport, owing to their low BER.

As summarized in Table 8.2, MWNT HT, with the largest aspect ratio as well as the lowest bulk electrical resistivity and Raman D/G ratio, displayed the best electrical performance, followed by PR-19 HT nanocomposites and P-55 composites.

The complex electrical permittivity of the nanocomposites ($\varepsilon = \varepsilon' - \varepsilon''$), as measured with the Agilent

16453A Dielectric Material Test Fixture is displayed in Figure 8.5. $\varepsilon'/\varepsilon_0$ represents the dimensionless real relative permittivity or dielectric energy storage, and $\varepsilon''/\varepsilon_0$ the dimensionless imaginary relative permittivity or polarization loss. The permittivity of vacuum is 8.85418 × 10^{-12} F/m. The real and imaginary permittivity of the nanocomposites displayed a generally decreasing behavior with respect to frequency (Figure 8.5). For batch-mixed (BM) composites, the MWNT HT nanocomposites exhibited an overall permittivity of (175 ± 60) – (93 ± 34) *j* at 100 MHz, decreasing to (15 ± 11) – (21 ± 3) *j* at 1.5 GHz. PR-19 HT nanocomposite permittivity decreased from (100 ± 65) – (63 ± 46) *j* at 100 MHz to (17 ± 7) – (18 ± 12) *j* at 1.5 GHz (n = 4 in both cases). In contrast, the permittivity of P-55 composites was (28 ± 17) – (12 ± 10) *j* (n = 4) at 100 MHz and (12 ± 5) – (3 ± 2) *j* (n = 4) at 1.5 GHz showing less frequency dependency. Likewise, for the twin-screw extruded composites, the permittivity of MWNT HT composites decreased from (140 ± 9) – (98 ± 25) *j* at 100 MHz to (29 ± 4) – (34 ± 3) *j* at 1.5 GHz, and for the PR-19 HT nanocomposites decreased from (105 ± 6) – (62 ± 24) *j* at 100 MHz to (14 ± 2) – (23 ± 2) *j* at 1.5 GHz (n = 4 in all cases). However, the twin-screw extruded P-55 composites were almost frequency insensitive at about (5 ± 4) – (4 ± 3) *j*.

The conductivity and permittivity are respectively the DC and AC "diffusivities" for electrical transport (Ramo et al., 1984). Hence, as the DC electrical conductivity decreases, the batch- mixed composites, MWNT HT (33.5 ± 5.6 × 10^{0} S/m), PR-19 HT (20.4 ± 3.3 × 10^{0} S/m) and P-55 (5.0 ± 1.8 × 10^{0} S/m), displayed a decreasing AC permittivity magnitude of 1.8 × 10^{-9} F/m, 1.0 × 10^{-9} F/m and 2.7 × 10^{-10} F/m, respectively at 100 MHz. A similar decreasing trend was observed for the twin-screw extruded composites. Thus, the more conductive the composite is, the greater the dielectric storage and loss that the composite displayed, and establishes that the capacitive behavior of the network also depends on the intrinsic transport properties of the particles themselves. As in the case of conductivity, the dielectric storage and loss of the composite also depend on the processing technique, because the AC permittivity magnitude of the twin- screw extruded composites displayed slightly lower values than those displayed by the batch-mixed ones, particularly for P-55 at 7.7 × 10^{-11} F/m (100 MHz).

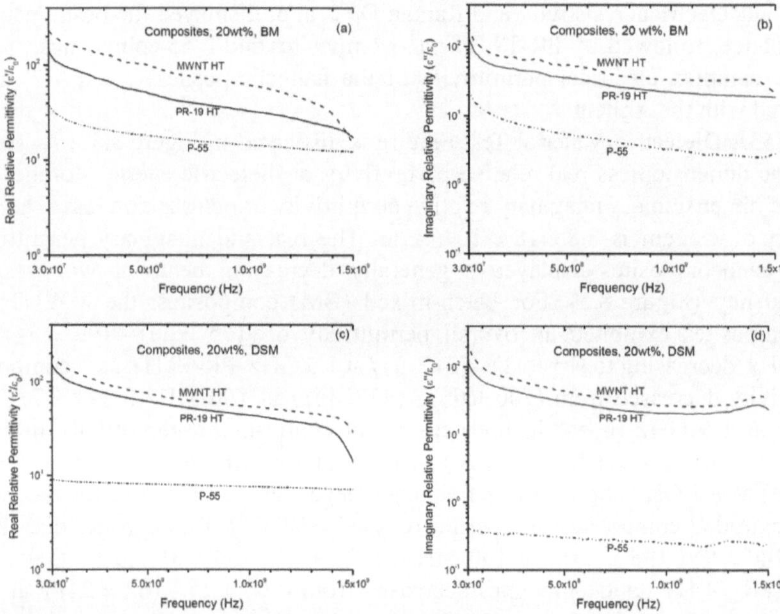

FIGURE 8.5 Relative real and imaginary electrical permittivity of representative composites made by: (a, b) batch mixing (BM) and (c, d) continuous twin-screw extrusion (DSM).

8.3.3 ELECTROSTATIC DISSIPATION AND ELECTROMAGNETIC SHIELDING

The electrostatic discharge (ESD) characteristic of the nanocomposites was also measured. All types of composites at 20 wt% were ESD dissipative and complied with the Mil-B-81705C requirements since they all were in the percolated regime. The decay time for each nanocomposite type is shown in Table 8.1 for 1% of cutoff. Only, twin-screw extruded P-55 composites displayed a slightly higher, but still dissipative, decay time of 1.0 s. This is consistent with its low electrical conductivity of only 0.8 S/m, as it is known that the electrostatic decay time increases as the conductivity of the material decreases (Pratt, 2000).

TABLE 8.1 Static decay-times for electrostatic dissipation (ESD) of batch-mixed and twin-screw extruded composites, measured at 1% cutoff and 50% relative humidity.

20 wt% Composites	Batch Mixing (BM) Decay time (sec)	Twin-screw Extrusion (DSM) Decay time (sec)
PR-19 HT	0.01	0.01
MWNT HT	0.01	0.01
P-55	0.01	1.00

The shielding ability of a homogeneous material is a function of its intrinsic impedance. For an EM plane wave, is the ratio of the magnitude of the electric field to that of the magnetic field that passes through the material (Paul, 1992). The electrical conductivity σ, electrical permittivity ε, and magnetic permeability μ, are related to the intrinsic impedance (Paul, 1992).

For non-magnetic conductive materials (i.e., $\mu = \mu_0 = 4\pi \times 10^{-7}$ H/m and $\sigma \gg \omega\varepsilon$), the intrinsic impedance η of such materials is primarily dependent on the DC electrical conductivity σ (Paul, 1992). The overall polarizability of the material can be expressed as a complex electrical permittivity, which accounts for the dielectric storage and losses (Ramo et al., 1984; Nanni and Valentini, 2011). Thus, in nanocomposites, whose conductivity is not high compared to pure metals (ie., 101 S/m vs.107 S/m), the polarization loss ε'' also plays a role in determining the EM SE, primarily in the absorptive component of shielding, and can be viewed as another entropic effect (Ramo et al., 1984).

Figure 8.6 (a) and (b) display the EM SE of the composites prepared by batch mixing (BM) and twin-screw extrusion (DSM), respectively. PR-19 HT nanocomposites displayed 24.7 and 16 dB of total shielding (at 1.5 GHz) when processed by BM and DSM, respectively. MWNT HT nanocomposites displayed larger values of 25.3 dB and 17.7 dB by BM and DSM techniques, respectively. In contrast, for P-55 composites, batch mixing (BM) provided 12.2 dB of shielding at 1.5 GHz, and much lower value of 2.3 dB from twin-screw extrusion (DSM). These results are consistent with the electrical conductivity and permittivity measurements, because with either processing technique, the greater the shielding, the higher the "lossy" electrical properties of the composite (i.e., conductivity and dielectric loss). Figure 8.7 displays a clear interrelationship between the EM SE and the in-plane conductivity of the composites, showing that the in-plane conductivity of the composite specimen is the one that determines how much of the electromagnetic plane-wave is reduced by electromagnetic polarization effects (Matumura et al., 2007). Uniformity at the macroscopic level is also desired in the composites so that at the wavelength-scale, the material appears "homogeneous" to the radiation and helps avoid radiation leakage through non-uniformities in the composite material. The last feature is particularly important because the wavelengths used in the present EM SE analysis range from 0.2 m (i.e., 1.5 GHz) to 10.0 m (i.e., 30 MHz). These wavelengths are at least three orders of magnitude larger than the cluster size (< 5×10^{-5} m). This is not the case of carbon-fiber composites whose macroscopic dimensions may be similar to the wavelength of the radiation. Therefore, rules applicable to homogeneous and/or uniform materials may not be completely suitable in this case.

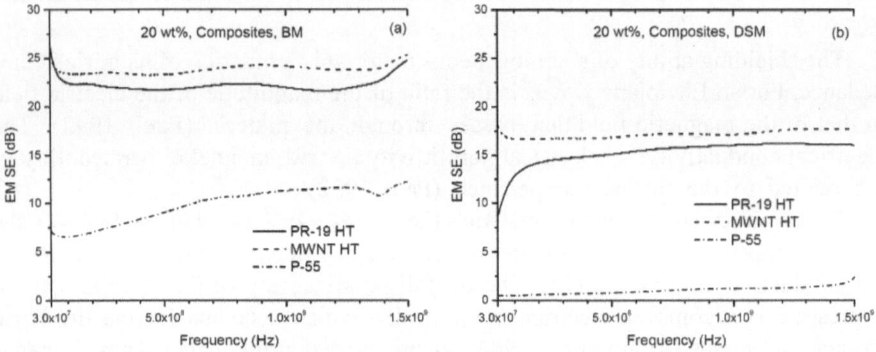

FIGURE 8.6 Electromagnetic shielding effectiveness (EM SE) of representative 20 wt% composites (a) batch-mixed and (b) twin-screw extruded.

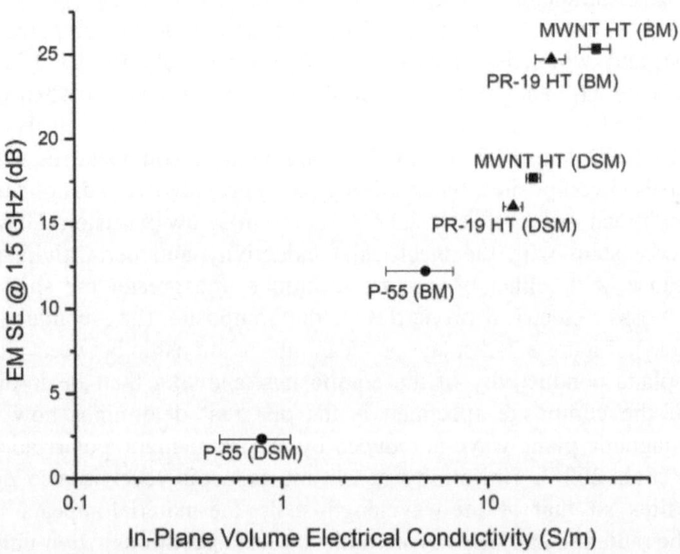

FIGURE 8.7 Relationshipelation between the EM SE (dB @ 1.5 GHz) and the in-plane electrical conductivity of different composites.

8.3.4 SHIELDING MECHANISMS

The absorptive and reflective shielding components the composites are displayed in Table 8.2. For both processing techniques, the absorptive component of shielding is the dominant one. In contrast, for P-55 composites, the reflection component of the batch-mixed composite is comparable to the absorptive one with EM SEA being very small at high frequencies, whereas in the twin-screw extruded ones reflection is the dominant component.

TABLE 8.2 Intrinsic properties of three types of particles as compared with the conductivity of their batch-mixed and twin-screw extruded composites at 20 wt% particle content. Ranges represent 95% confidence intervals.

Particle	Raman D/G ratio	Bulk electrical resistivity (Ω·cm)	Aspect Ratio (L/D)	In-plane electrical conductivity (S/m) Composites at 20 wt%	
				BM	DSM
MWNT HT	0.782 ± 0.105	0.028 ± 0.015	111–180	33.5 ± 5.6	16.7 ± 1.3
PR-19 HT	0.961 ± 0.044	0.031 ± 0.010	63–108	20.4 ± 3.3	13.3 ± 1.4
P-55	2.915 ± 0.079	0.116 ± 0.036	<150	5.0 ± 1.8	0.8 ± 0.3

In applications where a radiating source (e.g., antennas, heat sinks, high speed circuitry) must be kept from interfering with an external electronic circuit, or when an electronic circuit couples to other circuits in the same enclosure, an absorptive shielded enclosure is preferred. Reflective enclosures can cause large fields to build up internally, which can enhance unwanted coupling between circuits inside the enclosure. For batch-mixed composites, at 1.5 GHz, EM SEA values of 18.5, 18.7 and 7.2 dB were measured respectively for PR-19 HT, MWNT HT and P-55; whereas reflection values of 6.2 and 6.6 and 5.0 dB were respectively obtained. Similarly, absorption values of 11.2 and 12.8 and 0.8 dB were obtained for their twin-screw extruded counterparts. These results indicate that these types of nanocomposites are suitable materials for applications where absorptive shielding is required over the VHF and UHF frequency bands. Figure 8.8 shows the measured reflective (EM SER) and calculated absorptive (EM SEA) components of shielding for batch-mixed and twin-screw extruded composites.

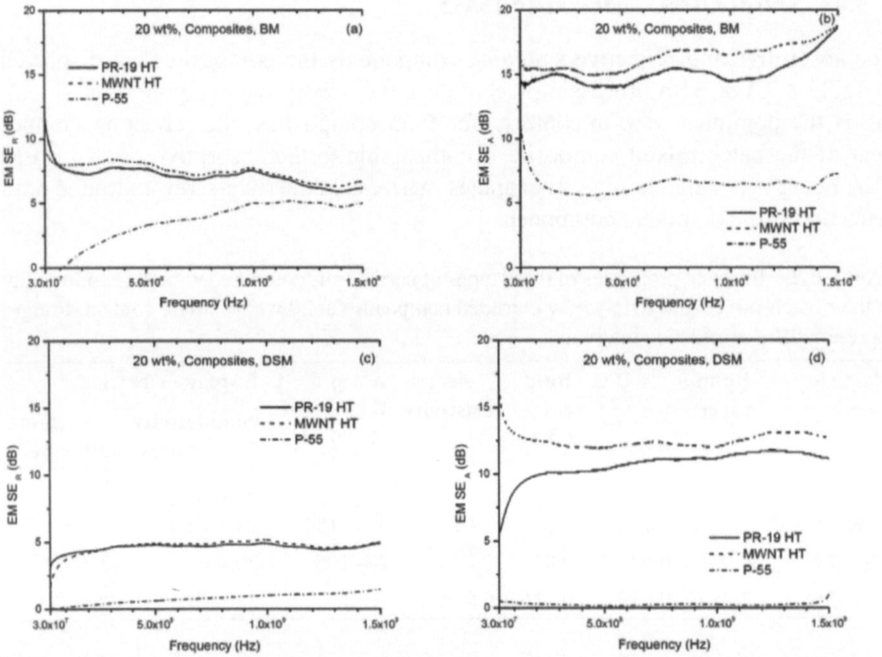

FIGURE 8.8 Measured reflective (EM SER) and calculated absorptive (EM SEA) components of shielding for (a, b) batch-mixed and (c, d) twin-screw extruded composites.

8.3.5 MECHANICAL PROPERTIES

Table 8.3 displays the tensile data for the different composites prepared in this study. At 20 wt%, the batch-mixed PR-19 HT nanocomposites displayed a tensile modulus of 638 ± 105 MPa, strength of 16.0 ± 2.5 MPa and strain-to-failure of 128 ± 49.0% (n = 6). Likewise, batch-mixed MWNT HT nancomposites possessed a tensile modulus of 696 ± 110 MPa, tensile strength of 17.7 ± 1.6 MPa and strain-to-failure of 114 ± 43.4% (n = 6), which displayed no significant change with respect to the twin-screw extruded nanocomposites. Pure LLDPE has a modulus of 325 ± 85 MPa, strength of 30 ± 5 MPa and strain-to-failure of 685 ± 105% (n = 6). Thus, as compared with pure LLDPE, the overall modulus of the nanocomposites was about twice, and they retained about 50% and 20% of the LLDPE strength and elongation-at-break, respectively. This reduction in strength and ductility is consistent with that reported in nanocomposite literature studies (Goh, 2011; Kasaliwal et al., 2011).

TABLE 8.3 Tensile properties for batch-mixed and twin-screw extruded composites at 20 wt% particle content. Ranges represent 95% confidence intervals.

	PR-19 HT (BM)	PR-19 HT (DSM)	MWNT HT (BM)	MWNT HT (DSM)	P-55 (BM)	P-55 (DSM)
Apparent Modulus (MPa)	683.0 ± 105.0	597.0 ± 41.0	696.0±110.0	638.0 ± 95.0	1079.0 ±88.0	896.0 ± 66.0
Yield Stress (MPa)	16.5 ±1.8	14.6 ± 1.0	18.3 ± 1.8	17.5 ± 1.4	20.0 ± 1.5	21.0 ± 1.4
Tensile Strength (MPa)	16.0 ± 2.5	14.7 ± 1.1	17.7 ± 1.6	17.8 ± 1.1	8.0 ± 1.4	7.6 ± 1.3
Elongation at break (%)	128.0 ± 49.0	179.0 ± 11.5	114.0 ± 43.4	123.0 ± 37.0	32.0 ± 8.0	31.0 ± 6.8

Although mechanical performance of composites can depend on the processing conditions, the batch-mixed and the twin-screw extruded nanocomposites developed in this study did not show any significant difference. Also, we note that while the nanocomposites were generally less flexible as compared to the pure LLDPE, they still retained a fairly high elongation-at-break of about 140%, compared to about only 30% displayed by short P-55 carbon fiber composites. Coupled with an EM SE of about 22–24 dB, PR-19 HT and MWNT HT nanocomposites indicate a potential in applications that require significant mechanical flexibility.

8.4 CONCLUSIONS

The electromagnetic shielding effectiveness (EM SE) values of the composites, at 20 wt%, displayed an increasing dependence with the in-plane electrical conductivity and permittivity of their composites, consistent with the electromagnetic shielding theory. Thus, over the 30 MHz–1.5 GHz frequency range, the 20 wt% batch-mixed MWNT HT (33.5 ± 5.6 S/m) and the PR-19 HT (20.4 ± 3.3 S/m) nanocomposites displayed a shielding performance of about 24 dB and 22 dB, respectively. In fact, they surpassed the EMC requirement of 20 dB for electronic systems over the Federal Communication Commission radiation emission limits for residential digital devices (i.e., at 40 dBμV/m for frequencies greater than 30

MHz; Paul, 1992). SEM analysis showed that the size of the agglomerates of both nanoparticles (PR-19 HT and MWNT HT) did not display any significant change after sonication, neither did the electrical conductivity. Therefore, it was discontinued after initial studies. For the three different particles here studied, discontinuous batch mixing, generally led to retaining larger electrical networks, and so to better shielding performance. On the other hand, for a given processing technique, a direct dependence of the intrinsic transport properties of the carbon-based particles was observed on composite properties. Thus, MWHT HT, whose aspect ratio and graphitic crystallinity were the highest of the studied set, displayed the largest conductivity, permittivity and shielding effectiveness in its composite form. Both types of nanocomposites displayed a predominantly absorptive shielding (18 dB); whereas, twin-screw extruded P-55 composites possessed an EM SE of only 1.3 dB. Thus, coupled with their flexibility both types of nanocomposites hold promise as light-weight, high-performance alternative materials for electromagnetic compatibility applications, where high levels of absorptive shielding are required.

ACKNOWLEDGEMENTS

This work made use of ERC Shared Facilities supported by the National Science Foundation under Award Number EEC-9731680. We also thank Prof. Todd Hubing for allowing the use of networks for EM SE measurements.

KEYWORDS

- **Electromagnetic shielding**
- **LLDPE**
- **Carbon Nanoparticle**
- **electrical conductivity**

REFERENCES

Al-Saleh, M. H.; Gelves, G. A.; Sundararaj, U. Copper nanowire/polystyrene nanocomposites: Lower percolation threshold and higher EMI shielding. *Comp. Part A: Appl. Sci. and Manufac.* **2011**, *42*, 92–97.

Al-Saleh, M. H.; Sundararaj, U. Electromagnetic interference shielding mechanisms of CNT/polymer composites. *Carbon* **2009**, 1738–1746.

Al-Saleh, M. H.; Sundararaj, U. Electrically conductive carbon nanofiber/polyethylene composite: effect of melt mixing conditions. *Polym. Adv. Technol.* **2011**, *22*, 246–253.

Bhattacharya, S.; Gupta, R. K.; Bhattacharya, S. N. The rheology of polymeric nanocomsposites. In *Polymer nanocomposite handbook*; Gupta, R.K.; Kennel, E.; Kim, K-J., eds; CRC Press Taylor and Francis Group: United States of America, 2010; pp. 151–203.

Dani, A.; Ogale, A. A. Electrical percolation behavior of short-fiber composites: experimental characterization and modeling. *Comp. Sci. Tech.* **1996**, *56*, 911–920.

Dani, A.; Ogale, A. A. Percolation in short-fiber composites: Cluster statistics and critical exponents. *Comp. Sci. Tech.* **1997**, *57*, 1355–1361.

Gelves, G.; Lin, B.; Sundararaj, U.; Haber, J. Low electrical percolation threshold of silver and copper nanowires in polystyrene composites. *Adv. Func. Mater.* **2006**, *16*, 2423–2430.

Goh, S. H. Mechanical properties of polymer-grafted carbon nanotube composites. In *Polymer-carbon nanotube composites, Preparation, properties and applications*; McNally, T.; Pötschke, P., eds; Woodhead Publishing: Cambridge, UK, 2011; pp. 347–375.

Janda, N. B.; Keith, J. M.; King, J. A.; Perger, W. F.; Oxby, T. J. Shielding-Effectiveness Modeling of Carbon-Fiber/Nylon-6,6 composites. *J. Appl. Polym. Sci.* **2005**, *96*, 62–69.

Jimenez, G. A.; Jana, S. C. Electrically conductive polymer nanocomposites of polymethylmethacrylate and carbon nanofibers prepared by chaotic mixing. *Comp. Part A: Appl. Sci. Manuf.* **2007**, *38*, 983–93.

Kasaliwal, G.; Villmow, T; Pegel, S; Pötschke, P. Influence of material and processing parameters on carbon nanotube dispersion in polymer melts. In *Polymer-carbon nanotube composites, preparation, properties and applications*; McNally, T.; Pötschke, P., eds; Woodhead Publishing: Cambridge, UK, 2011; pp. 92–132.

Keith, J. M.; King, J. A.; Miller, M. G.; Tomson, A. M. Thermal conductivity of carbon fiber/liquid crystal polymer composites. *J. Appl. Polym. Sci.* **2006**, *102*, 5456–5462.

King, J. A.; Johnson, B. A.; Via, M. D.; Ciarkowski, C. J. Effects of carbon fillers in thermally conductive polypropylene based resins. *Polym. Comp.* **2010**, *31*, 497–506.

King, J. A.; Morrison, F. A.; Keith, J. M.; Miller, M. G.; Smith, R. C.; Cruz, M.; Neuhalfen, A. M.; Barton, R. L. Electrical conductivity and rheology of carbon-filled liquid crystal composites. *J. Appl. Polym. Sci.* **2006**, *101*, 2680–2688.

King, J. A.; Via, M. D.; Keith, J. M.; Morrison, F. A. Effects of carbon fillers on rheology of polypropylene-based resins. *J. Comp. Mater.* **2009**, *43*, 3073–3089.

Kum, C. K.; Sung, Y.; Han, M. S.; Kim, W. N.; Lee, H. S.; Lee, S.; Joo, J. Effect of morphology on the electrical and mechanical properties of the polycarbonate/multi-walled carbon nanotube composites. *Macromol. Res.* **2006**, *14*, 456–460.

Lee, S.; Da, S.; Ogale, A. A.; Kim, M. Effect of heat treatment of carbon nanofibers on polypropylene nanocomposites. *J. Phy. Chem. Solid.* **2008**, *69*, 1407–1410.

Lee, S.; Kim, M.; Ogale, A. A. Influence of carbon nanofiber structure on properties of linear low density polyethylene composites. *Polym. Eng. Sci.* **2010**, *50*, 93–99.

Lee, S.; Kim, T.; Ogale, A. A.; Kim, M. Surface and structure modification of carbon nanofibers. *Synthetic Metals* **2007**, *157*, 644–650.

Ling, Q.; Sun, J.; Zhao, Q.; Zhou, Q. Microwave absorbing properties of linear low density polyethylene/ethylene–octene copolymer composites filled with short carbon fiber; *Mater. Sci. Eng., B* **2009**, *162*, 162–166.

Lozano, K.; Yang, S.; Zeng, Q. Rheological analysis of vapor-grown carbon nanofiber- reinforced polyethylene composites. *J. Appl. Polym. Sci.* **2004**, *93*, 155–162.

Mathur, R. B.; Pande, S.; Singh, B. P.; Dhami, T. L. Electrical and mechanical properties of multi-walled carbon nanotubes reinforced pmma and ps composites. *Polym. Comps.* **2008**, *29*, 717–727.

Matumura, K.; Kagawa, Y; Baba, K. Light transmitting electromagnetic wave shielding composite materials using electromagnetic wave polarizing effect. *J. Appl. Phys.* **2007**, *101*, 014912.

Morcon, M.; Simon, G. Polyolefin-carbon nanotube composites. In *Polymer-carbon nanotube composites, Preparation, properties and applications*; McNally, T.; Pötschke, P., eds; Woodhead Publishing: Cambridge, UK, 2011; pp. 511–544.

Nanni, F.; Valentini, M. Electromagnetic properties of polymer-carbon nanotube composites. In *Polymer-carbon nanotube composites, preparation, properties and applications*; McNally, T.; Pötschke, P., eds; Woodhead Publishing: Cambridge, UK, 2011; pp. 329–346.

Park, S. H.; Theilmann, P.; Yang, K.; Rao, A. M.; Bandaru, P. R. The influence of coiled nanostructure on the enhancement of dielectric constant and electromagnetic shielding efficiency in polymer composites. *Appl. Phys. Lett.* **2010**, *96*, 043115.

Paul C. R. *Introduction to electromagnetic compatibility*; John Wiley & Sons, Inc., United States of America, 1992.

Pegel, S; Villmow, T; Pegel, S; Pötschke, P. Quantification of dispersion and distribution of carbon nanotubes in polymer composites using microscopy techniques. In *Polymer-carbon nanotube composites, preparation, properties and applications*; McNally, T.; Pötschke, P., eds; Woodhead Publishing: Cambridge, UK, 2011; pp. 265–294.

Pratt T.H. *Electrostatic ignitions of fires and explosions*; American Institute of Chemical Enegineers, New York, 2000.

Ramo, S.; Whinnery, J. R.; Van Duzer, T. *Fields and waves in communication electronics*; John Wiley & Sons, Inc., Unite States of America, 1984.

Singjai, P.; Changsarn, S.; Thongtem, S. Electrical resistivity of bulk multi-walled carbon nanotubes synthesized by an infusion chemical vapor deposition method. *Mater. Sci. Eng., A* **2007**, *443*, 42–46.

Villacorta, B.; Ogale, A. A. In *Effect of Ultra-High Thermal Treatment of Carbon Nanofibers on EMI SE of LLDPE Nanocomposites;* ANTEC 2011, 2011, *11*, 0093.

Villacorta, B.; Ogale, A.; Hubing, T. Effect of Heat Treatment of Carbon Nanofibers on the Electromagnetic Shielding Effectiveness of Linear Low Density Polyethylene Nanocomposites. Accepted in *Polym. Eng. Sci.* **2012**, DOI#23276.

Villmow T.; Kretzschmar B.; Pötschke, P. *In Influence of screw configuration and mixing conditions in twin-screw extrusion on dispersion of MWNT in PCL composites*. ANTEC 2011, 2011, *11*, 0509.

Wang, S. F.; Ogale, A. A. Continuum space simulation and experimental characterization of electrical percolation behavior of particulate composites. *Comp. Sci. Technol.* **1993**, *46*, 93–103.

Xiao, K. Q.; Zhang, L. C.; Zarudi, I. Mechanical and rheological properties of carbon nanotube reinforced polyethylene composites. *Comp. Sci. Technol.* **2007**, *67*, 177–182.

Yang, D.; Rochette, J.; Sacher, E. Functionalization of Multiwalled Carbon Nanotubes by Mild Aqueous Sonication. *J. Phys. Chem. B* **2005a,** *109*, 7788–7794.

Yang, S.; Lozano, K.; Lomeli, A.; Foltz, H.D.; Jones, R. Electromagnetic interference shielding effectiveness of carbon nanofiber/LCP composites; *Composites Part A: Appl. Sci. Manuf.* **2005b**, *36*, 691–697.

CHAPTER 9

EFFECT OF DIFFERENT TYPES OF GRAPHENE AND GRAPHENE LOADINGS ON THE PROPERTIES OF NITRILE RUBBER NBR COMPOUNDS

ANNA MARZEC and MARIAN ZABORSKI

Institute of Polymer and Dye Technology, Technical University of Lodz ul. Stefanowskiego 12/16 90-924 Lodz, Poland; E-mail: anna.marzec@p.lodz.pl, marian.zaborski@p.lodz.pl

CONTENTS

ABSTRACT

Compounds of acrylonitrile-butadiene rubber (NBR) with graphene nanoplatelets were prepared. Graphene with different specific surface areas (300, 500, and 750 m^2/g) was used as a filler at 5, 10, and 15 phr loadings. NBR rubber compounds were vulcanized using a conventional sulfur-based *cross-linking system*. The effect of graphene nanoflakes concentrations as well as specific surface area on the vulcanization/curing kinetics, strain-stress behavior, cross-link density and UV stability of composites were investigated. The morphology of the graphene nanoplatelets and their dispersion in the elastomer matrix were estimated using scanning electron microscopy. Nitrile-butadiene rubbers showed significant improvement in mechanical properties even when a small amount of around 5 phr of graphene nanoplatelets was incorporated. It was also found that NBR filled with graphene exhibited enhanced UV stability in comparison to unfilled rubber material.

9.1 INTRODUCTION

Graphene, a new class of two-dimensional carbon nanostructure has attracted tremendous attention from both the experimental and theoretical scientific communities in recent years. This unique nanostructure holds great promise for potential applications in many technological fields such as nanoelectronics, sensors, nanocomposites, batteries, super capacitors and hydrogen storage (Kuilla et al., 2010; Srivastava et al., 2011; Wolf, 2014; Xu et al., 2014;Yong et al., 2012). Graphene has very high mechanical, thermal and electrical properties suitable for thermally and electrically conducting reinforced nanocomposites, electronic circuits, sensors and transparent and flexible electrodes for displays and solar cells, and so on. (Bai et al., 2011; Ozbas et al., 2012; Potts at al., 2011; Zhan et al., 2012). Especially, in the area of polymer nanocomposites, graphene is considered to be one of the most promising modifiers because of its excellent properties. Graphene as a nanofiller in polymer matrixes may be preferred over other conventional nanofillers, for example, Na-MMT, LDH, CNT, owing to its high surface area, aspect ratio, TS, thermal and electrical conductivity. For these reasons, graphene is considered the best carbon nanofiller among the traditional fillers available (Li et al., 2012; Moahmoud, 2011). Although incorporation of graphene into a thermoplastic matrix such as poly(methyl methacrylate), polycarbonate, poly(ethylene oxide) has been very successfull, application of graphene nanoplatelets in relatively non-polar polymers as rubber has not been extensively reported (Khan et al., 2010; Pang et al., 2010; Wu et al., 2010). In the current study, graphene/NBR composites with various contents of graphene were prepared, and their properties acrylonitrile-butadiene rubber (NBR) were studied.

The influence of graphene particles concentrations as well as specific surface area on the vulcanization/curing kinetics, strain-stress behavior, cross-link density and UV stability of composites were investigated.

9.2 MATERIALS

Acrylonitrile butadiene rubber NBR (Perbunan 28-45F) containing 28 wt % of acrylonitrile was obtained from Lanxess. Its Mooney viscosity was (ML1+4(100 °C):45). It was cured with sulfur in the presence of 2-mercaptobenzothiazole (MBT), zinc oxide with and stearic acid. We used three types of graphene nanoplatelets with specific surface area: 300 m^2/g (G-300), 500 m^2/g (G-500), and 750 m^2/g (G-750) from XG Science Inc. (USA).

9.3 MEASUREMENT METHODS

Rubber compounds with the general formulations given in Table 9.1 were prepared using a laboratory two-roll. The cure characteristics, including the optimum cure time, taken as the time to reach a 90% maximal cure level (T_{90}) and the delta torque of each compound, were determined at 160 °C using an ALPHA PMDR 2000. The tensile properties of vulcanizates were determined according to ISO-37 with a universal machine Zwick 1435. The cross-link density of vulcanizates was determined by equilibrium swelling in toluene, based on the Flory-Rehner (Flory and Rehner, 1943) equation using the Huggins parameter of elastomer–solvent interaction μ = 0.381 + 0.671Vr (equation 9.1).

$$v_e = -\frac{\ln(1-V_r)V_r + \mu V_r^2}{V_0(V_r^{\frac{1}{3}} - \frac{V_r}{2})} \tag{9.1}$$

where: v_e – cross-link density, v_r – volume fraction of elastomer in swollen gel, v_0 – molar volume of solvent [mol/cm^3]. The morphology of the graphene particles and the dispersion in the elastomeric matrix containing amounts of graphene was estimated using scanning electron microscopy with a LEO 1530 SEM microscope. The NBR vulcanizates were broken down in liquid nitrogen and the fracture surfaces of the vulcanizate were examined. Prior to the measurements, the samples were coated with carbon. Resistance to ageing under UV radiation was tested by exposing the vulcanizates to a UV source (Atlas UV200) for a total of 5 days (140 h). The test timeline was as follows: day segment – UV power 0.7 W/m^2; temperature 60 °C; duration 8 h; night segment – UV power 0.0 W/m^2; temperature 50 °C; duration 4 h. The vulcanizates subjected to UV ageing were tested for tensile strength and cross-link density.

TABLE 9.1 Rubber formulations

The mixture composition	*phr	*phr	*phr	*phr
NBR rubber	100	100	100	100
Zinc oxide	5	5	5	5
Sulfur	2	2	2	2
Mercaptobenzothiazole	2	2	2	2
Stearic acid	1	1	1	1
Graphene**	–	5	10	15

*phr = parts per hundred of rubber by weight
**Graphene with different specific surface area: 300 m^2/g (G-300), 500 m^2/g (G-500), 750 m^2/g (G-750)

9.4 RESULTS AND DISCUSSION

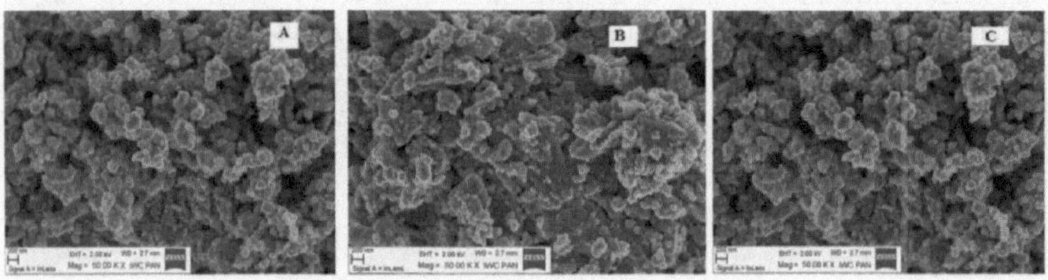

FIGURE 9.1 SEM image of graphene particles with different specific surface area: (a) 300 m^2/g (G-300), (b) 500 m^2/g (G-500), (c) 750 m^2/g (G-750).

Figure 9.1 shows the SEM images of the morphology of the graphene with specific surface area: 300 m^2/g (A), 500 m^2/g (B), and 750 m^2/g (C). The greatest irregular shape of graphene platelets characterized by an area of 300 m^2/g. For graphenes with 500 m^2/g and 750 m^2/g, particles are smaller, more focused, and have demonstrated a greater tendency to agglomerate.

TABLE 9.2 Rheometric properties NBR/Graphene blends

Sample name	Graphene loading phr	Minimum torque	Maximum torque	ΔM (M_{max} – M_{min})	T_5 min	Scorch time t_2 min	T_{90} min
NBR	0	0.5	8.6	8.1	1.1	0.8	6.8
NBR/G-300	5	0.6	11.0	10.4	2.0	1.5	8.6
	10	0.7	9.8	9.1	2.6	1.7	16.3
	15	1.0	12.8	11.8	2.3	1.6	15.8
NBR/G-500	5	0.6	10.1	9.5	2.5	1.8	9.7
	10	0.7	9.8	8.9	2.9	1.8	16.3
	15	1.0	13.4	12.4	2.7	2.0	19.2
NBR/G-750	5	0.6	10.6	10.0	2.6	1.8	13.7
	10	0.8	10.1	9.7	3.4	2.1	18.0
	15	1.2	13.8	12.6	3.3	2.4	21.3

$M_{max.}$ – Minimum torque [dNm], $M_{min.}$ – Minimum torque [dNm], ΔM – torque increase [dNm], T_{90} – vulcanization time [min]

The effects of graphenes on the properties of rubber mixtures were evaluated with respect to the curing kinetics and tensile strength of the vulcanizates (Table 9.2). Rheometric tests showed that adding graphene to the rubber mixture had significantly affected the vulcanization time (T_{90}), torque increase (ΔM), and the initial viscosity values (M_{min}) in comparison to the compounds without filler. The largest torque value was observed for samples containing 15 phr of filler. The use of these fillers also increased the scorch time of compounds. Increase in the surface area of fillers as well as increase in their proportion in the mixture affected the prolonged scorch times. The same tendency for the time of curing (T_{90}) was observed. Longest curing was characterized by a blend containing 15 phr of graphene with specific surface area 750 m^2/g (T_{90} = 21.3 min). The increase in curing time together with the increase in surface area graphenes, may be associated with the absorption curing substance on the surface of fillers. The vulcanization process produced samples, which were then subjected to tensile strength tests. The presence of filler in the composites increased the modules at value 100 %, 200 %, 300 % stress (Table 9.3).

The values of the modules (SE_{100}, SE_{200}, and SE_{300}) increased with increasing specific surface area and number of added graphene. The same trends were observed when testing the tensile strength (TS), the highest value reported for vulcanizate NBR/G-750/15 phr (13.8 MPa). It observed that the increase in surface area reduces the value of Eb and cross-link density for each type of grapheme. The distribution of graphene nanoplatelets in NBR composite seems to be inhomogeneous (Figure 9.2 A–J). The SEM images revealed agglomerates in the vulcanizates filled with all typs of graphenes. The reason for this is the hydrophilic nature of graphenes, incompatible with the hydrophobic matrix. Next, the functional properties of the vulcanizates, after undergoing the 140 h of UV ageing, were re-examined. After ageing, all samples exhibited lower TS and EB values and a higher curing density. However, the vulcanizates containing graphenes exhibited less UV degradation compared to pure sample. Based on the values (Figure 9.2), it can be conclude that graphene enhanced UV stability of NBR composites, especially at 15 phr loading.

TABLE 9.3 Effects of graphene on selected properties of NBR

Sample		Sample condition	SE_{100} [MPa]	SE_{200} [MPa]	SE_{300} [MPa]	TS [MPa]	Eb [%]	v_T*10^{-5} [mol/cm^3]
NBR		non-aged	1.20	1.58	1.88	3.82	499	5.91
		after UV	1.33	-	-	1.54	189	6.13
300 m^2/g	5	non-aged	1.47	2.12	2.99	8.86	533	6.18
		after UV	1.56	2.23	3.16	4.52	399	6.62
	10	non-aged	1.48	2.31	3.47	8.05	526	5.74
		after UV	1.58	2.42	3.64	7.85	505	5.87
	15	non-aged	1.92	3.29	5.17	11.4	543	6.12
		after UV	2.10	3.68	5.75	10.4	474	6.41

500 m^2/g	5	non-aged	1.35	1.85	2.61	8.01	507	5.82
		after UV	1.55	2.14	3.02	4.08	412	6.22
	10	non-aged	1.54	2.48	410	9.92	505	6.24
		after UV	1.60	2.52	4.08	8.18	451	6.54
	15	non-aged	1.79	3.12	5.31	12.5	518	5.86
		after UV	1.94	3.47	5.98	11.7	457	6.26
750 m^2/g	5	non-aged	1.48	2.14	3.23	8.16	463	5.82
		after UV	1.51	2.16	3.20	5.12	402	6.20
	10	non-aged	1.65	2.69	4.39	11.3	480	4.98
		after UV	1.68	2.65	4.30	7.25	422	5.64
	15	non-aged	1.92	3.52	6.42	13.8	481	5.66
		after UV	1.96	3.95	7.69	12.0	403	6.36

SE_{100}, SE_{200}, S_{300} – stress at 100, 200, 300% elongation [MPa], *TS* – tensile strength [MPa], *Eb* – relative elongation at break [%], v_T – curing density of vulcanizates calculated using the Flory Rehner equation [mol/cm³]

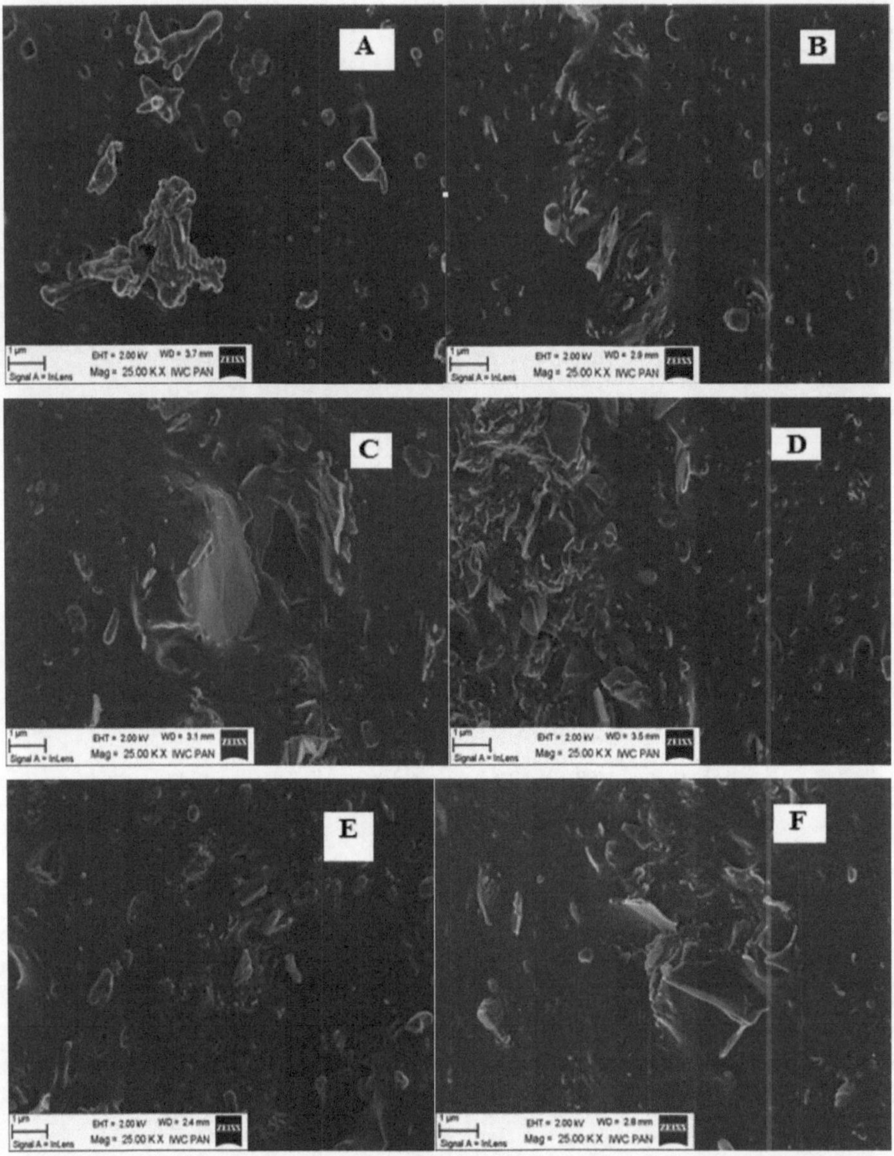

FIGURE 9.2 SEM image of vulcanizate: (A) NBR, (B) NBR/G-300/5, (C) NBR/G-300/10 (D) NBR/G-300/15, (E) NBR/G-500/5, (F) NBR/G-500/10, (G) NBR/G-500/15, (H) NBR/G-750/5, (I) NBR/G-750/10, (J) NBR/G-750/15

9.5 CONCLUSIONS

Graphene-based polymer nanocomposites exhibit superior mechanical properties compared to the neat composites. Nitrile-butadiene rubbers showed significant improvement in mechanical properties even when a small amount of around 5 phr of graphene nanoplatelets was incorporated. Improvement in the mechanical properties of graphene-filled polymer composites is not only dependent on the properties of graphene, but also on the amount of fillers in the host polymer matrixes. It was also found that NBR compounds filled with graphene exhibited enhanced UV stability in comparison to unfilled rubber material. The degree of graphene influence on vulcanization kinetics, cross-link density and elongation at break, and UV stability was dependent on the type and amount of filler and as well as specific surface area.

KEYWORDS

- **Acrylonitrile-butadiene rubber**
- **Graphene**
- **UV stability**

REFERENCES

Bai, X.; Wan, Ch.; Zhang, Y.; Zhai, Y. Carbon 2011, 49, 1608–1613.

Flory, P.J.; Rehner, J. J. Chem. Phys., 1943, 11, 52–56.

Khan, U.; May, P.; O'Neill, A.; Coleman, J. N. Carbon 2012, 48, 4035–4041.

Kuilla, T.; Bhadra, S.; Yao, D.; Kim, N. H.; Bose, S.; Lee, J. H. Prog. Polym. Sci. 2010, 35, 1350–1375.

Li, Y.; Wang, Q.; Wang, T.; Pan, G. J. Mater. Sci. 2012, 47, 730–738.

Mahmoud, W. E. Eur. Polym. J. 2011, 47, 1534–1540.

Ozbas, B.; O'Neill, Ch. D.; Register, R. A.; Aksay, I. A.; Prud'homme, R. K.; Douglas, H. J. Polym. Sci. Pol. Phys. 2012, 50, 910–916.

Pang, H.; Chen, T.; Zhang, G; Zeng, B.; Li, Z. Mater. Lett. 2010, 64, 2226–2229.

Potts, J. R.; Dreyer, D. R.; Bielawski, Ch. W; Ruoff, R. S. Polym. 2011, 52, 5–25.

Srivastava, R. K.; Srivastava, S.; Narayanan, T. N. ; Mahlotra, B. D.; Vajtai, R.; Ajayan, P. M.; Srivastava, A. ACS Nano. 2011, 6,168–175.

Wu, X.; Liu, P. Macromol. Res. 2010, 18, 1008–1012.

Zhan, Y.; Lavorgna, M.; Buonocore, G.; Xia, H. J. Mater. Chem., 2012, 22, 10464–10468.

Yong, Y. C. ; Dong, X. C. ; Chan-Park, M. B. ; Song, H.; Chen, P. ;2012, ACS Nano, 6, 2394–2400.

Wolf, E. L. Springer International Publishing, 2014, 69–77

Xu, Y., Liu; Z. W.; Xu, Y. L.; Zhang, Y. Y.; Wu, J. L.; 2014,9, 415–420.

CHAPTER 10

SYNTHESIS AND CHARACTERIZATION OF SILVER NANOPARTICLE SOLS IN PRESENCE OF DIFFERENT POLYMERIC STABILIZERS

I. SINHA[1*], MANJEET SINGH[2], and R. K. MANDAL[3]

[1]Department of Applied Chemistry, Institute of Technology, Banaras Hindu University, Varanasi 221005, India; [2]Center for Applied Chemistry, Central University of Jharkhand, Brambe, Ranchi 835205, India; [3]Department of Metallurgical Engineering, Institute of Technology, Banaras Hindu University, Varanasi 221005, India, E-mail: isinha.apc@itbhu.ac.in

CONTENTS

ABSTRACT

We review our recent work on synthesis of the silver nanoparticles using glucose as the reductant. Non-toxic or biocompatible polymer molecules such as poly (vinyl pyrrolidone) (PVP), poly(vinyl alcohol) (PVA) and starch are used to stabilize the sols. Synthesis conditions are varied by adding different amounts of NaOH resulting in interesting changes in pH of the sol formed. NaOH accelerates glucose oxidation to gluconic acid by Ag^+. This decreases the pH of the sols, which affects the reduction and the aggregation kinetics. Such kinetic control in presence of different stabilizers leads to formation of interesting anisotropic as well as fractal aggregate silver nanostructures. Coherently scattered domain size obtained from X-ray diffraction (XRD) of the nanoparticles formed gives us important information regarding the kinetics of nucleation. Small-angle x-ray scattering (SAXS) analysis, necessary for quantitative understanding of different aspects of aggregation behavior, is coupled with the localized surface plasmon resonance (LSPR) behavior information and TEM imaging to delineate various nanostructure characteristics.

10.1 INTRODUCTION

Silver nanoparticles (AgNPs) have attracted extensive attention due to their localized surface plasmon resonance (LSPR) characteristics and large effective light scattering cross section. In case of aggregation of such particles, electromagnetic interaction through dipole coupling broadens and red shifts the LSPR. Sensor applications are either based on such aggregation induced LSPR changes or LSPR shifts due to changes in the local refractive index of the medium. Aggregation sensors have enabled highly sensitive and low cost colorimetric assays for detection of small molecules, DNA, proteins, toxic metal ions, and pollutants. Conversely, interparticle junctions (or hot spots) in AgNPs aggregate structures are also particularly favorable for surface enhancing Raman scattering (SERS). However, uncontrolled aggregation of the destabilized NPs is problematic for bioanalytical applications of SERS spectroscopy. The formation of aggregate nanostructures can be controlled by stabilizers adsorbed on to the nanoparticle surfaces (Rycenga et al., 2011).

The most frequently used chemical reduction synthesis route for preparing stable AgNP suspensions utilize stabilization by ligands or polymers, especially solvent-soluble polymers, either natural or synthetic, with some affinity for metals. Stabilizers influence both the reduction rate of metal ions and the aggregation process of metal atoms. The use of synthetic polymers and different types of organic solvents render nanoparticles that may not be biocompatible. Therefore, in view of the biocompatibility requirement, polymers, such as PVP (polyvinyl pyrrolidone) and PVA (polyvinyl alcohol) have been widely used in the synthesis of AgNPs as a stabilizer. Due to presence of polar functionality and polymer backbone, these polymers can provide stability to the AgNPs by steric stabilization even in an aqueous medium. In contrast to the homopolymers PVP and PVA, a polyschacharide starch has been used in recent years for synthesis of AgNPs. Starch has been shown to act as a reductant as well as capping material and yields highly stable and water-soluble nanoparticle suspensions. The ability of all these polymers, to stabilize nanoparticles depends very much on the nanoparticles environment, that is, pH, solvent refractive index and ionic strength.

The present communication reviews our recent work and compares the stabilizing ability of the three biocompatible polymer stabilizers PVP, PVA and starch on AgNPs synthesis in an aqueous medium (Singh et al., 2009; Singh et al., 2011a; Singh et al., 2011b; Singh et al., 2011c). Except the polymer stabilizer, other synthesis conditions are kept the same to enable comparison between the studies. Glucose is used as the reductant in the presence of varying amounts of NaOH. In an aqueous medium, besides the effect on pH, NaOH addition may also influence the reduction kinetics and change the stabilizing ability of polymers. Thus, the formation of different aggregate nanostructures is possible. Small-angle X-ray scattering (SAXS) was used to study the nano aggregates formed in the sol state and to understand LSPR absorbance changes from this perspective. It is important to appreciate that in view of the length scale of the nanoparticle aggregate structures, which affects or induces LSPR changes, small-angle X-ray scattering (SAXS) is the most appropriate non-invasive technique for such investigations.

10.2 THE AGNPS SYNTHESIS PROTOCOL

Briefly, the common AgNPs preparation method followed is as given below. The details of the methods followed for each polymer stabilizer are given in references (Rycenga et al., 2011; Singh et al., 2009; Singh et al., 2011a; Singh et al., 2011b; Singh et al., 2011c). Ten milliliter aqueous solution of $AgNO_3$ (1 M) and 1 wt% of the polymer were mixed and heated up to 60 °C under continuous stirring until almost clear solution was obtained. Different reductant solutions were prepared by adding different amounts of NaOH to the aqueous solution of glucose (10 ml, 2 M). These solutions were added to different $AgNO_3$ solutions by fast injection. The color of the final solution immediately changed to dark brown in all cases indicating the formation of silver nanoparticles. The reaction was continued for 10 min under rapid stirring.

10.3 COMPARISON BETWEEN STARCH, PVA AND PVP STABILIZED AGNPS SOLS

Table 10.1 gives the typical variation of λ_{max} (the wavelength of maximum LSPR absorbance) and coherently scattered domain size with the amount of NaOH and the pH of the (starch, PVA and PVP stabilized) AgNPs sols as reported in references (Singh et al., 2009; Rycenga et al., 2011; Singh et al., 2011a; Singh et al., 2011b;

Singh et al., 2011c). In all cases, the X-ray diffraction (XRD) patterns of AgNPs show formation of FCC silver. Coherently scattered domain sizes are calculated by full width at half maximum (FWHM) using Scherrer formula after making necessary corrections for instrumental broadening. The broadening of the LSPR curves is quantified by the width of the curve at its half peak height or full width at half maximum (FWHM). The respective values of $\Delta\lambda_{FWHM}$ for different samples are shown in the fifth column of Table 10.1. As it is well known, the optical absorption spectra of noble metal NPs, owing to their LSPR absorbance, shift to longer wavelengths with increasing particle size. Further, for spherical nanoparticles $\Delta\lambda_{hw}$ increases with the broadening of the particle size distribution. Alternatively, shape anisotropy causes the splitting of LSPR peak into more peaks due to the different oscillations that can arise from different orientations of the particle with respect to the electric field of the incident electromagnetic radiation. For instance, in case of nanorods, oscillations of electrons along major (longitudinal plasmon band) and minor axes (transverse plasmon band) give rise to corresponding LSPR peaks. Broadening of the LSPR absorbance curve may therefore be due to either polydisperse size distribution or due to polydisperse shape and size distribution. Such broadening is evident from the $\Delta\lambda_{hw}$ values and the shape of the LSPR absorbance curve for AgNPs sol samples PP2, PA2 and ST1. In the PVP stabilized sample PP2 we, in fact, observe both longitudinal and transverse modes of LSPR absorbance signifying anisotropic shape formation (Singh et al., 2011a).

TABLE 10.1 AgNPs sol characterization

NP Sample		NaOH conditions (in mmoles)	pH	λ_{max} (nm)	$\Delta\lambda_{FWHM}$ (nm)	Coherently scattered domain size (nm)	SAXS inference
Starch stabilized	ST0	0	9.4	398	102	–	Spherical surface fractal
	ST1	5	1.5	423	149	25	Spherical surface fractal and small rods/oblate particles
	ST2	10	6.0	389	101	18	Spherical surface fractal

PVA stabilized	PA0	0	7.9	454	268	36	Bimodal size distribution
	PA1	0.05	4.4	445	204	17	Bimodal size distribution
	PA2	0.25	4.0	473	295	42	Bimodal size distribution (AgNPs embedded PVA cross-linked fine nanostructures)
PVP stabilized	PP0	0	8.4	436	191	26	Spherical shapes and triangular nanoplate like nanostructures
	PP1	0.1	6.2	421	115	19	Trimodal mass fractal nanostructureos
	PP2	0.5	3.3	442 & 750	225	27	Polygonal nanoplates and nanorods

The TEM micrograph of sample PP2 (Figure 10.1) clearly shows the formation of hexagonal/triangular nanoplates like structures. Similar is the case with sample PA2, where AgNPs embedded PVA, cross-linked fine nanostructures (Figure 10.2) are formed (Singh et al., 2011b). Similarly, in starch stabilized sample ST1 we observe broadening of the LSPR absorbance curve. Further, TEM micrographs also indicate the formation of oblate/rod shaped/spherical AgNPs (Singh et al., 2011c). Now we try to understand the mechanism of such anisotropic aggregation.

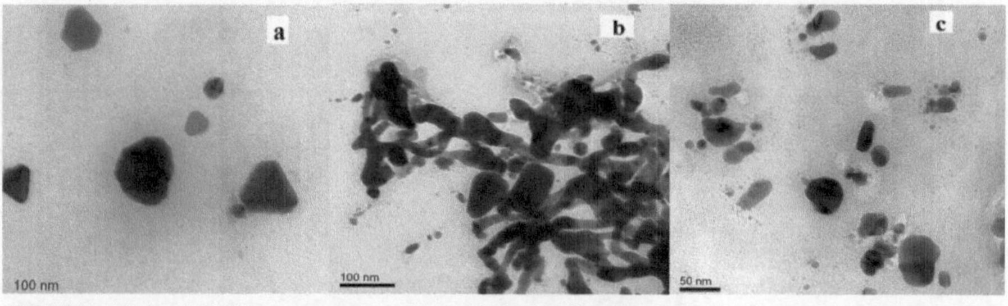

FIGURE 10.1 TEM micrographs of (a) PVP (PP2), (b) PVA (PA2) and (c) starch (ST1) stabilized AgNPs.

The Guinier–Porod empirical model proposed by Hammouda (2010) was used to analyze the small angle scattering data. In both PVP and PVA stabilized AgNPs sol SAXS profiles, we observe two distinct power law regimes. The size of the scattering object is inversely proportional to q (the scattering vector magnitude) while the slope of the linear fit to the regime gives the dimensionality of the object. For instance, if the ln–ln plot shows clear demarcation into two regions such that the low-q or intermediate-q exponent value is ~2 and this is in conjunction with a high-q region with D~3–4, then this may indicate lamellae or plate like object. We found that the SAXS profiles of PVP and PVA stabilized samples can be explained either by a bimodal size distribution or the presence of anisotropic nanostructures such as nanoplates containing different dimensions in a structure. Therefore, we are able to arrive at proper conclusions only by combining the possible SAXS models with the LSPR and TEM analysis. Hence, LSPR and TEM evidence pointed out that the SAXS profiles of PP2 and PA2 represent anisotropic nanostructures. The starch stabilized AgNPs sols SAXS profile also shows two or more distinct power law regimes. However, the dimensionality trend is opposite to that found for PVP and PVA stabilized samples. The larger objects in the sol have proper spherical surface fractal dimensions, while at the smaller structural level the SAXS profile indicates sparse mass fractal or even rod like structures (sample ST1). Such a SAXS scattering profile is also typical of those displayed by unimers. Starch stabilization of nanoparticle surfaces may be similar to that of block copolymer nanoreactor approaches. That is, the amylose hydrophobic interior associates with the nanoparticle surface while the amylose hydrophilic surface affords solubility and steric stabilization.

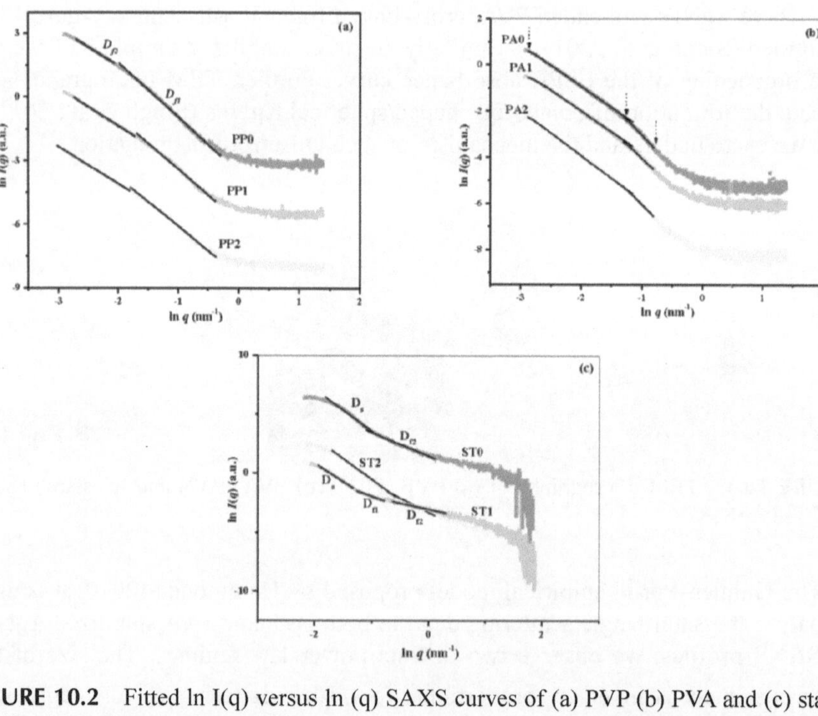

FIGURE 10.2 Fitted ln I(q) versus ln (q) SAXS curves of (a) PVP (b) PVA and (c) starch stabilized AgNPs sols. Fits according to empirical Guinier–Porod model (Hammouda, 2010).

One common feature is that we observe some amount of anisotropic aggregate structure formation at the low pH of the sol. The decrease in pH with an increase in amount of NaOH added is because it facilitates the open-chain form of glucose, which is oxidized to gluconic acid. This is the trend observed for the PVA, and the PVP stabilized sols. Here, it is important to keep in perspective the earlier reported glucose reduction of the Tollen's reagent under alkaline condition. In that case, also gluconic acid is formed; however, no shape control could be obtained with this technique. Further, the binding strength of metal complexes with sugar-type ligands, which contain hydroxyl or carbonyl oxygen donor atoms only, is weak in neutral or acidic aqueous solutions. Since the pH is the least for samples PP2, PA2 and ST1, clearly, gluconic acid does not contribute to the shape direction.

Among various techniques of shape controlled syntheses of metal nanostructures, kinetic control of the reduction rate offers a simple alternative for obtaining anisotropic nanostructures. For a face-centered cubic (fcc) noble metal, truncated nanocubes (or cubooctahedra) and multiple twinned particles (MTPs) have the lowest free energy and are therefore favored by thermodynamics (Wiley et al., 2004; Xiong et al., 2005). To obtain a shape other than the thermodynamic ones, the kinetics of nucleation must be carefully controlled (Rycenga et al., 2011). In our investigation, the coherently scattered domain size is an indication of the reduction kinetics. Faster reduction kinetics would mean high nucleation rate and smaller coherently scattered domain size. We note that for different polymeric stabilizers considered, anisotropic aggregate nanostructures (PP2, PA2 and ST1) result only when the coherently scattered domain size is the highest. Thus, the sample PP0 also shows the formation of nanoplates as well spherical nanoparticles. Samples ST1, PP2 and PA2 are also in an acidic pH regime. In case of PVP stabilization, it appears that both alkaline (PP0) and acidic pH values facilitate slower reduction kinetics. However, the control over anisotropic shapes is much better in the lower pH regime.

10.4 CONCLUSIONS

It is apparent from the above discussion that polymer stabilization at lower pH conditions significantly affects the kinetics of AgNPs nucleation and growth. Preparation of polymer-stabilized nanoparticles by a chemical method involves two processes: reduction of metal ions to zerovalent atoms and coordination of the stabilizing polymer to metal nanoparticles. We note that the synthesis protocol for all these cases involves the controlled addition of the reductant to the aqueous solution of polymer and metal ion. It is apparent that the kinetics of reduction depends on the nature and the strength of polymer-metal ion interaction (Toshima and Yonezawa, 1998). It seems that this polymer-metal ion interaction is strongest in case of PVP stabilization (Washio et al., 2006; Xiong et al., 2006). Further, low pH conditions apparently facilitate this kind of polymer-metal ion interaction. The subsequently reduced metal atoms maintain an interaction with the stabilizing polymer. These interactive forces exercise kinetic control over both nucleation and growth: seeds with stacking faults can nucleate and then grow into anisotropic shapes.

KEYWORDS

- **Silver nanoparticles**
- **Sols**
- **Starch**
- **PVA**
- **PVP**

REFERENCES

Hammouda B. J. Appl. Crystallogr. 2010, 43, 1474–1478.

Rycenga M.; Cobley C. M.; Zeng J.; Li W.; Moran C. H.; Zhang Q.; Qin D.; Xia Y. Chem. Rev. 2011, 111, 3669–3712.

Singh M.; Sinha I.; Mandal R. K. Mater. Let. 2009, 63, 425–427.

Singh M.; Singh A. K.; Mandal R. K.; Sinha I. Colloids and Surfaces A: Physicochem. Eng. Aspects 2011a, 390, 167–172.

Singh M.; Sinha I.; Singh A. K.; Mandal R. K. J. Nanopart. Res. 2011b, 13, 4387–4394.

Singh M.; Sinha I.; Singh A. K.; Mandal R. K. J. Nanopart. Res. 2011c, 13, 69–76.

Toshima N.; Yonezawa T. New J. Chem. 1998, 1179–1201.

Washio I.; Xiong Y.; Li Y.; Xia Y. Adv. Mater. 2006, 18, 1745–1749.

Wiley B.; Herricks T.; Sun Y.; Xia Y. Nano Lett. 2004, 4, 1733–1739.

Xiong Y.; Chen J.; Wiley B.; Xia Y.; Aloni S.; Yin Y. J. Am. Chem. Soc. 2005, 127, 7332–7333.

Xiong Y.; Washio I.; Chen J.; Cai H.; Li Z.-Y.; Xia Y. Langmuir 2006, 22, 8563–8570.

CHAPTER 11

SYNTHESIS OF AG-CU ALLOY NANOPARTICLES IN PRESENCE OF STARCH, PVP AND PVA AS POLYMERIC STABILIZERS: STRUCTURAL AND LSPR STUDIES

MANJEET SINGH[1*], I. SINHA[2], and R. K. MANDAL[2]

[1]Center for Applied Chemistry, Central University of Jharkhand, Ranchi, Jharkhand-835205, India

[2]Institute of Technology, Banaras Hindu University, Varanasi-221005, India;
*E-mail: msitbhu@gmail.com

CONTENTS

ABSTRACT

We report the aqueous phase chemical reduction synthesis of Ag–Cu alloy nanoparticles in presence of three different polymer stabilizers starch, PVP (Poly vinyl pyrrolidone) and PVA (Poly vinyl alcohol). We study the effect of varying proportions of Ag and Cu salts precursor on the extent of solid solubility using this route of synthesis. The localized surface plasmon resonance (LSPR) absorbances of sols are analyzed from the perspective of nanoparticles composition and their structural features. Small-angle X-ray scattering (SAXS) analysis of the Ag–Cu nanoparticles sols has been done extensively and shows the formation of mass fractal aggregate nanostructures in all cases. It has been observed that the starch stabilized Ag-Cu alloy nanoparticles shows the presence of LSPR maxima corresponding to the Ag-rich and Cu-rich phases, whereas PVP and PVA stabilized Ag-Cu alloy nanoparticles show LSPR maxima corresponding to Ag-rich phases only. The extent of solid solubility is different for each polymeric stabilizer. X-ray diffraction study (XRD) analysis shows the presence of multiple phases of Ag-Cu nanoparticles owing to the different levels of solid solubility. The solid solubility of Cu in Ag has been found to be ~10at% in starch, ~18at% in PVP and ~8at% in PVA and solid solubility of Ag in Cu 1.5at% in starch, 2at% in PVP and PVA.

11.1 INTRODUCTION

Bimetallic alloy nanoparticles have improved optical, electronic, and catalytic properties different from those of the component metals. The Ag–Cu nanoparticles have been used to improve the properties of lead-free interconnects, dental amalgams, and so on (Chung et al., 2008; Hongjin et al., 2005). For example, current commercial micron-size silver–copper particles are not suitable for fine pitch interconnect assembly. Better methods for synthesis of finer (<1 μm) alloy particles need to be developed. Other important feature of the alloy system is the localized surface plasmon resonance (LSPR) behavior, which can be tuned by changing the composition of the component metal. In this context, an important issue pertains to the possibility of enhancing solid solubility of components in such small finite systems (nanoparticles) (Cable and Schaak, 2005; Eccles et al., 2010) at room temperature. The Ag–Cu alloy system showing eutectic behavior and wide range of the miscibility gap is a best-suited system for this study. Owing to the same crystal structure (F.C.C.) of Ag and Cu, and size mismatch of less than 15%, Hume-Rothery rule predicts the formation of solid solution under metastable condition. Such condition can be achieved by liquid phase synthesis route of nanoparticles or alloy particles. There are few reports on the Ag–Cu nanoparticles synthesized by chemical reduction route (Hongjin et al., 2005; Singh et al., 2009a). Hongjin et al. have reported the synthesis and the LSPR behavior of phase-separated Ag–Cu nanoparticles sols. In their investigation, copper acetate hydrate and silver nitrate were added sequentially for preparation of Ag–Cu nanoparticles with poly (vinylpyrrolidone) (PVP) as a stabilizer in ethylene glycol. In a recent study, Cu shell on Ag-rich core and Ag shell on Cu-rich core bimetallic nanoparticles have been synthesized by the conventional polyol route (Tsuji et al., 2010). We had earlier shown that Ag–Cu ultra-fine particles with varied degrees of solid solubility could also be prepared in an aqueous medium (Singh et al., 2009b). Since the polymer stabilizer plays a very important role in directing the growth and consequently the solid solubility, different types of stabilizers give a different level of solid solubility. In this work, effect of three polymeric stabilizers, starch, PVP, and PVA on solid solubility, were studied. Also the effects of these polymers on the LSPR behavior of Ag–Cu nanoparticles were studied.

11.2 EXPERIMENTAL

$AgNO_3$ and Cu $(NO_3)_2{\cdot}3H_2O$ were taken in a definite ratio, and their aqueous solution prepared in 20 mL Millipore water. One gram of each polymer, starch, PVP, and PVA were added separately to prepare three types of polymer mixed salt solution and stirred until clear solution was obtained. Hydrazine hydrate (1 mL, 99–100%) and NaOH (1 mL, 1 M) were added in these solutions under rapid stirring. A black precipitate was obtained immediately. Stirring was continued for the next 15 min to ensure completion of reaction. Different samples were made by changing the molar ratios of $AgNO_3$ and $Cu(NO_3)_2{\cdot}3H_2O$. The precipitate was washed with Millipore water after filtration. In case of starch stabilization, nanoparticles sol was obtained, whereas in case of PVP (Singh et al., 2009b) and PVA stabilization powder was obtained after filtration. In all the three cases, precipitate was dried and the powder thus obtained was used for X-ray diffraction analysis. UV-Visible analysis (Cary Bio 100 UV-Visible spectrophotometer) was done by dispersing the powder in water. Small-Angle X-ray Scattering (SAXS) and TEM (Tecnai 20G^2) analysis was carried out by taking a drop of dispersion of the powder obtained. Atomic force microscope (AFM) analysis was also carried out by taking a pinch of powder on the sample holder in the tapping mode.

11.3 RESULTS AND DISCUSSION

The X-ray diffraction (XRD) of all samples were done at a scan speed of 0.25°/min. Extensive overlapping of Ag, Cu (of different levels of solid solubility) peaks, broadening and splitting of peaks are observed. The major overlapping peaks are deconvoluted using standard routines to determine the peak positions and line widths. The lattice parameters of the identified phases are compared with the experimental data of Linde (Linde, 1966). Alloying, results in change of the lattice parameter from that of the pure element to one intermediate. Coherently scattered domain sizes are calculated by full width at half maximum (FWHM) using Scherrer formula after making necessary corrections for instrumental broadening. The XRD patterns for various PVA stabilized NPs samples are shown in Figure 11.1. Table 11.1 summarizes the quantitative details of the XRD analysis.

TABLE 11.1 Characteristics of the phases present, corresponding lattice parameters and the coherently scattered domain size of the PVA stabilized Ag–Cu alloy NPs

Sample number (Ag:Cu) Molar ratio	Phases observed (%)	d-spacing (A°)	Lattice parameter	XRD domain size (nm)
A0 (Ag)	fcc Ag	2.362 (111)	4.09	13
A1 (3:1)	2H-Ag	2.396(0002)		
	fcc Ag	2.374(111)	4.09	12
	fcc Ag97.90Cu2.1	2.355 (111) 2.039 (200)	4.078	
A2 (2:1)	2H-Ag	2.372 (0002)		
	fcc Ag	2.365 (111)	4.09	11
	fcc Ag97.90 Cu2.1	2.355 (111) 2.039 (200)	4.078	15
A3 (1:1)	2H- Ag	2.385 (0002)		
	fcc Ag	2.362 (111)	4.09	12
	fcc Ag92.5Cu7.5	2.343 (111)	4.058	20
	fcc Ag0.9Cu99.1	2.091 (111)	3.621	–
A4 (1:2)	2H-Ag	2.395 (0002)		
	fcc Ag	2.365 (111)	4.09	12
	fcc Ag94.4Cu5.6	2.349 (111)	4.068	15
	fcc Ag2.0Cu98.0	2.094 (111) 1.812 (200)	3.626	
A5 (1:3)	2H-Ag	2.531 (10$\bar{1}$0) 2.395 (0002)		
	fcc Ag	2.365 (111)	4.09	13
	fcc Ag98.66Cu1.34	2.355 (111)	4.078	13
	fcc Ag2.09Cu97.91	2.093 (111) 1.812 (200)	3.625	-

A6 (1:5)	2H-Ag fcc Ag fcc Ag94.4Cu5.6 fcc Ag2.09Cu97.91	2.531 (10$\bar{1}$0) 2.374 (0002) 2.362 (111) 2.349 (111) 2.093 (111) 1.812 (200)	 4.09 4.068 3.625	 15 –

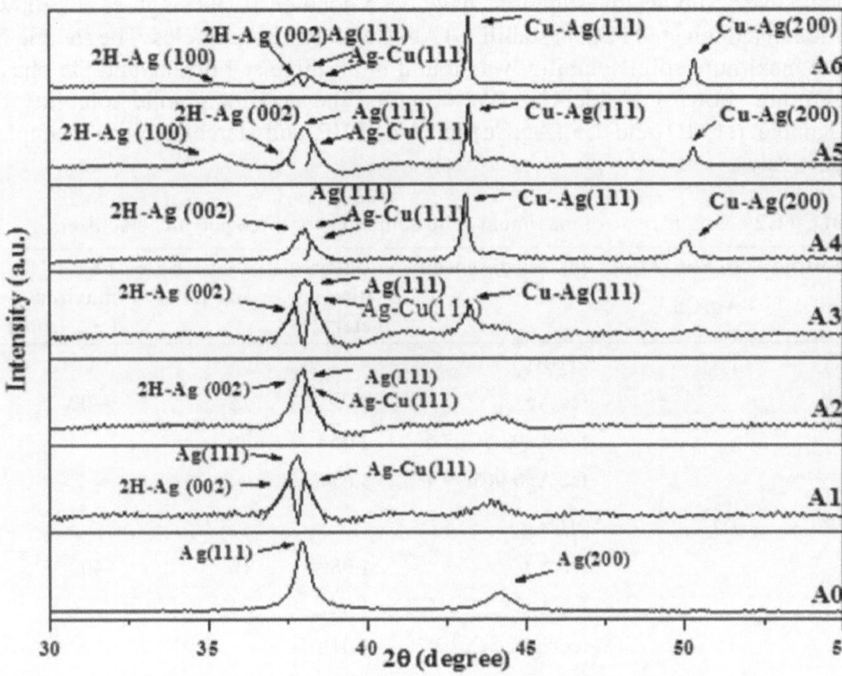

FIGURE 11.1 XRD patterns of PVA stabilized Ag–Cu alloy NPs

The marginal reduction in the lattice parameter is due to the small alloying of fcc (Ag) phase in A1 by (~2.1%) Cu. Two different fcc (Ag) (111) lines are indexed in the XRD for sample A2 (Figure11.1). One of them corresponds to the nearly pure fcc (Ag) (111) line earlier observed in A1. The other fcc (Ag) (111) line (a = 4.078 Å) occurs owing to increased solid solubility of Cu in the fcc (Ag) phase corresponding to the alloy phase. Similarly different levels of solid solubility are found in all the samples (Figure11.1). The maximum solid solubility of Cu in Ag is found in A3. The solubility for this sample shows ~7.5 at.% Cu in Ag (Fig. 11.1).We also observe that this peak in the XRD spectrum overlaps with several other closely spaced peaks, which suggest the presence of other alloy compositions with Cu solid solubility in Ag ranging from 3 to 7.5 at.%. The domain size of all phases, found in the samples considered, is in the nanometer range (Table 11.1).

Except in A0, in all other samples, the formation of the hexagonal 2H-(Ag) phase was observed. In the XRD of A1, A2, A3 and A4 only the (0002) line of the 2H-(Ag) phase is observed. While for A5 and A6 both (10 0) and (0002) lines of the 2H-(Ag) phases are found. The d-value of the (0002) plane in the ICDD [PDF card No. 01-071-5025] is 2.395 Å but in our investigation, it comes at slightly lower value for some samples (Table 11.1) that can be attributed to the alloying of hexagonal silver. Similar investigations have been done on PVP (Singh et al., 2009b) and starch (Singh et al., 2011) stabilized Ag–Cu alloy nanoparticles. The sample for which maximum solid solubility was found in all three stabilizers, and the phases present are shown in Table 11.2. We observed the maximum solid solubility for the sample 1:1, 1:1 and 1:5 (Ag:Cu) in PVA, PVP, and starch stabilized samples, respectively.

TABLE 11.2 Comparison of maximum solid solubility for three polymer stabilizers

Stabilizers	Molar ratio Ag:Cu	Phases observed	Lattice parameter	Domain size (nm)	LSPR maxima (λ_{max} (nm))
PVA	1:1	2H-Ag			
		fcc Ag	4.09	12	~400
		fcc Ag92.5Cu7.5	4.058	20	
		fcc Ag0.9Cu99.1	3.621		
PVP	1:1	2H-Ag			
		fcc Ag	4.089	10	~400
		fcc Ag88.3 Cu11.7	4.042	21	
		fccAg82.3Cu17.7	4.018		
		Fcc Ag1.8Cu98.2	3.626		
Starch	1:5	2H-Ag			~420 and 588
		fccAg94.32Cu5.68	4.063	15	
		fccAg89.49Cu10.51	4.046	21	
		fcc-Cu	3.600		

The last column of the Table 11.2 indicates the LSPR maxima position of the dispersed powder in water determined by UV-Visible spectra. Appearance of peak corresponding to only (Ag)-rich phase (~420 nm) in PVA and PVP polymer stabilizers indicates that the size of the (Cu)-rich particles (~588 nm) are more than the

LSPR limit (>100 nm). (Cu)-rich phase only appears in the case of starch stabilization. This indicates that the starch stabilized particles are finer than the PVA and PVP stabilized particles. SAXS analysis can only be performed in case of starch stabilized nanoparticles sol because in the other two cases, particles formed are too bigger to be investigated in SAXS. SAXS results (Singh et al., 2011) show the formation of mass fractal aggregates in all cases as well as the bimodal distribution of the particles. TEM photograph also indicates the bimodal size distribution (Singh et al, 2011) of particles, one corresponding to (Ag)-rich phase and the other corresponding to (Cu)-rich phase.

To investigate particles morphology, AFM photograph was taken under tapping mode, of samples A2 as shown in Figure 11.2. From the photograph it is clearly seen that particles have sizes greater than 100 nm in conformity with the UV-Vis analysis. Since the photograph is taken in powder form some aggregated particle is also visible in addition to the isolated particles.

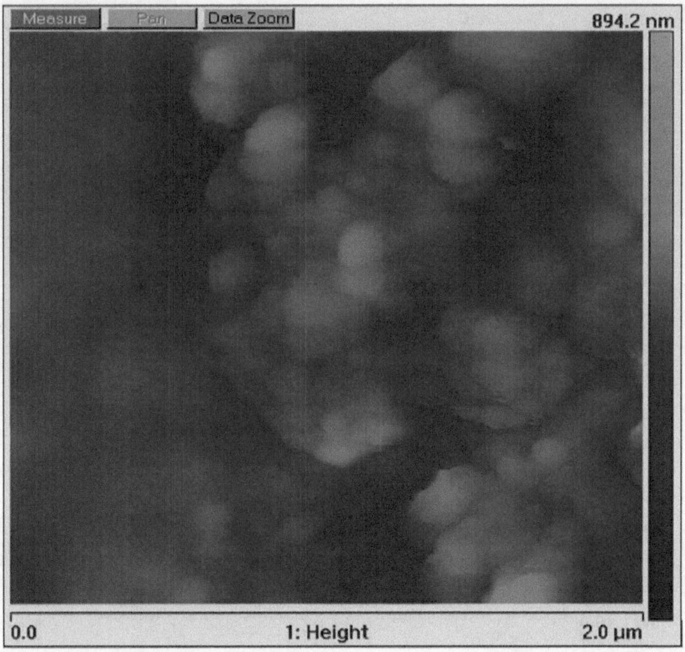

FIGURE 11.2 AFM photo graph of sample A2

11.4 CONCLUSIONS

The Ag-Cu alloy nanoparticles have been synthesized by chemical reduction route in presence of three stabilizers, starch, PVA, and PVP. The synthesized Ag-Cu alloy particles displayed different degrees of solid solubility for each stabilizer. The starch stabilized Ag–Cu nanoparticles display two distinct absorption peaks in UV-Visible spectra. These are attributed to Ag-rich and Cu-rich solid solutions. The nanoparticles suspension has been obtained only in starch stabilization, whereas nanopowder has been obtained in PVP and PVA stabilization. In case of starch stabilized alloy nanoparticles, SAXS studies were undertaken to understand the LSPR results of the sol synthesized. SAXS studies revealed mass fractal aggregate formation in the starch stabilized Ag–Cu alloy NPs. The maximum solid solubility of Cu in Ag was found to be ~10at% in starch, ~18at% in PVP, and ~7.5at% in PVA. The maximum solid solubility of Ag in Cu was found to be ~1.5at% in starch and ~2at% in PVP and PVA. In addition to the formation of Ag-rich FCC phases, Ag (rich)-2H phase was also observed.

KEYWORDS

- **Poly vinyl pyrrolidone**
- **Poly vinyl alcohol**
- **Localized surface plasmon resonance**

REFERENCES

Cable R.E., Schaak R.E. Low-temperature solution synthesis of nanocrystalline binary intermetallic compounds using the polyol process. Chem. Mater. 2005;17;6835–6841.

Chung K, Hsiao L, Lin Y, Duh J. Morphology and electrochemical behavior of Ag–Cu nanoparticle-doped amalgams. Acta. Biomater. 2008; 4;717–724.

Eccles J.W.L, Bangert U., Bromfield M., Christian P., Harvey A.J. Do nanomaterials form truly homogeneous alloys? J. Appl. Phys. 2010;107;104325-1-6.

Hongjin J, Kyoung-sik M, Wong C.P. Synthesis of Ag–Cu alloy nanoparticles for leadfree interconnect materials, advanced packaging materials: processes, properties and interfaces. Proceedings International Symposium; 2005, p. 173–177, 16–18 March.

Linde RK. Lattice parameters of metastable silver–copper alloys. J. Appl. Phys. 1966;37;934.

Singh M., Sinha I., Mandal R.K. Role of pH in the green synthesis of silver nanoparticles. Mater. Lett. 2009a; 63;425–427.

Singh M., Sinha I., Mandal R.K. Synthesis of nanostructured Ag–Cu alloy ultrafine particles. Mater. Lett. 2009b; 63;2243–2245.

Singh M., Sinha I., Singh A.K., Mandal R.K. LSPR and SAXS studies of starch stabilized Ag–Cu alloy nanoparticles. Colloids and Surfaces A: Physicochem. Eng. Aspects 2011;384;668–674.

Tsuji M., Hikino S., Tanabe R., Matsunagaa M., Sano Yoshiyuki. Syntheses of Ag/Cu alloy and Ag/Cu alloy core Cu shell nanoparticles using a polyol method. Cryst. Eng. Commun. 2010;12;3900–3908.

CHAPTER 12

REMOVAL OF Cr (VI) FROM AQUEOUS SOLUTION BY SUPER ABSORBENT POLY [N, N-DAPB/N, N-DMAAM/PNAAC] HYDROGELS

YATIN N. PATEL and MANISH P. PATEL*

Department of Chemistry, Sardar Patel University, Vallabh Vidyanagar-388120, Gujarat, India

*Corresponding author: Tel: +91 2692 226856, E-mail: patelmanish1069@yahoo.com

CONTENTS

ABSTRACT

Fast swelling super absorbent poly [*N, N*-DAPB/*N, N*-DMAAm/PNAAc] hydrogels (A_1–A_5) were synthesized via microwave irradiated free radical solution polymerization of DAPB, DMAAm and PNAAc using *N, N'*-methylenebisacrylamide (MBAm) as crosslinker. According to swelling experiments, hydrogel A_2 shows higher percentage of swelling compared to other hydrogels. The hydrogel has been characterized by FT-IR, TGA and SEM analysis. A batch system was applied to study the adsorption of Cr (VI) from aqueous solutions by super absorbent hydrogels. The experiments of Cr (VI) adsorption were performed at different initial ion concentrations, pH of the solution, adsorbent dosage, and contact time. Moreover, the optimum pH for the adsorption was found to be 1.0 with the corresponding adsorbent dosage level of 20 mg. Subsequently, the equilibrium was achieved for the 200 mg/L solution of Cr (VI) with 120 min of contact time. Furthermore, Langmuir and Freundlich adsorption isotherm was applied. The hydrogel A2 shows better results after desorption up to five cycles.

12.1 INTRODUCTION

In natural waters, the range of chromium concentration is quite large, from 5.2 to 208,000 mg/L (Richard and Bourg, 199). Nevertheless, for most natural waters, the chromium concentration is below the 50 μg/L value recommended for drinking water by the World Health Organization or the US Environmental Protection Agency (USEPA, 1996; WHO, 1997).

Cr (VI) is the most toxic form, being carcinogenic and mutagenic to living organisms (IARC, 1987). In addition, it leads to liver damage, pulmonary congestion, and causes skin irritation resulting in ulcer formation (Cieslak-Golonka, 1996). Trivalent Trivalent Cr (III) is about 300 times less toxic than Cr (VI) and due to the fact that it has limited hydroxide solubility, it is less mobile and fewer bio available (Krishnan and Ayyappan, 2006). Cr (III) is essential to animals and plants and plays an important role in sugar and fat metabolism, although in excess can cause allergic skin reactions and cancer (Kotas and Stasicka, 2000). The use of Chromium chemicals in several industrial processes (leather tanning, mining of chrome ore, production of steel and alloys, dyes and pigment manufacturing, glass industry, wood preservation, textile industry, film and photography, metal cleaning, plating and electroplating, etc.) leads to contamination of natural waters mainly due to improper disposal methods (Sarin et al., 2006).

Because of its high toxicity, Cr (VI) must be substantially removed from the wastewater before being discharged into the aquatic system. Different technologies are available: chemical precipitation, coagulation, ion exchange, membrane technologies, adsorption, and so on. The most common conventional method for Cr (VI) removal is a reduction to Cr (III) at pH2.0 and precipitation of Cr $(OH)_3$ by increasing pH to 9–10 using lime. The disadvantage of precipitation is the disposal of the solid waste. Nowadays, adsorption becomes by far the most versatile and widely used technology, and activated carbon, the most commonly used sorbent. However, the use of activated carbon is expensive, so there has been increased interest in the last years in the use of other adsorbent materials. The removal of chromium from wastewaters using bacteria, fungi, algae, and different plants has been reported, but although these material show good performance under laboratory conditions, their use for large scale effluents treatments may not be suitable due to their relatively poor natural abundance (Gao et al., 2008). The use of adsorbents showed higher potential for chromium removal from aqueous solution (Abdel-Halim and Al-Deyab, 2011; Cheng et al., 2011; Lo et al., 2011; Roy et al., 2011; Liu et al., 2012; Tang et al., 2012).

Hydrogels are water insoluble, three-dimensional networks of polymeric chains, which are crosslinked by chemical or physical bonding, possess the ability to absorb a large amount of water and swell, while maintaining their three-dimensional structure. The swelling behavior of hydrogels may be characterized by water adsorption (Saraydin et al., 2001). Hydrogels are finding increased use in metal ion sorption. In our earlier studies, we have established that hydrogels are very effective in partitioning of metal ions from the solution phase. Partitioning capacity of hydrogels is related to the capacity to absorb water. The presence of functional groups like –COOH and $–CONH_2$ not only improves water uptake capacity of the hydrogels, but these also act as efficient anchors for active molecular species and metal ions.

The purpose of this work was aimed to gain a new super absorbent hydrogel, which is cost-effective, safe and able to carry out almost complete removal of heavy-metal ions from wastewater. In recent years, many researchers along with our research group have worked on the removal of heavy-metal ions, and textile dyes using different hydrogels having different functional groups (Dadhaniya et al., 2009; Rivas et al., 2009; Patel et al., 2011; Patel et al., 2012). There is no reference available for the removal of Cr (VI) by super absorbent poly [*N, N*-DAPB/*N, N*-DMAAm/PNAAc] hydrogels from aqueous solution. We have synthesized this hydrogel keeping an objective to improve their water swelling capacity and removal of Cr (VI) from aqueous solution by using DMAAm and PNAAc. In view of the health problems caused by Cr (VI), the present study was performed to evaluate super absorbent poly [*N, N*-DAPB/*N, N*-DMAAm/PNAAc] hydrogels A_2 as adsorbent for the removal of Cr (VI) from polluted water by systematic evaluation of the parameters involved such as P^H, initial ion concentration, contact time, and adsorbent dosage.

12.2 MATERIALS AND METHODS

12.2.1 MATERIALS

Acrylic acid (AAc), pyrrolidine, allyl bromide, diethyl ether, *N, N'*- methylenebisacrylamide (MBAm) crosslinker, potassium persulfate (KPS) initiator, sodium hydroxide, and potassium dichromate ($K_2Cr_2O_7$) used were of analytical grade. *N, N*-dimethyl acrylamide monomer was supplied by Aldrich (Steinheim, Germany). Double distilled water was used in hydrogel synthesis and metal removal studies. The stock solution of monomer partially neutralized acrylic acid was prepared by dropping 40 ml of 40% sodium hydroxide (W/V) into 60 ml of magnetically stirred acrylic acid with ice cooling (Bajpai et al., 2006).

12.2.2 INSTRUMENTATION

The functional groups of the synthesized hydrogel were confirmed using a Perkin-Elmer spectrum GX FTIR spectrophotometer with 400–4000 cm^{-1} range *via* KBr pellet. METTLER Thermo Gravimetric Analyzer (TGA) was used to determine the thermal behavior of the hydrogel. A Hitachi S-3000N Scanning Electron Microscope was used to observe the surface morphology of the hydrogel. Shimadzu-18A UV-Visible spectrophotometer was used to determine the equilibrium concentration of metal ion.

12.2.3 METHODS

Synthesis of *N*-diallylpyrrolidinium bromide monomer was synthesized by adopting a described procedure in our previous work (Liu et al., 2012).

Synthesis of super absorbent hydrogels

The super absorbent poly [*N, N*-DAPB/*N, N*-DMAAm/PNAAc] hydrogels (A_1–A_5) are prepared via microwave irradiated free radical cyclopolymerization in presence of MBAm as a crosslinker and KPS as an initiator. Monomers, crosslinker, and initiator were mixed together in the appropriate molar ratio and dissolved in 10 mL of double distilled water in a 100 mL single neck round bottom flask under magnetic stirring. This mixture was stirred to give homogeneous solution. The flask was put in a microwave and irradiated at 280 W for 1 min. Then the flask was put in an oven at 60 °C for 3 h to reduce the residual monomer content in the gel. The polymerization reactions were carried out by varying the concentrations of monomers in the feed. The hydrogels (A_1–A_5) were immersed in double distilled water and kept for one day by changing the water at every 4 h intervals. The swollen hydrogel was immersed in acetone for 4 h by replacing fresh acetone twice. Finally the hydrogels were dried in air forced oven at 60 °C to constant weight. The hydrogels sample code, composition detail and swelling characteristics are presented in Table 12.1.

TABLE 12.1 Composition detail and swelling characteristic of studied super absorbent hydrogels (A_1–A_5)

Hydrogel Code	*N, N*-DAPB Mole	*N, N*-DMAAm Mole	PNAAc Mole	MBAm Mole	KPS Mole	% S (in D.D Water)
A_1	1	1	1	0.0032	0.0018	13,520
A_2	1	1	2	0.0032	0.0018	25,954
A_3	2	1	1	0.0032	0.0018	5789
A_4	0.5	2	0.5	0.0032	0.0018	9103
A_5	1	0.5	1	0.0032	0.0018	7588

12.2.4 SWELLING MEASUREMENT

In order to determine the percentage of equilibrium swelling (%S) of the super absorbent hydrogels (A_1–A_5), the gravimetric method was employed. An accurately weighed hydrogel was immersed in a beaker containing 100 mL of double distilled water. After attaining the equilibrium, the hydrogel was taken out, and the excess water was removed by filter paper and then weighed accurately using an REPTECH electronic balance (Modal RA-123 with 0.001 g accuracy).

The percentage of equilibrium swelling of hydrogel was calculated using the following Eq. (12.1).

$$\%S = \frac{W_{eq} - W_0}{W_0} \times 100 \tag{12.1}$$

where W_{eq} is the weight of swollen gel at equilibrium and W_0 is the weight of dry gel.

12.2.5 ADSORPTION EXPERIMENTS

Adsorption of Cr (VI) from aqueous solutions was investigated in batch experiments. The effects of several parameters, such as pH of the solution, contact time, adsorbent dosage, and initial metal ion concentration on adsorption were studied. Through the entire adsorption experiments, the stock solution of the metal ion was diluted to the required concentration. All the adsorption isotherm experiments were carried out at room temperature. The adsorbent amount was kept constant at 20 mg in a 50 mL solution of desired concentration of metal ion at varying pH in a 100 mL stopper the conical flask. The pH of the solution in the experiments was adjusted with HCl and NaOH solutions. The reaction mixture was agitated at 150 rpm for 3 h in a Scigenics Orbitek mechanical shaker. Subsequently, the aqueous solution containing the adsorbent was filtered out to separate the adsorbent.

The final concentrations of the metal ion which remained in the solution were determined by UV-Visible spectrophotometer using a Shimadzu-18A UV-Visible spectrophotometer, using precalibrated curves. The amount of metal ions adsorbed on the hydrogel was calculated, based on the difference of the metal ions concentration in the aqueous solution before and after adsorption, according to the following Eq. (12.2) (Kalyani et al., 2005). The amount of metal taken up by the adsorbent was calculated by applying Eq. (12.2).

$$q_e = \frac{C_0 - C_e}{m} \times V \tag{12.2}$$

Where q_e is the amount of metal taken up by the adsorbent (mg/g), C_0 is the initial metal ion concentration put in contact with adsorbent (mg/L), C_e is the equilibrium metal ion concentration (mg/L) after the batch adsorption procedure, m is the adsorbent mass (g) and V is the volume of the metal ion solution (L). The percentage of the removal of Cr (VI) can be calculated from the following Eq. (12.3).

$$\%removal = \frac{C_0 - C_e}{C_e} \times 100 \tag{12.3}$$

12.3 RESULTS AND DISCUSSION

The batch adsorption and swelling experiment show that hydrogel A_2 has a higher Cr (VI) removal efficiency from 200 mg/L Cr (VI) solution, and it shows higher percentage of swelling in double distilled water. Therefore, hydrogel A_2 was selected for the adsorption of Cr (VI) from aqueous solution. The hydrogel A_2 has been characterized by FT-IR, TGA and SEM analysis.

12.3.1 ADSORPTION EXPERIMENTS

FT-IR analysis of hydrogel A_2

A representative FT-IR spectrum of hydrogel A_2 is presented in Figure12.1. The FT-IR analysis of the hydrogel showed the presence of peaks corresponding to the functional groups of monomeric units, that is, *N, N*-DAPB, *N, N*-DMAAm and PNAAc in the copolymeric hydrogel chain. A broad peak corresponding to –COO–H of acrylic acid was observed at 3439 cm^{-1}. The peaks corresponding to C–H stretching were observed at 2924 cm^{-1}. Two peaks were observed at 1722 cm^{-1} corresponding to the C–O stretching of acrylic acid and 1627 cm^{-1} corresponding to the carbonyl stretching of dimethyl acrylamide units. Two peaks were observed at 1405 cm^{-1} corresponding to C–N stretching of DAPB and 1265 cm^{-1} corresponding to the

C–N stretching of dimethyl acrylamide. The IR spectra also show a small peak at 1458 cm^{-1} due to the CH_2 bending.

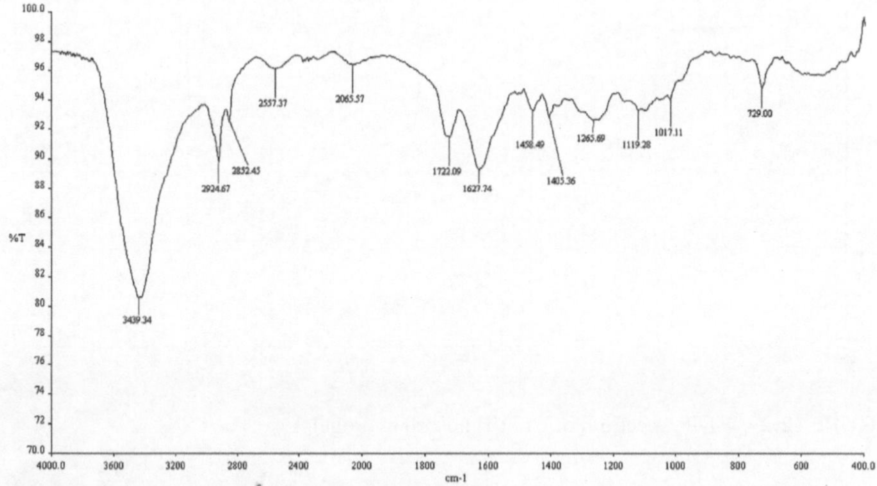

FIGURE 12.1 FT-IR spectrum of hydrogel A_2

FT-IR analysis of Cr (VI) adsorb hydrogel A_2

Figure 12.2 represents FT-IR spectrum of Cr (VI) adsorbed hydrogel A_2. In the binding of metal ions with hydrogel some new bands are observed and certain shift has been also observed. The additional peaks appeared at 947, 904 and 884 cm^{-1}. The appearance of these peaks may be due to the M–O (metal-oxygen) bond. The hydrogels contain many ionic groups in its backbone, which can bind with the cationic Cr (VI) ions and form the M–O bonds. Therefore, the insertion of the metal ion into the hydrogel network may be responsible for the absorption band shift at higher frequency. The peaks observed at 1405 and 1458 cm^{-1} in FT-IR spectrum of hydrogel A_2 is shifted to lowering the frequency, 1403 and 1453 cm^{-1} respectively. The peaks observed at 1627, 1722, 2924, and 3439 cm^{-1} in FT-IR spectrum of hydrogel A_2 is shifted to higher frequency, 1630, 1726, 2927, and 3440 cm^{-1} respectively.

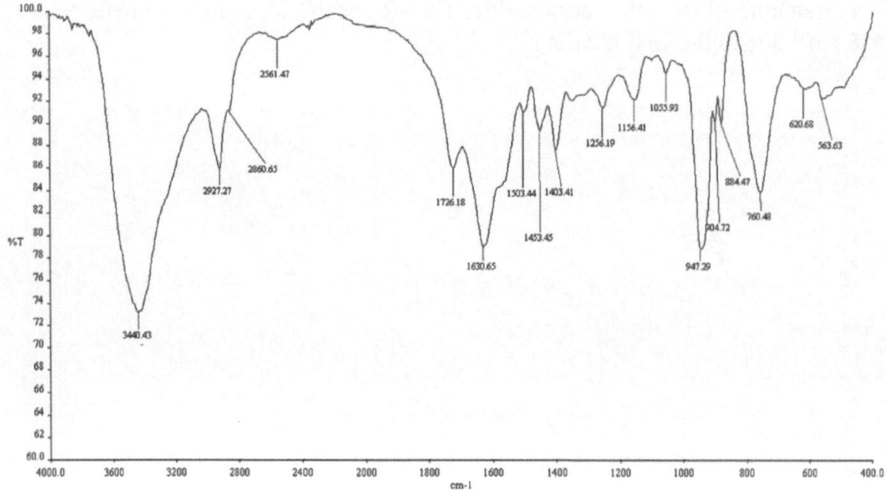

FIGURE 12.2 FT-IR spectrum of Cr (VI) adsorb hydrogel A_2

TGA analysis of hydrogel A_2 and Cr (VI) adsorb hydrogel A_2

The TGA thermogram was taken for the hydrogels with and without metal ion absorption. The relative thermal stabilities of both the different hydrogels were assessed by comparing the weight loss in the temperature range 50–600 °C. The TGA data furnished in Figure 12.3 shows the nature of the thermograms of the hydrogel and the hydrogel with metal ion. From the TGA thermograms, it is observed that the Cr (VI) adsorbed A_2 hydrogel has much higher thermal stability as compared to A_2 hydrogel. Up to 400 °C, there is a negligible weight loss in the Cr (VI) adsorbed hydrogel. The percentage of weight loss of A_2 hydrogel at 600 °C is 87.81%, while that of Cr (VI) adsorbed A_2 hydrogel at 600 °C is 33.48%. The high thermal stability of Cr (VI) adsorbed A_2 hydrogel may be due to the insertion of metal ion into the hydrogel network. The metal ion makes the hydrogel network more thermally stable as compared to virgin A_2 hydrogel. Hence, Cr (VI) adsorbed A_2 hydrogel shows only 33.48% weight loss as compared to 87.81% weight of A_2 hydrogel.

The difference in the morphology of A_2 hydrogel with and without adsorbed metal ion was observed by SEM analysis. Figure 12.4 (a) and (b) represent the SEM microphotographs of hydrogel A_2 and Cr (VI) adsorb hydrogel A2. The adsorbed metal ions on the surface of hydrogels are clearly seen in Figure 12.4 (b), and we cannot see any pores, that is, there may be a trapping of pores by the metal ions.

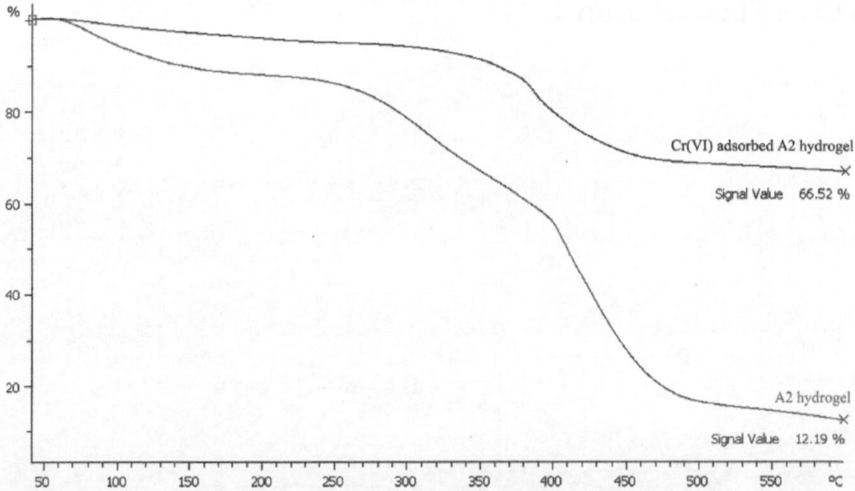

FIGURE 12.3 TGA curve of hydrogel A_2 and Cr (VI) adsorb hydrogel A_2

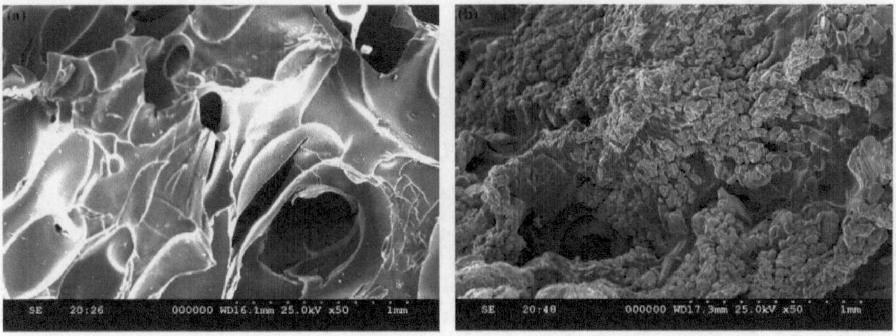

(a) (b)

FIGURE 12.4 (a) SEM analysis of hydrogel A_2 (b) Cr (VI) adsorb hydrogel A_2

12.3.2 SWELLING STUDIES

A fundamental relationship exists between the swelling of a polymer in a solvent and the nature of the polymer and the solvent. The percentage of equilibrium swelling is the most important parameter about swelling studies. The %S was calculated from Eq. (12.1) (Karadag et al., 2001; Saraydina et al., 2001).

The water intake of initially dry hydrogels was followed for a short time. Swelling isotherm of poly [*N, N*-DAPB/*N, N*-DMAAm/PNAAc] hydrogels (A_1–A_5) is shown in Figure 12.5. It can be seen that %S increases with time until a certain point, then it becomes constant.

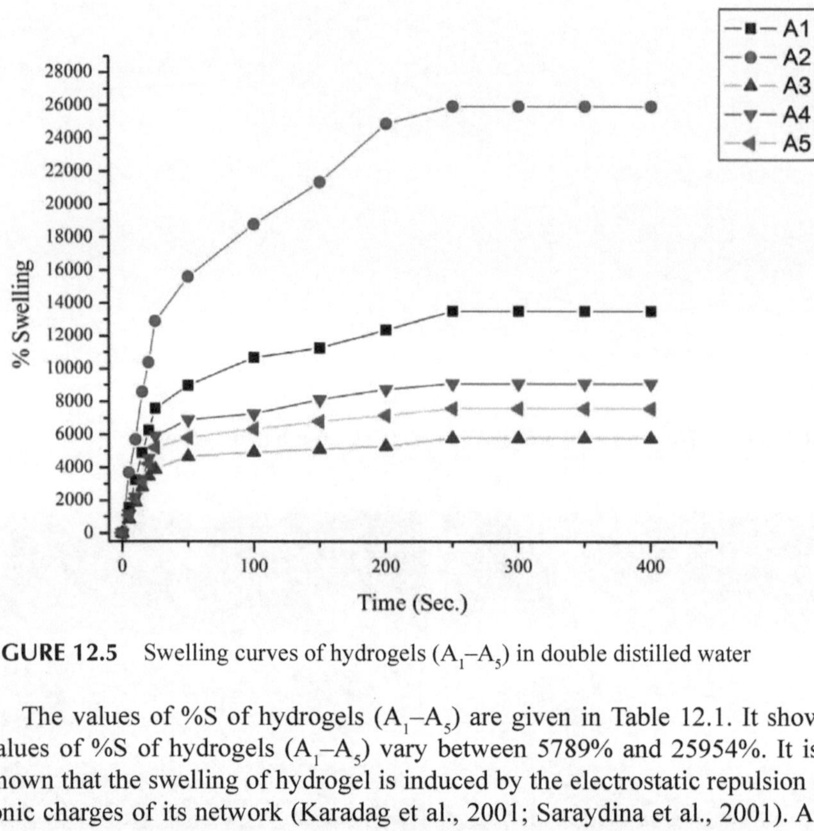

FIGURE 12.5 Swelling curves of hydrogels (A_1–A_5) in double distilled water

The values of %S of hydrogels (A_1–A_5) are given in Table 12.1. It shows the values of %S of hydrogels (A_1–A_5) vary between 5789% and 25954%. It is well known that the swelling of hydrogel is induced by the electrostatic repulsion of the ionic charges of its network (Karadag et al., 2001; Saraydina et al., 2001). Acrylic acid contains many ionic units (–COOH) and *N, N*-DMAAm contains many ionic units (–$N(CH_3)_2$). At the highest concentration of *N, N*-DMAAm and acrylic acid in the hydrogel, an increase in the equilibrium swelling is observed.

12.1.3 EFFECT OF pH

The pH of the solution plays an important role in the adsorption process and particularly on the adsorption capacity as the sorbent surface creates positive or negative charges on its surface depending on the solution pH. In order to evaluate the influence of pH on sorption of Cr (VI), the experiments were carried out with the pH range of 1–13, which are shown in Figure12.6.

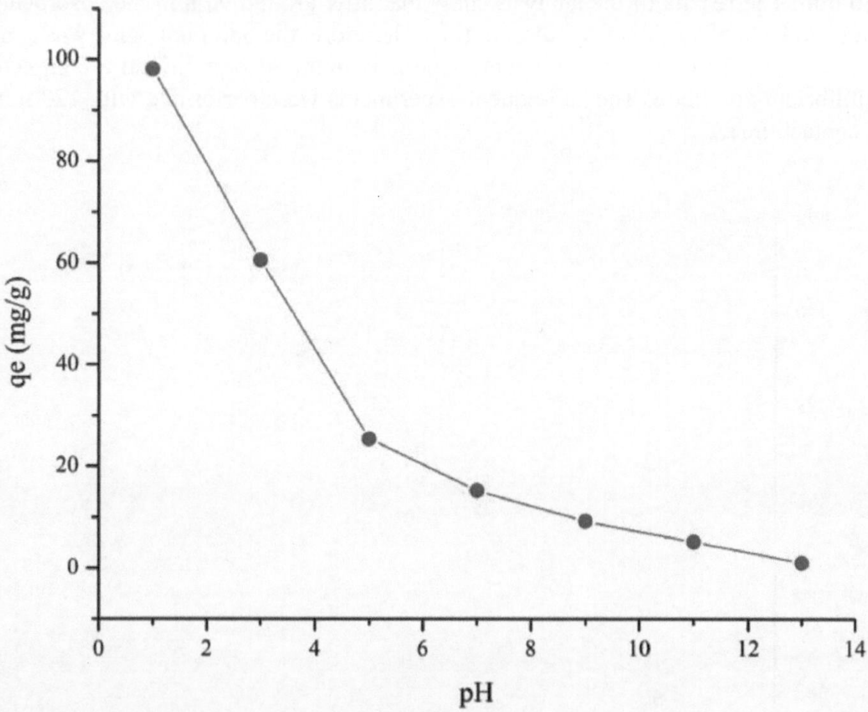

FIGURE 12.6 The effect of pH on adsorption of Cr (VI) by hydrogel A_2

For this purpose, 20 mg of hydrogel A_2 was added to 200 mg/L Cr (VI) solution, where the pH was adjusted using hydrochloric acid and sodium hydroxide. These samples were stirred for 3 h at 150 rpm. The adsorption of Cr (VI) ions by hydrogel A_2 decreases with an increase in pH of the solution. The maximum adsorption capacity of the hydrogel occurred at pH 1. We know that the hydrogel A_2 is independent over the entire pH range even though there is a variation of an adsorption capacity q_e with respect to pH. It is due to a pH dependent protic equilibrium in which, at acidic pH, the predominant species is $HCrO_4^-$ and is predominant (Spinelli et al., 2004). As this hydrogel A_2 is not influenced by pH, the decrease in the adsorption capacity at basic pH is attributed to the fact that, to neutralize the CrO_4^{-2} species, two quaternary nitrogens are necessary, while only one is needed to neutralize an $HCrO_4^-$ species.

1.3.4 EFFECT OF CONTACT TIME

Adsorption reactions for adsorbents were carried out using 50 mL solution containing 200 mg/L of Cr (VI) at pH1.0 and 20 mg of adsorbent at room temperature. Figure12.7 shows Cr (VI) uptake capacity at different time intervals, 10 min to

120 min. The results of the analysis show that after around 90 min the adsorbents removed 98% of Cr (VI) from the solution. Therefore, the optimum value was considered to be 90 min, but to guarantee the maximum adsorption and a complete equilibrium condition. The subsequent experiments were performed with 120 min of contact time.

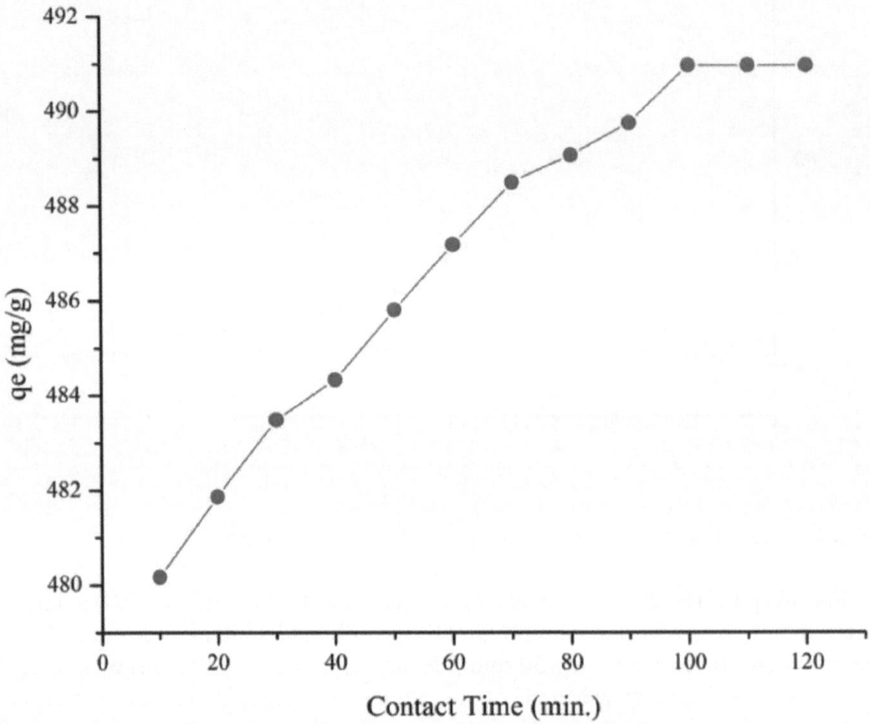

FIGURE 12.7 The effect of contact time on adsorption of Cr (VI) by hydrogel A_2

12.3.5 EFFECT OF AMOUNT OF ADSORBENT

To investigate the effect of the adsorbent dosage on the adsorption, different amount of adsorbent was tested, where the weight between 4 mg and 20 mg was used with 200 mg/L of Cr (VI) solution at pH 1.0 at room temperature. The samples were stirred for 120 min. The results are shown in Figure12.8. The adsorption capacity dropped from 988.68 to 490.90 mg/g for Cr (VI) by increasing the adsorbent amount from 4 mg to 20 mg, respectively. The decrease in adsorption capacity by increasing the adsorbent amount is basically due to the sites remaining unsaturated during the adsorption process.

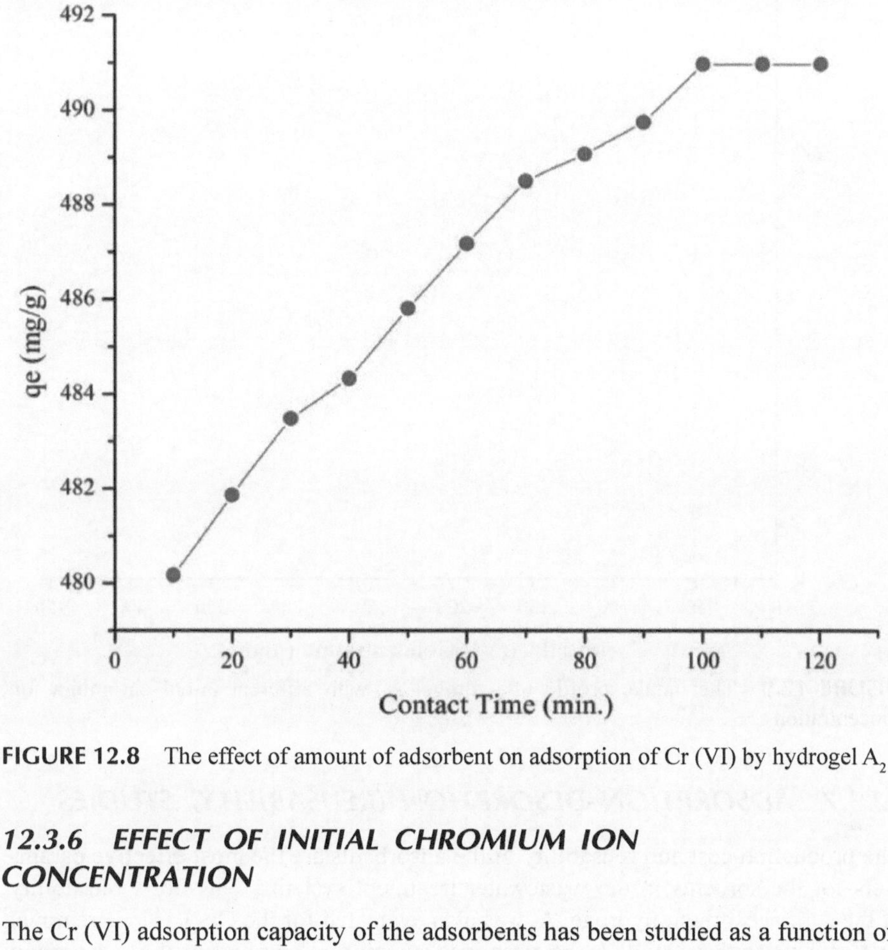

FIGURE 12.8 The effect of amount of adsorbent on adsorption of Cr (VI) by hydrogel A_2

12.3.6 EFFECT OF INITIAL CHROMIUM ION CONCENTRATION

The Cr (VI) adsorption capacity of the adsorbents has been studied as a function of the initial concentration. The Cr (VI) ion concentrations were between 50 mg/L and 400 mg/L at pH 1.0 and the experiments were performed at room temperature, and the Cr (VI) ion concentration was measured after 3 h of contact. After 400 mg/L concentration, Cr (VI) removal efficiency of hydrogel A_2 decreases and therefore, 50 mg/L to 400 mg/L initial Cr (VI) concentration range was selected. It was found that an increase in Cr (VI) ion concentration leads to an increase in adsorption capacity of Cr (VI) by adsorbent as shown in Figure12.9. This increase in adsorption capacity in relation to the Cr (VI) ion concentration could be explained with the high driving force for mass transfer.

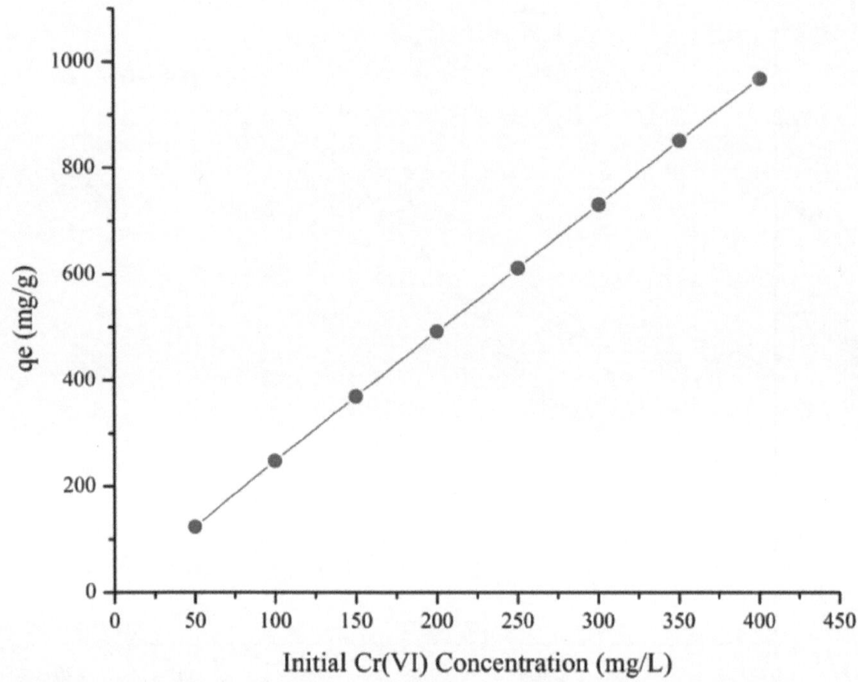

FIGURE 12.9 The uptake profile of hydrogel A_2 with different initial chromium ion concentration

12.3.7 ADSORPTION–DESORPTION (REUSABILITY) STUDIES

The production cost and reusability of the adsorbents are the most effective parameters for the sorbents in the wastewater treatment systems. Therefore, reusability of the super absorbent hydrogel A_2 was also evaluated for the Cr (VI) ions removal studies. During these studies, the optimized super absorbent hydrogel A_2 adsorption condition, (initial ion concentration), which can provide maximum ion removal, was used.

In this part of the study, the super absorbent hydrogel A_2, adsorbed Cr (VI) ions, was treated with a 0.2M NaOH solution to remove the adsorbed Cr (VI) ions for the investigation of the reusability of the hydrogel A_2 in this ion adsorption study. This characteristic is very important for the economical perspective of the use of super absorbent hydrogel as an adsorbent for heavy metal removal. The initial Cr (VI) ion concentration was 200 mg/L. Desorption time was fixed as 2 h throughout the adsorption period. Desorption ratio was calculated by using the following Eq. (12.4)

$$Desorption\,ratio = \frac{Amount\ of\ desorbed\ Cr\ (VI)\ ions\ into\ the\ desorption\ medium}{Amount\ of\ adsorbed\ Cr\ (VI)\ ions\ onto\ the\ hydrogel} \times 100 \quad (12.4)$$

Adsorption–desorption studies were replicated for five times and desorbed super absorbent hydrogel A_2 were washed with distilled water before applying the following adsorption steps. The obtained results were shown in Figure12.10. As can be seen in the Figure12.10, the adsorption capacity for Cr (VI) ions was approximately 490 mg/g adsorbent in the first step; these values decreased to approximately 460 mg/g adsorbent for the Cr (VI) ions in the following steps. In spite of these decreases in the adsorption of Cr (VI) ions, no excessive changes were observed after the second step. Furthermore, almost all of the adsorbed Cr (VI) ions could be removed from the hydrogel according to the found desorption ratios.

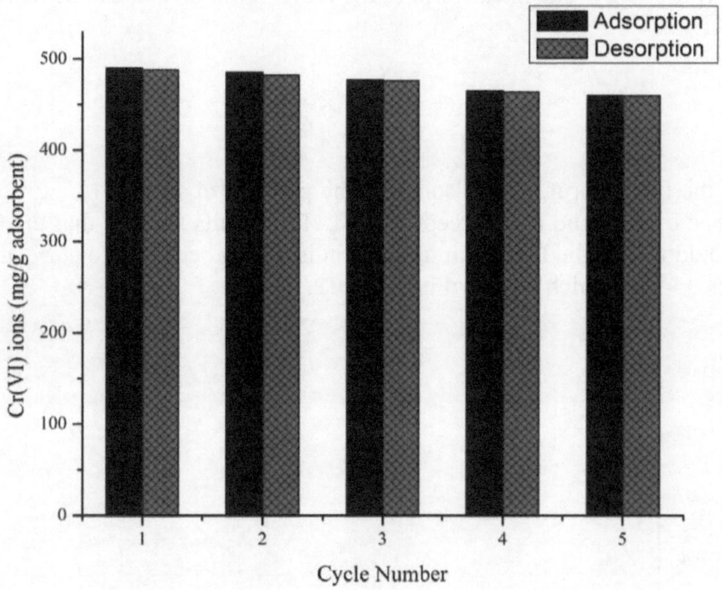

FIGURE 12.10 Adsorption-desorption values for Cr (VI) ions with hydrogel A_2

12.3.8 ADSORPTION ISOTHERMS

The relationship between the amount of a substance adsorbed per unit mass of adsorbent at constant temperature and its concentration in the equilibrium solution is called the adsorption isotherm. The more common models used to investigate the adsorption isotherm are Langmuir and Freundlich equations (Langmuir, 1918; Freundlich, 1939). The experimental results of this study were fitted with these two models. The equilibrium adsorption capacity of the adsorbent was calculated by the following Eq. (12.2). The equilibrium adsorption of Cr (VI) ion solutions by hydrogel A_2 was measured (50 mL of 50 mg/L–400 mg/L Solutions) after equilibrium time.

Langmuir isotherm model

The most widely used Langmuir equation, which is valid for monolayer sorption on a surface with finite number of identical sites, is given by Eq. (12.5).

$$q_e = \frac{q_m K_L C_e}{(1+K_L C_e)} \tag{12.5}$$

where q_e is the equilibrium adsorption capacity of metal ion on the adsorbent (mg/g);C_e, the equilibrium metal ion concentration in solution (mg/L); q_m, the maximum capacity of the adsorbent (mg/g);and K_L, the Langmuir adsorption constant (L/mg) related to free energy. The linear form of the Langmuir equation can be represented as follows:

$$\frac{1}{q_e} = \frac{1}{q_m} + \frac{1}{q_m K_L} \times \frac{1}{C_e} \tag{12.6}$$

By using the linear form of this isotherm, the plot of $1/q_e$ versus $1/C_e$ gives a line with a slope of $1/q_m$ and an intercept of $1/K_{L.}$ The results showed that the Cr (VI) uptake would follow the Langmuir adsorption isotherm because the correlation coefficient is 0.9955, which is shown in Figure12.11.

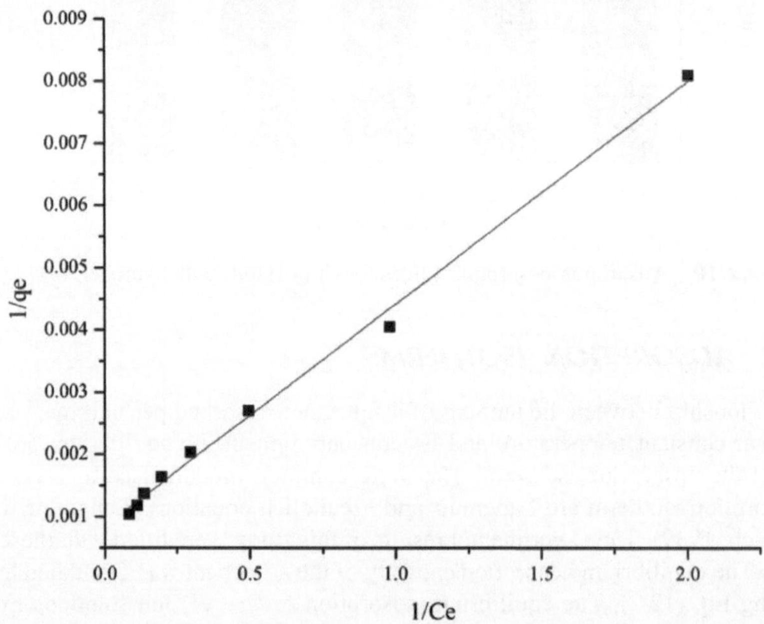

FIGURE 12.11 Langmuir adsorption isotherm

Freundlich isotherm model

Freundlich equation is derived to model the multilayer adsorption and for the adsorption on heterogeneous surface. The empirical equation used to describe the Freundlich isotherm is given by Eq. (12.7).

$$q_e = K_F C_e^{1/n} \tag{12.7}$$

The linear form of the Freundlich equation can be represented as follows:

$$\log q_e = \log K_F + \frac{1}{n} \log C_e \tag{12.8}$$

Where q_e and C_e are defined as above, K_F is the Freundlich constant (L/mg), which indicates the relative adsorption capacity of the adsorbent related to the bonding energy, and n is the heterogeneity factor representing the deviation from linearity of adsorption and is also known as Freundlich coefficient. According to Eq. (12.8), the plot of the log q_e versus log C_e gives a straight line and the K_F and n values can be calculated from the intercept and slope of the straight line, respectively. The results showed that Cr (VI) uptake by hydrogel would follow the Freundlich adsorption isotherm, because the correlation coefficient is 0.9842, which is shown in Figure12.12. The Langmuir and Freundlich both isotherm fitting to the experimental data have been seen in the biosorption study of Cr (VI) from aqueous solution (Ahmad et al., 2005; Abbas et al., 2008).

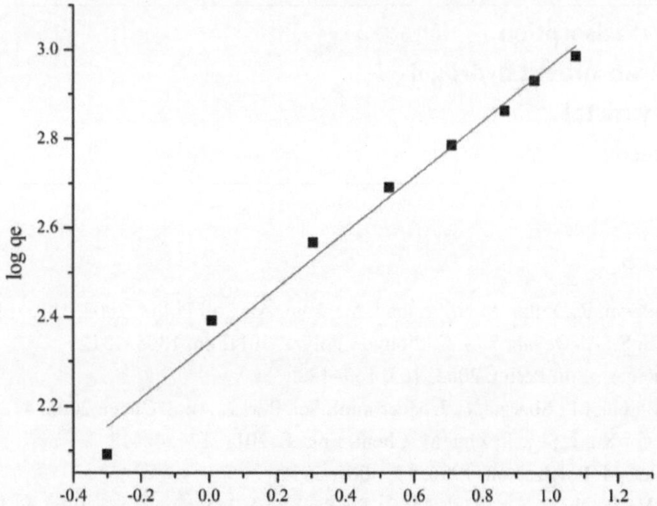

FIGURE 12.12 Freundlich adsorption isotherm

12.3 CONCLUSIONS

The fast swelling poly [*N, N*-DAPB/*N, N*-DMAAm/PNAAc] super absorbent hydrogels are synthesized via microwave irradiated free radical solution polymerization. Hydrogels show higher percentage of swelling in double distilled water in a short time period. The effect of various parameters on the adsorption of Cr (VI) from aqueous solution has been investigated. The optimum condition for Cr (VI) removal from aqueous solution (initial Cr (VI) concentration: 200 mg/L, adsorbent dosage: 20 mg, pH: 1.0, contact time: 120 min) was concluded. The Langmuir and Freundlich isotherm models show good fit to the equilibrium data with correlation coefficient 0.9955 and 0.9842 respectively. Desorption experiments show that even after five adsorption-desorption cycles, hydrogel can be reused without significant losses of its initial properties.

ACKNOWLEDGEMENT

Appreciation is expressed for the studies in the Sophisticated Instrument Center for Applied Research and Testing [SICART], Vallabh Vidyanagar for FTIR analysis, Department of Materials Science, Vallabh Vidyanagar for SEM and TGA analysis. The Authors are grateful to the University Grant Commission, New Delhi for providing the Research Fellowship and are also thankful to the Department of Chemistry, Sardar Patel University for providing the research facilities.

KEYWORDS

- **Cr (VI) adsorption**
- **Super absorbent hydrogel**
- **Heavy metal**
- **Isotherms**

REFERENCES

Abbas, M.; Nadeem, R.; Zafar, M.N.; Arshad, M. Water Air Soil Pollut. 2008, 191, 139–148.

Abdel-Halim, E. S.; Al-Deyab, S. S. Carbohydr. Polym. 2011, 86, 1306–1312.

Ahmad, R. Water Air Soil Pollut. 2005, 163, 169–183.

Bajpai, S. K.; Bajpai, M.; Sharma, L. J. Macromol. Sci. Pure & Appl. Chem. 2006, 43, 1323–1337.

Cheng, Q.; Li, C.; Xu, L.; Li, J.; Zhai M. Chem. Eng. J., 2011, 173, 42–48.

Cieslak-Golonka, M. Polyhedron. 1996, 15, 3667–3918.

Dadhaniya, P. V.; Patel, A. M.; Patel, M. P.; Patel, R.G. J. Macromol. Sci. Pure & Appl. Chem. 2009, 46, 447–454.

Freundlich, H.; Heller, W. J. Amer. Chem. Soc. 1939, 61, 2228–2230.

Gao, H.; Liu, Y.; Zeng, G.; Xu, W.; Li, T.; Xia, W. J. Hazard. Mater. 2008, 150, 446–452.

IARC (International Agency for Research on Cancer). IARC monographs on the evaluation of carcinogenic risks to humans: overall evaluation of carcinogenicity. An updating of IARC Monographs, vols. 1–42, Supplement 7, 1987, WHO, Lyon, France.

Kalyani, S.; Priya, J. A.; Rao, P. S.; Krishnaiah, A. Sep. Sci. Technol. 2005, 40, 1483–1495.

Karadag, E.; Saraydın, D.; Guven, O. Macromol. Mater. Eng. 2001, 286, 34–42.

Kotas, J.; Stasicka, Z. Environ. Pollut. 2000, 107, 263–283.

Krishnani, K. K.; Ayyappan, S. Rev. Environ. Contam. Toxicol. 2006, 188, 59–84.

Langmuir, I. J. Amer. Chem. Soc. 1918, 40, 1361–1403.

Liu, Y.; Kang, Y.; Huang, D.; Wang, A. J. Chem. Technol. Biotechnol. 2012, 87, 1010–1016.

Lo, I. M. C.; Yin, K.; Tang, S. C. N. J. Environ. Sci. 2011, 23, 1004–1010.

Patel, A. M.; Patel, R. G.; Patel, M. P. J. Macromol. Sci. Pure & Appl. Chem. 2011, 48, 339–347.

Patel, Y. N.; Patel, M. P. J. Macromol. Sci., Pure & Appl. Chem. 2012, 49, 490–501.

Richard, F.; Bourg, A. Water Res. 1991, 25, 807–816.

Rivas, B. L.; Maureira, A.; Guzman, C.; Mondaca, M. A. J. Appl. Polym. Sci. 2009, 111, 78–86.

Roy, P. K.; Swami, V.; Kumar, D.; Rajagopal, C. J. Appl. Polym. Sci. 2011, 122, 2415–2423.

Saraydin, D.; Karadag, E.; Guven, O. J. Appl. Polym. Sci. 2001, 79, 1809–1815.

Saraydina, D.; Karadag, E.; Caldirana, Y.; Guven, O. Radiat. Phys. Chem. 2001, 60, 203–210.

Sarin, V.; Sarvinder Singh, T.; Pant, K. K. Bioresour. Techonol. 2006, 97, 1986–1993.

Spinelli, V. A.; Laranjeira, M. C. M.; Favere, V.T. React. Func. Polyms. 2004, 61, 347–352.

Tang, S. C. N.; Lo, I. M. C.; Mak, M. S. H. Water Air Soil Pollut. 2012, 223, 1713–1722.

USEPA. USEPA Drinking Water Regulations and Health Advisories. 1996, Washington, DC. EPA 822-B-96-002.

WHO. Guidelines for drinking water quality. Health criteria and other supporting information, vol. 1, 2nd ed., 1997, World Health Organization, Geneva, Switzerland.

CHAPTER 13

AN IDEAL HYDROGEL FOR REMOVAL OF METHYLENE BLUE FROM AQUEOUS SOLUTION

P. SOUDA and LISA SREEJITH*

Soft Materials Research Laboratory, Department of Chemistry, National Institute of Technology, Calicut, India

*Corresponding author: Tel: 09447203130; E-mail: lisa@nitc.ac.in

CONTENTS

ABSTRACT

Poly(acrylate- acrylic acid-co maleic acid) (PAAAM) hydrogel in form of rods are prepared by heating the aqueous solution of acrylic acid(AA) and maleic acid (MA) and crosslinked with varying quantities of crosslinker N,N'-methylenebisacrylamide (NMBA). The introduction of MA having two carboxylic acid groups enables hydrogel with good swelling property and excellent Methylene Blue (MB) sorption capacity. The sorption experiments were carried out to optimize various parameters viz, the amount of crosslinker in the hydrogel, initial MB concentration, contact time, and pH that influence the adsorption. Desorption was carried in acidic solution and more than 70% MB could be removed, recommending the PAAAM hydrogel had the potential for reuse.

13.1 INTRODUCTION

New methods are being developed to correct the imbalances of the water system caused by effluents as removal of organic matter from sewage and industrial waste is posing a big problem (Demirbas, 2011; DoğrueL et al., 2013). Different types of materials such as activated carbon filters, organically modified synthetic porous silica, dentric polymers, and hydrogels (Saraydin et al., 1998; Kasgoz, 2006) are nowadays used for removing heavy metals and dyes from industrial wastes. Hydrogels are the cross-linked polymers which can absorb a large amount of water without dissolving in it. Softness, smartness, and the capacity to store water make the hydrogels unique materials. Since it can absorb a large amount of water, they are otherwise called "holding water agents" or "super absorbents". The capacity of the hydrogel to absorb water is enormous and could be 1000 times the mass of polymer. The ability of hydrogel to absorb water is due to the hydrophilic functional groups attached to its back bone, while its water resistivity is due to the presence of crosslinks between the chains (Okay, 2009). There are many applications for the hydrogels, which includes drug delivery (Drury and Mooney, 2003), preparation of contact lens (Hyon et al., 1994), and *waste water purification (Cavus* and Gurdag, 2009; Carvalho et al., 2010)

Polyelectrolyte hydrogels are formed by crosslinking the polymer chains, which contain a number of ionisable groups. The ionisable functional groups are responsible for the intelligent behavior and adsorption capacity of hydrogels. Methylene blue (MB) has wide-range applications, which include coloring paper, temporary hair colorant, dying cottons, and wools. Although not strongly hazardous, it can cause heart beat increase, vomiting, shock, and cyanosis on inhalation. Removal of MB from polluted water is a challenge due to the difficulty of treating industrial effluents by conventional methods.

Hydrogels having free carboxylic acid or carboxylate groups have the capacity to be complex with the cations such as metal ions or organic dyes. Acrylic acid (AA) is a monocarboxylic acid which polymerize very easily and form poly acrylic acid with many applications including biomedical (Kim and Lee,1999), heavy metal removal (Katime and Rodríguez, 2001; ; Tang et al., 2009; Guclu et al., 2010) and dye removal (Paulino et al., 2006; Zendehdel et al., *2010*). In this paper, we report the preparation of poly (acrylate-acrylic acid-co-maleic acid) hydrogel, referred herein as PAAAM, and for its suitability for removal of MB as a function of pH; the crosslinker concentration, treatment time and initial concentration of MB were investigated in detail. The comparatively moderate cost and the ease of preparation make the hydrogels a promising candidature for the said application. The adsorption capacity of the hydrogel seems to be affected by these parameters as studied by Systronics Double beam UV-Vis absorption spectrophotometer.

13.2 EXPERIMENTAL

13.2.1 MATERIALS

The monomer acrylic acid (AA), comonomer maleic acid (MA), initiator ammonium per sulfate (APS) and crosslinker N,N'-methylenebisacrylamide (NMBA) of analytical grade were purchased from Hi Media, Mumbai, India. The chemical structures of these materials were as given in scheme 1(a–e). Potassium hydroxide (KOH) used for neutralization of AA was procured from Merck, India. Methylene blue (MB) for adsorption study was purchased from Spectrum chemicals.

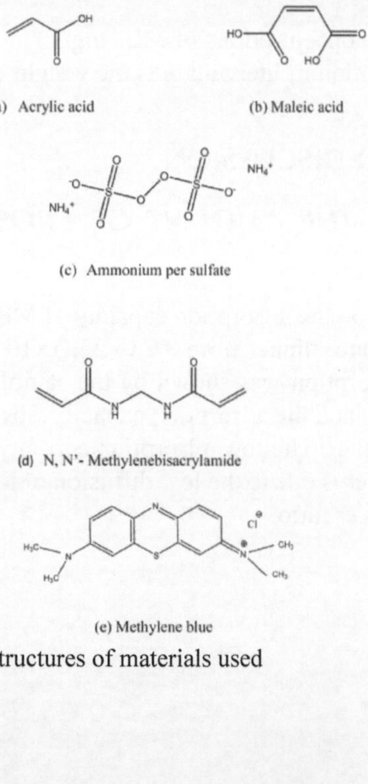

SCHEME 13.1 Chemical structures of materials used

13.2.2 APPARATUS AND PROCEDURE

The poly (acrylate-acrylic acid-co maleic acid) (PAAAM) hydrogels were synthesized by copolymerization of AA and MA in aqueous medium using NMBA as crosslinker and APS as an initiator. MA and NMBA were dissolved in double distilled water to a total volume of 5 mL and then mixed with APS (0.015g) and 50% neutralized acrylic acid. Here AA:MA mole ratio is fixed at 80:20 and the NMBA concentration varied from 0.032 mmol to 0.22 mmol. The resulting solution was transferred to a PVC straw of diameter 0.3 cm and kept in an electric oven at 60 °C for a period of 2 h (Bajpai et al., 2006). The resulting transparent gels were cut into small cylindrical pieces. They were then washed with double distilled water to remove the unreacted monomers and dried at 40 °C till they attained constant weight. The hydrogels formed were highly transparent, very soft, elastic, and cylindrical in shape. When a low amount of crosslinker (below 0. 032mmol) of NMBA was used the hydrogel had not formed. So the minimum amount of crosslinker for the formation of PAAAM hydrogel was optimized to 0. 032mmol.

Adsorption of MB by the hydrogel was studied by introducing approximately 0.01 g of dry hydrogel in 20 mL of MB and was left to adsorb for 2 days. Then, the gel was removed from the solution, and the remaining MB was monitored by UV-Vis absorption spectrophotometer at 660 nm.

The amount of adsorbed MB was calculated as,

$$\text{Adsorption capacity (mg/g) } Q_e = ((C_0 - C_t)\, V)/m \tag{13.1}$$

Where C_0 and C_t are the concentrations of MB (mg/L) at the beginning and at time *t*, V is the volume of solution in liter and *m* is the weight of dried hydrogel in gm.

13.3 RESULTS AND DISCUSSION

13.3.1 EFFECT OF THE AMOUNT OF CROSSLINKER ON REMOVAL OF MB

The effect of crosslinker on the adsorption capacity of MB was determined by varying the concentration of crosslinker from 3.2 to 22.7($\times 10^{-2}$ mmol) as shown in Figure 13.1. Maximum adsorption was shown by the sample having crosslinker concentration of 0.032 mmol and the adsorption capacity was decreased by the increase in the amount of crosslinker. The low adsorption capacity of the hydrogel having a high amount of crosslinker is due to the low diffusion of MB in this hydrogel, which is having a high crosslinker ratio.

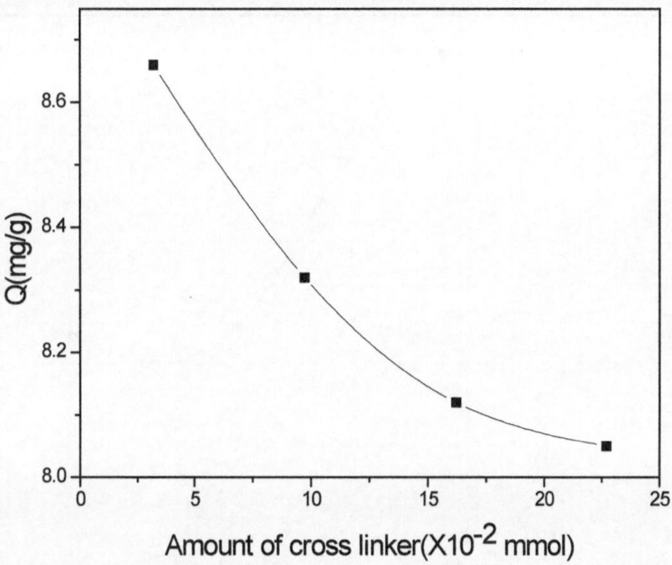

FIGURE 13.1 Effect of a crosslinker amount on the adsorption of MB on to hydrogel (pH = 5.7, temp = 25 °C, C_0 = 9 ppm).

13.3.2 EFFECT OF pH

To evaluate the effect of pH on the adsorption of MB on to hydrogel, the adsorption experiments were carried out in solutions having different pH values at 25 °C. MB having pH values between 2 and 11 were prepared from MB solution and buffer solutions, and the adsorption capacity were determined. As shown in Figure 13.2, the adsorption capacity of hydrogel was continuously increased in the pH range of 3–6. MB is a cationic dye, which will adsorb on the surface of the hydrogel that is having a high amount of negatively charged carboxylate ions. When pH increases above pka values of AA and MA(Pka of AA = 4.25, Pka2 of MA = 6.07) the –COOH groups completely ionize, and the surface of hydrogel will be covered by $-COO^-$ groups, which will favor the adsorption of MB. At higher pH, there is a sharp decrease in the adsorption due to the screening effect of the counter ions which will shield the carboxylate ions (Koskun and Delibas , 2011).

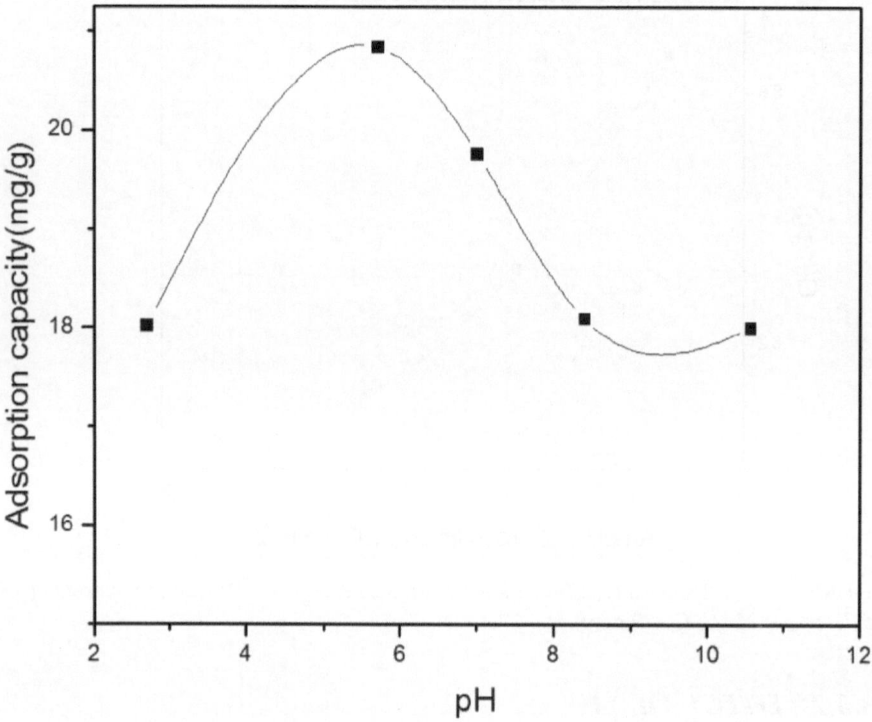

FIGURE 13.2 Effect of pH on the adsorption of MB on to hydrogel (temp = 25 °C, C_0 = 20 ppm, [NMBA] = 0.032 mmol).

13.3.3 EFFECT OF INITIAL MB CONCENTRATION

Free gel was added to definite concentrations of the MB solutions (2–20 mg/l) at room temperature and was noted for its adsorption. It is clear from Figure13.3 that the dye adsorption increases sharply with an increase in the initial dye concentration. When C_0 was reached at 5 ppm and 10 ppm, the Q_e was reached at 10.04 and 20.81 respectively, which were much higher than reported Qe values of other adsorbents (Table 13.3). Equilibrium adsorption isotherm is an important criterion to determine the mechanism of dye adsorption on hydrogel. The Langmuir and Freundlich models are widely used to examine the adsorption isotherms. Freundlich isotherm models are based on the assumption that the surface of the adsorbent is not homogeneous. The experimental data in Figure 13.3 was also analyzed with the Freundlich isotherm model, which describes a heterogeneous system with multilayer adsorption. The linear form of Freundlich isotherm equation

$$\ln Q_e = \ln k + \left(\frac{1}{n}\right) \ln C_e \tag{13.2}$$

with k is Freundlich constant related to the adsorption capacity and $\frac{1}{n}$ is related to the adsorption intensity of the adsorbent (Zhang and Wang , 2010).

Freundlich isotherm curve (Figure 13.4) and the related constants *k* and *n* and R^2 for MB (Table 13.1) shows that the regression coefficient value is well fit in the Freundlich isotherm curve, indicating the adsorbent surface is not homogeneous.

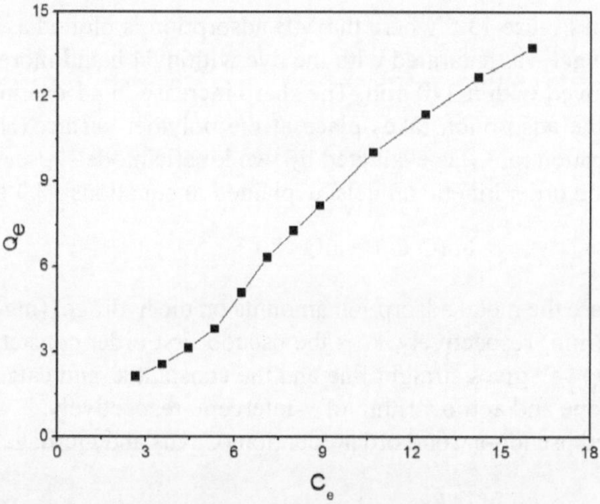

FIGURE 13.3 Effect of initial MB concentration on the adsorption of MB on to the hydrogel. (pH = 5.7, temp = 25 °C, [NMBA] = 0.032 mmol).

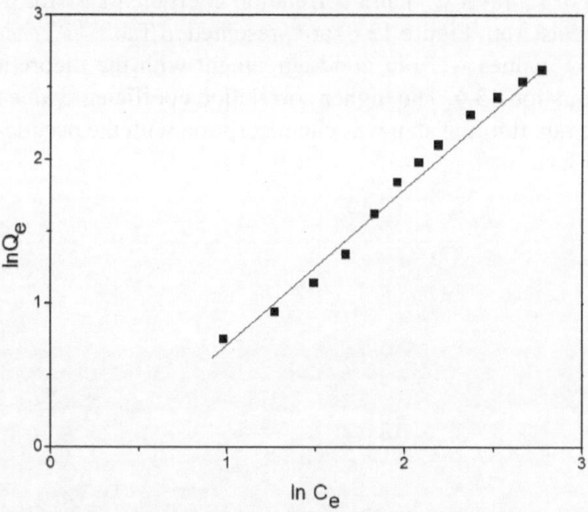

FIGURE 13.4 Freundlich isotherm curve of MB adsorption.

TABLE 13.1 Constants of Freundlich isotherm models for MB adsorption

k	n	R^2
0.63	0.88	0.98625

13.3.4 EFFECT OF TIME

It was noted from Figure 13.5 where the MB adsorption is plotted as a function of time that the free gel was saturated with the dye within 54 h and more than 70% of the dye was removed within 240 min. The sharp increase in adsorption rate, points to the fact that the adsorption takes place at the polymer surface (Hasaine et al., 2003). The adsorption rate was evaluated by two kinetic models, pseudo-first order and pseudo second order kinetic models, explained in equations 13.3 and 13.4

$$\ln (Q_e Q_t) = \ln Q_e \ k_1 t \tag{13.3}$$

where Q_e and Q_t are the metal adsorption amounts on the hydrogel (mg/g) at equilibrium and time t (min) respectively, k_1 is the pseudo first order constant (min^{-1}). By plotting $\ln(Q_e–Q_t)$ vs t gives straight line and the constant k_1 and calculated Q_e was obtained from slope and antilogarithm of y-intercept, respectively.

The linearized pseudo-second order equation (Cavus and Gurdag, 2009)

$$\frac{t}{Q_t} = \frac{1}{k_2 Q_e^2} + \frac{t}{Q_e} \tag{13.4}$$

where k_2 is the pseudo-second order rate constant for adsorption process in g/(mg. min).The values of k_2, $h(=k_2 Q_e^2)$ and correlation coefficients R^2 for pseudo-second order kinetic models(from Figure 13.6) are presented in Table 13.2. The experimentally calculated Q_e values were in good agreement with the theoretical Q_e values obtained from equation.13.4. The higher correlation coefficient value thus obtained indicates that the rate limiting step was chemisorption with the pseudo-second order kinetics (Table 13.3).

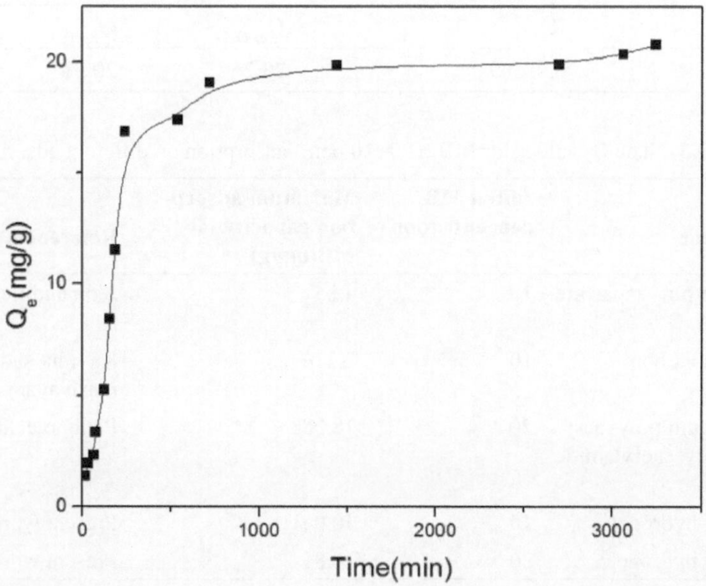

FIGURE 13.5 Adsorption rate MB by the hydrogel (temp = 25 °C, C_i = 20 ppm, [NMBA] = 0.032 mmol).

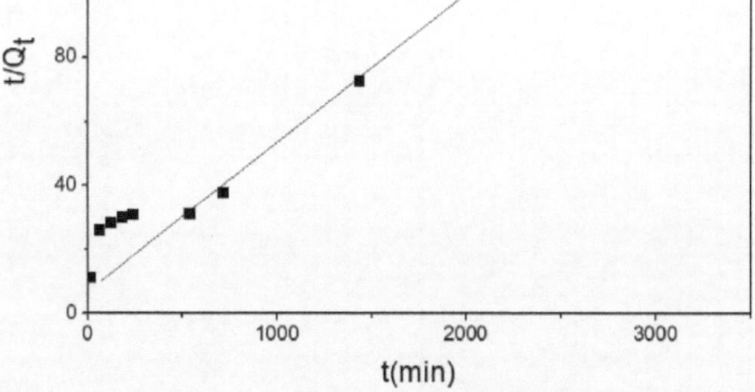

FIGURE 13.6 Pseudo-second-order kinetic plot for adsorption of MB on hydrogel.

TABLE 13.2 Kinetic parameters for the adsorption of MB on hydrogel

k_2	R^2	$Q_{e,thor}$	$Q_{e,exptl}$
0.00013	0.98569	23.26	20.81

TABLE 13.3 The Q_e values for MB (C_0 = 10 ppm) adsorption on different adsorbents

Adsorbent	Initial MB concentration	Maximum adsorption capacity for MB(mg/g)	References
Alginate/polyaspartate hydrogel	10	4.85	Jeon and Lei (2008)
Coir pith carbon	10	2.314	Kavitha and C. Namasivayam (2007)
Arabic gum/poly acrylate /poly acrylamide hydrogel	20	18.083	Paulino et al. (2006)
PAAAM hydrogel	10	10.04	Present work
PAAAM hydrogel	20	20.81	Present work

13.3.5 THERMAL STABILITY OF HYDROGEL

Figure 13.7 shows the TGA graph as a function of weight loss against temperature for the PAAAM hydrogel and for its resultant complex of PAAAM-MB. No weight loss was noticed up to 150 °C for the PAAAM gel, but a sharp weight loss was noticed at 184.86 *°C* – indicating the decomposition of hydrogel back bone. This sharp weight loss for the loaded gel (PAAAM.MB) is at 211 *°C points to the increased thermal stability of the gel after chelating with MB.*

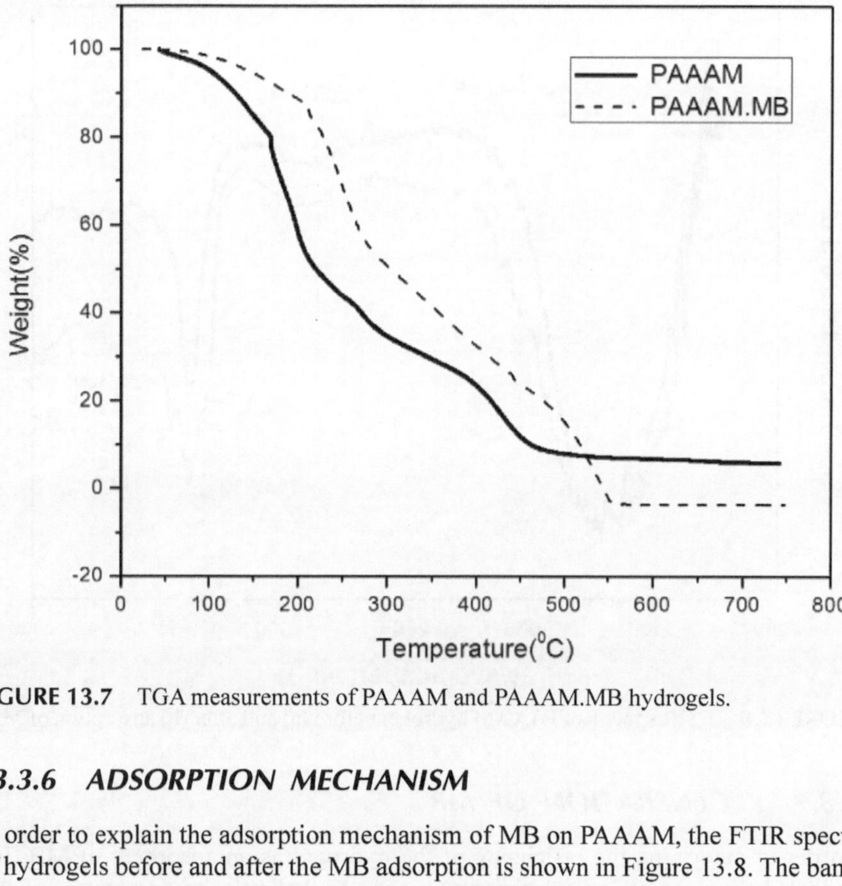

FIGURE 13.7 TGA measurements of PAAAM and PAAAM.MB hydrogels.

13.3.6 ADSORPTION MECHANISM

In order to explain the adsorption mechanism of MB on PAAAM, the FTIR spectra of hydrogels before and after the MB adsorption is shown in Figure 13.8. The bands of PAAAM at 3000–3500cm^{-1}(–OH stretching of AA and MA present in the polymer), 2059 2059 2059 cm^{-1}(C–N stretching of NMBA), 1636 1636 1636 cm^{-1} (–C = O stretching of a carboxylate group), 1561 1561 1561 cm^{-1} (C–O bending), 1363 1363 1363 cm^{-1}1281 cm^{-1}(C–O stretching) were observed. After MB adsorption, the peaks at 1561 cm^{-1}, 1363 cm^{-1}1281 cm^{-1}which were present in the PAAAM almost disappeared, showing indications of chelating of MB on COOH and COO^{-} groups on hydrogel.

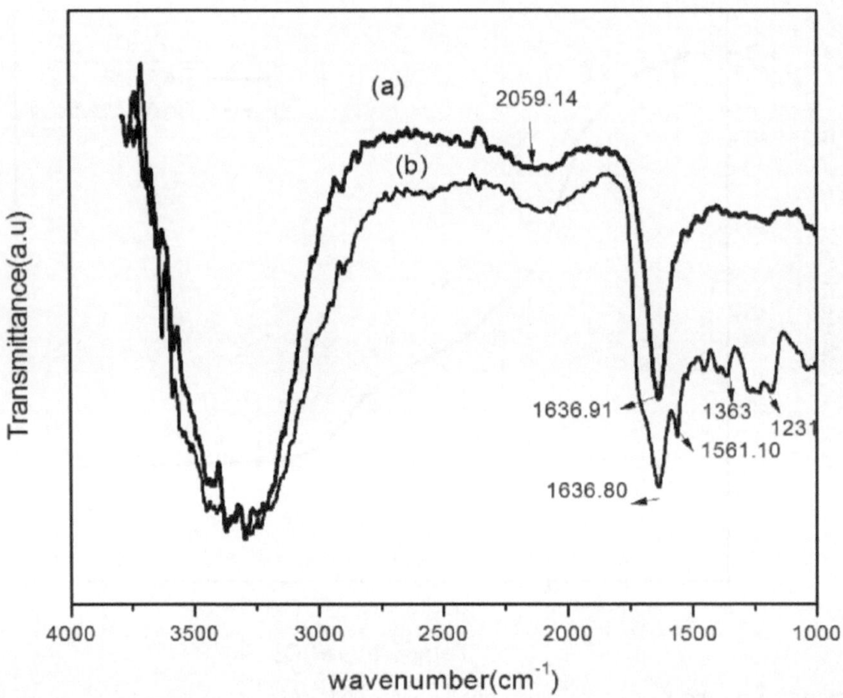

FIGURE 13.8 FTIR spectra of PAAAM hydrogel before (a) and after (b) adsorption of MB.

13.3.7 REGENERATION OF MB

In order to determine the efficiency of the hydrogel as an adsorbent of MB, we conducted the desorption experiments. Figure 13.9 shows the percentage of desorption at different desorption medias. The results showed that desorption in double distilled water (DW) and basic medium were low; it was 1.46% and 27.14%, respectively. Higher desorption is shown in acidic media, which is 74.3% in 0.1M acetic acid (AA) and highest desorption in 0.1M HCl which are 99%. At low pH value the binding site on the hydrogel will protonate, so the H^+ ion will compete with MB molecules, and thus the dye affinity of the hydrogel is decreased. This reveals the electrostatic interaction of MB molecules and hydrogels as explained from FTIR results. This regenerated hydrogel undergoes four, consecutive, adsorption-desorption cycles and the results are shown in Figure 13.10. Adsorption capacity of hydrogels decreases very slowly when regenerated, but the decrease is not significant.

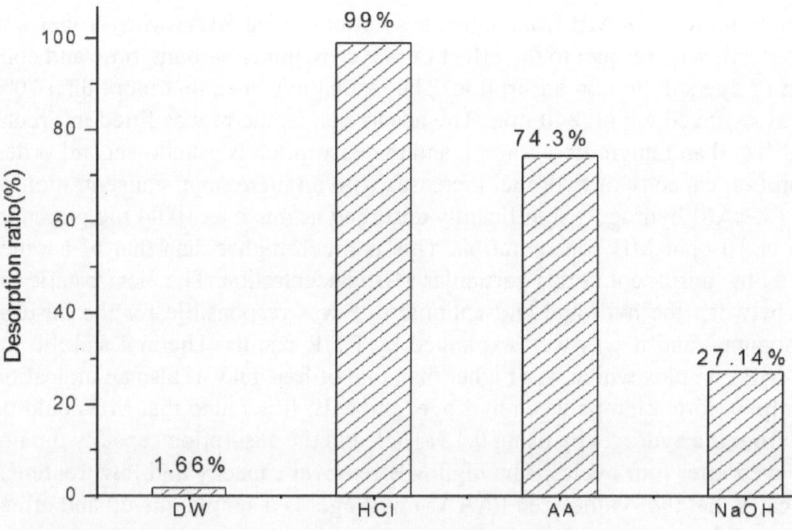

FIGURE 13.9 Desorption of MB at different desorbing media.

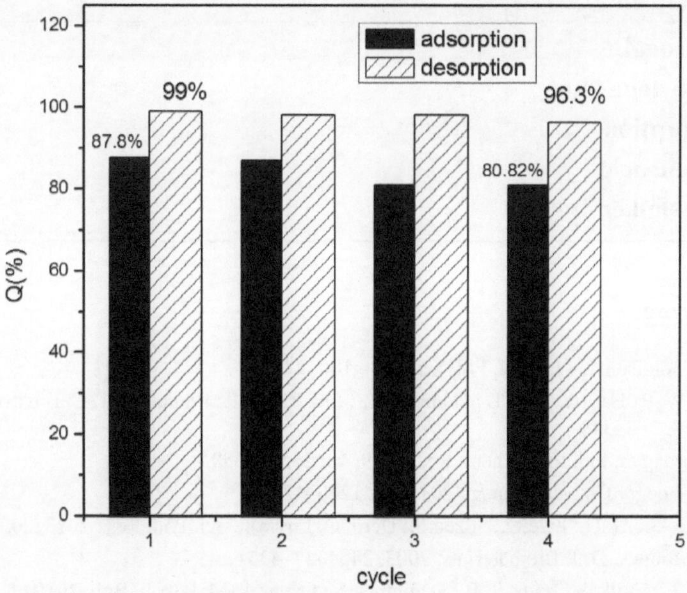

FIGURE 13.10 Desorption–desorption cycles.

13.4 CONCLUSIONS

The effective removal of MB from aqueous solutions using PAAAM hydrogel was studied in detail with respect to the effect of pH, crosslinker amount, time and concentration of dye solution on adsorption. The results indicated that more than 70% of MB was removed within 240 min. The adsorption isotherm was fitted in Freundlich isotherm than Langmuir isotherm, and the adsorption is pseudo-second-order. The adsorption capacity of hydrogel increases with an increase in concentration of dye. The PAAAM hydrogels significantly adsorbed as much as 10.04 mg/g methylene blue at 10 ppm MB concentration. This is much higher than that of another methylene blue adsorbent at this particular MB concentration. The electrostatic interaction between the hydrogel and cationic dye was responsible for the adsorption phenomena, and it was well explained by FTIR results. Thermal stability of hydrogel-MB complex which was higher than that of free gel was also an indication of the complex formation between hydrogel and MB. It is found that MB could be desorbed from the hydrogel by using 0.1 N HCl, and the adsorption capacity did not decrease even after four cycles. The high MB removal capacity and low treatment time indicates that the synthesized PAAAM hydrogel is a simple, novel, and effective sorbent with potent reusability.

KEYWORDS

- **Hydrogel**
- **Methylene blue**
- **Adsorption**
- **Maleic acid**
- **Crosslinker**

REFERENCES

Bajpai, S. K.; Johnson, S.; J. Appl. Pol. Sci. 2006,100, 2759.

Carvalho, H. W. P.; Batista, A. P. L.; Hammer, P. ; Luz, G. H. P.; Ramalho , T. C. Environ. Chem. Lett. 2010, 8, 343–348.

Cavus, S.; Gurdag G. Ind. Eng. Chem. Res. 2009, 48, 2652–2658.

Demirbas, A. Energy Conver. Manag.. 2011, 52, 1280–1287.

Doğruel, S.; Çokgör,E.U.; Ince,O.; Sözen,S.; Derin O. Environ. Sci. Poll. Res.. 2013, 20, 340–350.

Drury, J. L.; Mooney, D. J. Biomaterials. 2003, 24, 4337–4351.

Guclu, G.; Al, E.; Emik, S.; Iyim , T. B.; Ozgumus, S.; Ozyurek, M. Polym. Bull. 2010 65,333–346.

Hasaine, K.; Saadet, O.; Murat. O. Polymer, 2003, 44, 1785.

Hyon, S.; Cha, W.; Ikada, Y.; kita, M.; Ogura Y.;Honda, Y. J. Biom. Sci. 1994, 5, 5.

Jeon, Y. S.; Lei, J. ; Kim J.H. J. Ind. Eng. Chem. 2008, 14, 726–731.

Kasgoz, H. Polym. Bull. 2006, 56, 517–528.

Katime, I.; Rodríguez, E. J. Macr. Sci., Part A. 2001, A38 (5&6), 543–558.

Kavitha D.; Namasivayam, C. Biores.Tech., 2007, 98,14–21.

Kim, S. Y.; Lee, Y. M. J. Appl. Pol. Sci. 1999, 74, 1752–1761.

Koskun, R;. Delibas, A. Polym. Bull. 2012, 68, 1889–1903.

Okay, O. Hydrogel sensors and actuators, Springer, Heidelberg, 2009.

Paulino, T.; Guilherme, M. R.; Reis, A. V.; Campese, G. M.; Muniz, E. C.; Nozaki, J. J. Coll. Int. Sci. 2006, 55–62.

Saraydin, D.; Karadag, E.; Guven, O. J. Appl. Pol. Sci. 2001, 79, 1809–1815.

Tang, Q.; Sun, X.; Li , Q.; Lin, J. ; Wu, J. J Mater. Sci. 2009, 44,726–733.

Zendehdel, M.; Barati, A.; Alikhani, H.; Hekmat, A.; Iran. *J. Environ. Health. Sci. Eng. 2010, 7, 423–428.*

Zhang, J.; Wang, A. J. Chem. Eng. Data. 2010, 55, 2379.

CHAPTER 14

MODIFICATION OF PVC BY BLENDING

MAHUYA DAS[1*], RUPA BHATTACHARYYA[2], and DEBABRATA CHAKRABARTY[3]

[1]Department of Chemistry, JIS College of Engineering, Kalyani, Nadia, Pin – 741235, West Bengal, India; [2]Department of Chemistry, Narula Institute of Technology, 81, Nilgunj Road, Kolkata – 700109, West Bengal, India; [3]Department of Polymer Science & Technology, 92, A. P. C. Road, Kolkata – 700009, West Bengal, India

*E-mail: d_mahuya@yahoo.com

CONTENTS

ABSTRACT

Blending of polymers for property improvement or for economic advantage has gained considerable importance in the present scenario. Nevertheless, blends offer the possibility of customizing products to meet specific end needs. These advantages of polymer blending on performance and economy have accelerated R&D activities in the field of polymer blends in terms of academic and industrial interest. Poly(vinyl chloride) (PVC) is a stiff and rigid homopolymer which is widely used as a commodity plastic. Hence, for many applications, it is necessary to blend PVC with other polymers or copolymers in order to modify its properties. This chapter deals with blend formation of PVC with a wide range of elastomers and plastomers which aid in the property enhancement and subsequent improved application of PVC under different conditions.

14.1 INTRODUCTION

The scientific and commercial progress in the area of polymer blends during the past two decades has been tremendous and was driven by the realization that new molecules are not always required to meet needs for new materials, and that blending can usually be implemented more rapidly and economically than the development of new chemistry (Encyclopedia of Polym Sci & Engg., 1988). A continuous need for new materials is created by the increasing complexity of products that result from technological breakthrough and the desire to obtain improved materials for existing applications. Alongside the development of new improved base polymers and their copolymers, blend technology is one of the most widely used routes to satisfy this need. Polymer blends encompass many different kinds of materials containing two or more polymer components in blend or network form (Mathew and Deb, 1992). In principle, blending two polymers together in order to achieve a balance of properties not achievable with a single one is an obvious and well-founded practice. Polymer blends cover different kinds of materials containing two or more polymeric components and most of them exhibit phase separation to a greater or lesser extent. Thus, blends are characterized as a physical mixture of two or more components, which exhibit phase separation to a greater or lesser degree. The properties of the blends formed are largely dependent on the microstructure of the polymeric phases and hence their corresponding morphology.

The formation of a polymer blend aims in achieving certain specific properties, which are difficult to obtain from any one of the single components alone. Here, the certain specific properties refer to the properties of the polymers which are improved by formation of blends. Specific types of components of polymer blends modify some specific kinds of properties only. In other words, a large number of polymer properties cannot be improved simultaneously by blending two different polymers. Blend development is far less costly and to smaller extent time-consuming than the development of new polymers and therefore, blends are economically attractive. Nevertheless, blends offer the possibility of customizing products to meet specific end needs. These advantages of polymer blending on performance and economy have accelerated R&D activities in the field of polymer blends in terms of academic and industrial interest (Yen et al., 2003).

In this chapter, we consider the blending phenomenon of poly(vinyl chloride) (PVC) with the other polymers and elucidate how the modification of its properties overcomes the disadvantages related to both processing and properties. PVC belongs to a major class of engineering plastics that possess many unique properties that are suitable for a wide variety of technical and industrial products. It is available at a relatively low cost, is non-flammable and has good corrosion and chemical resistances. All these advantages have made PVC a unique commodity plastic. Presence of the chloride group turns PVC into a rigid polymer due to formation of hydrogen bonding which in turn gives rise to the inherent problems of poor impact strength and difficult processing of PVC. These problems are usually overcome by incorporation of suitable plasticizers and lubricants with subsequent sacrifice of the mechanical properties such as modulus and ultimate tensile strength to a great extent. The rigid PVC is made flexible by the incorporation of these plasticizers and lubricants. In this attempt, the impact modification of PVC is also achieved. However, in addition to these added lubricants and plasticizers, certain low-molecular-weight polymeric additives can also be incorporated to contribute to the internal plasticization and hence to the impact modification of the polymers. The process of blending with certain other polymeric materials which are referred to as impact modifiers and processing aids, offers a unique tool for overcoming the deficiencies of PVC as mentioned above, without much loss in the mechanicals. In association with the loss in mechanical properties, the thermal property of PVC is another significant factor in terms of its stability and processing. Dehydrochlorination of PVC that leads to the formation of conjugated polyene sequences is profound at higher temperature (>200 °C) which imparts in PVC a reddish brown colour and poses a major problem in high temperature applications. Thus, in order to amend PVC out of its inherent rigidities, blending imposes a most suitable solution for this purpose. Various polymeric materials are implemented and experimentation of blending technology in this regard revealed significant improvement in both mechanical and thermal properties. However, PVC can be blended with both elastomers and plastics to improve the physical, mechanical, and thermal properties as well as to modify the processing characteristics and cost reduction of the final product (Hafezi et al., 2007).

In the attempt to modify the properties of PVC, it has been blended with various other kinds of polymers, elastomeric or plastic in origin, to achieve a proper balance in physico mechanical and thermal properties, which are again influenced by the microstructural phase miscibility and phase compatibility. In such a context, we have analyzed the properties of the poly blends with one component as PVC and estimated the extent of modification, in point of fact of microstructure blend formation.

14.2 BLENDS OF PVC WITH VARIOUS POLYOLEFINS

The PVC/PE blend is now utilized in several applications such as blister packaging, low halogen-low smoke fire resistant electrical cable sheathing, and so on. Processing of the PVC/PE mixtures is not likely to yield products with the expected mechanical properties because of the poor adhesion of the phases due to the thermodynamic incompatibility. Raija Mikkonen and Autti Savolainen (Mikkonen and Savolainen, 1990) reported an experimental study investigating the rheological, morphological, and mechanical properties of a heterogeneous polymer blend system consisting of low-density polyethylene (LDPE) and plasticized PVC. The components were mixed using a single screw extruder, which was equipped with a special measuring head for the determination of rheological quantities. The dependencies of viscosity, die swell, and ultrasonic velocity on blend composition were qualitatively similar, exhibiting a minimum at about 70 wt% of PVC. The morphology of the blend system at this blending ratio was different from morphologies of the other blends. Tensile properties of blends, except elongations at break were not significantly inferior to those of the LDPE component. As a part of fundamental study of the behavior of mixed plastics during reprocessing and service, blends of LDPE and PVC showed that the Young's modulus increased steadily from pure LDPE to pure PVC, whereas both tensile strength and elongation at break passed through a minimum at about 5% PVC (Rajendran et al., 2007). Plasma treatment was applied to one of the components (LDPE) in order to affect the degree of compatibility. For this purpose, different monomers such as carbon tetrachloride and vinyl chloride were used. The surface energy results of blends prepared from untreated and treated LDPE-PVC showed considerable differences with appreciable increases for the latter indicating an increase in the work of adhesion as a result of the plasma surface modifications applied. The tensile test results and the measured surface energies were found to show a similar parallel behavior (Akovali et al., 1998).

In situ cross linking was performed in a PVC/HDPE (weight ratio 50/50 blend) using a cross-linking system composed of dicumyl peroxide, triallyl isocyanurate, and magnesium oxide. The results of selective solvent extraction and SEM studies reveal that a semi interpenetrated network of cross-linked HDPE and uncross-linked PVC was formed during the process and the mechanical properties of the blends thus improved. When powdery nitrile rubber was added simultaneously to the blend, it improved the phase dispersity of PVC and HDPE, and at the same time attracted the cross-linking agents to the interfacial layer. A good synergism between powdery NBR and the cross linking system thus occurred (Fang et al., 1997).

Stress–strain behavior coupled with fractography was used by Ivanov et al. (Ivanov et al., 2009) to investigate the weld line strength of 30/70 w/w poly(vinyl chloride)/high polyethylene density polyethylene (PVC/HDPE) blends. Weld lines are created by the impingement of two opposing fountain flows when molten material has been injected into the mould. In general, the modulus, yield stress and strain, and fracture strain of the blends studied by Jarusa et al. (Jarusa et al., 2000) decreased as the PVC molecular weight increased in each case. The engineering stress–strain curves of PVC/HDPE blends without a weld line are shown in Figure14.1, and that with a weld line is presented in Figure 14.2. The weld line strength determined in uniaxial tension depended upon the domain shape of the PVC phase at the weld line, with elongated domains causing to weld line weakness.

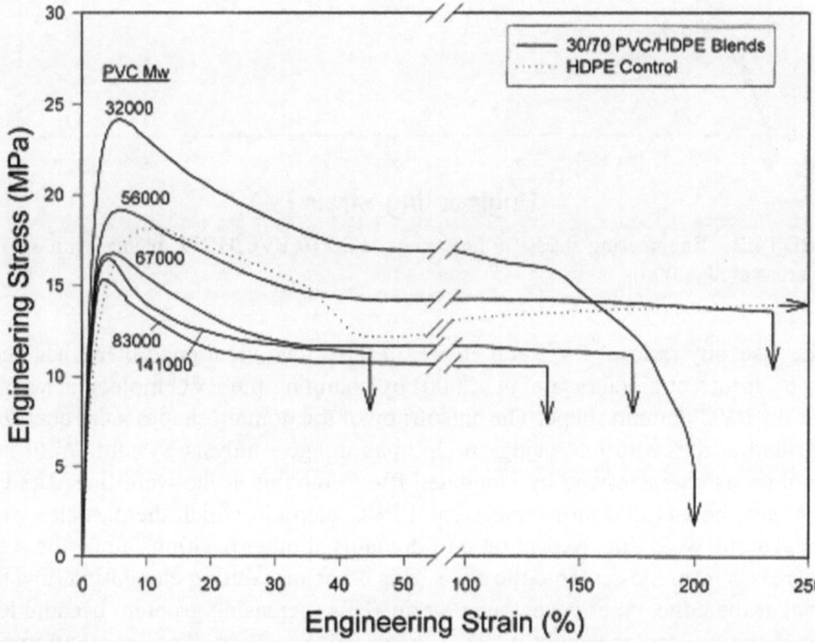

FIGURE 14.1 Engineering stress–strain curves of 30/70 PVC/HDPE blends without weld line (Jarusa et al., 2000)

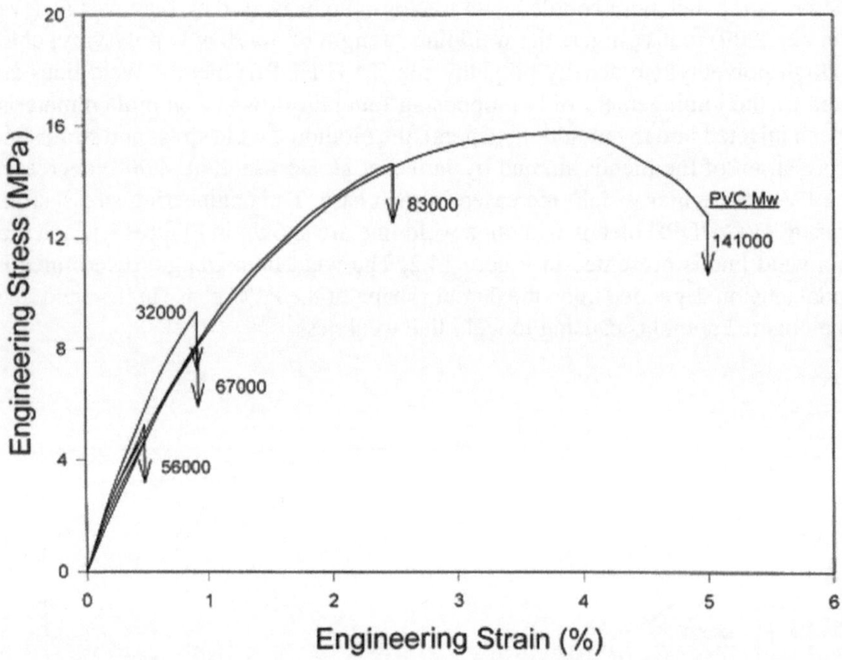

FIGURE 14.2 Engineering stress–strain curves of 30/70 PVC/HDPE blends with a weld time (Jarusa et al., 2000)

The viscosity ratio of PVC and HDPE in the above mentioned blend has been varied by Jarusa et al. (Jarusa et al., 2000) by changing the PVC molecular weight to alter the PVC domain shape. The anisotropy of the domain shapes were analyzed at bulk and at the weld line using an Optimas Image Analysis System. Weld line morphology is characterized by elongated PVC domains at the weld line. The exception was the 141,000 molecular weight PVC blend in which the particles were spherical at the weld line. Aspect ratio of domains at other positions along the weld line increases from the center to the edge. This is because during elongation flow the material at the edge experiences more strain. This increasing gradient became less severe as the molecular weight of PVC increased. Scanning electron micrographs of the fracture plane of those blend samples are shown in Figure14.3 (a–d). Fracture surface of all blends sample exhibits an oxbow shape which is represented by Figure 14.3 (a). The oxbow shape is created by one surface with a ridge in the center and a matching surface with a Center Valley. Fractography was performed on the surface with the ridge. In all cases, initiation occurred in the trough that surrounded the center ridge of the weld line.

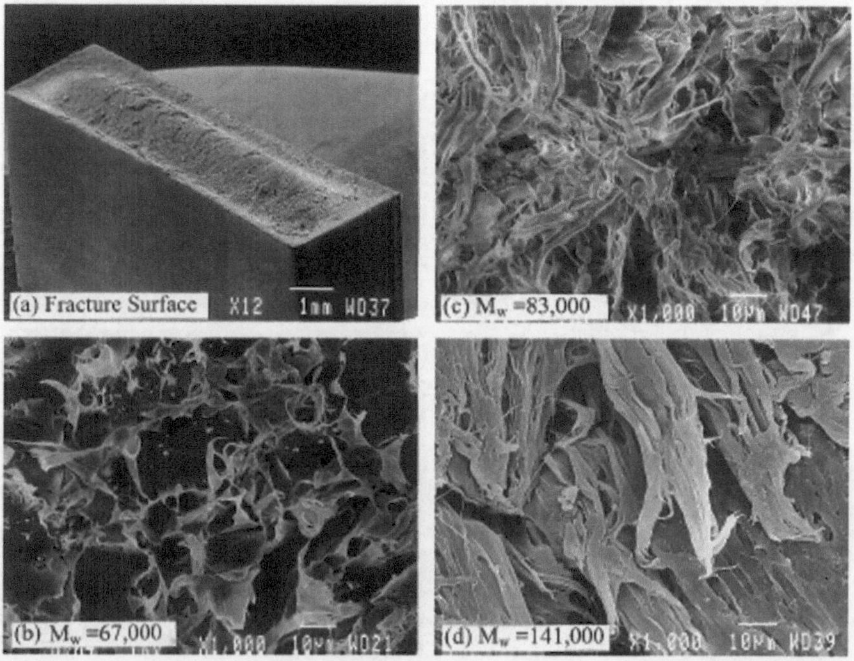

FIGURE 14.3 Typical fracture surface of the low-molecular-weight PVC blend showing the trough of the weld line where fracture initiated and magnifications of the initiation site for (a) Fracture surface; (b) PVC M_w. 67,000; (c) PVC M_w. 83,000; and (d) PVC M_w. 141,000 (Jarusa et al., 2000)

Macroscopic views of the region directly below the fracture plane are taken by scanning electron microscope and presented in Figure 14.4 (a)–(c). From the micrographs, clearly in the low-molecular-weight blends, fracture occurred along the weld line with no subsurface deformation and clear weld line profile. On the other hand, as the molecular weight is increased the fracture surface showed some distortion of the weld line profile with continuous and considerable deformation on a plane perpendicular to the stress at the site of the weld line, which is insignificant. The region with the highest degree of fibrillation, indicating the region of slowest crack growth, identified the fracture initiation site which was usually close to one end. The size of the initiation site depended strongly on PVC molecular weight. The fact is attributed to the brittle nature of blends with lower molecular weight PVC, which has undergone fracture at the weld line before a neck is formed.

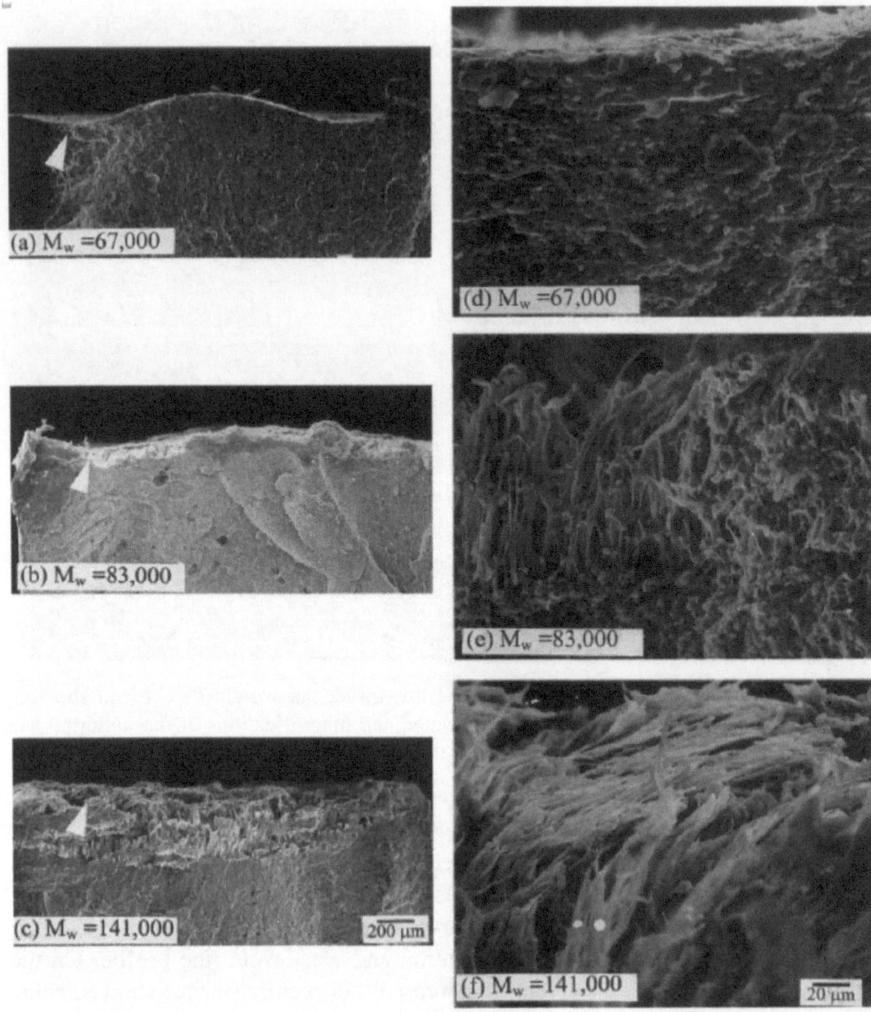

FIGURE 14.4 Macroscopic views of the morphology below the fracture plane: (a) PVC M_w. 67,000; (b) PVC M_w. 83,000; and (c) PVC M_w. 141,000: The arrows indicate the fracture initiation site. Higher magnification of the morphology below the initiation site: (d) PVC M_w. 67,000; (e) PVC M_w. 83,000; and (f) PVC M_w. 141,000 (Jarusa et al., 2000)

The elongation of the PVC phase at the weld line was found to be the cause of weak weld lines. The domain shape at the fracture initiation site was used in conjunction with a modified Nielsen approach to predict the ductile to brittle transition at the weld line. For the composition studied, a critical aspect ratio of the PVC phase of 1.24 was determined. Elongation of the PVC domains above this critical value led to brittle weld lines. The domain shape at the weld line was modeled by a Taylor analysis of droplet deformation in an elongational flow field. The calculations predicted that a viscosity ratio of 21 would produce a particle with an aspect ratio of 1.24. The observed weld line strength confirmed this prediction that blends with a viscosity ratio below 21 were brittle, and those with a viscosity ratio above 21 had been ductile weld lines.

In another study, the effect of the viscosity ratio on the morphology and mechanical properties of PVC/PE (50/50 mass ratio) blends was relevant. The melt viscosity of PVC, and PE was modified via changing the rotation speeds during the mixing process. The component having a lower melt viscosity constitutes the continuous phase in the blends. Phase inversion occurs when the melt viscosities of the two components are the same. The sample with a dual-continuous morphology has the best tensile strength and elongation at break, whereas, the sample in which PE forms the continuous phase has the greatest impact strength (Fang et al., 2000).

In order to overcome low-impact resistance and low processability PVC has been blended with polyolefin elastomers (POEs). These polymers are immiscible, and thus the resultant materials do not attain optimal properties. The polar nature of the PVC and nonpolar character of the polyolefins require that blends of this type be compatibilized in order to obtain a useful material that combines the strength of PVC with the impact resistance of POE (Fang et al., 2000). Chlorinated polyethylenes (CPEs) have been used as a compatibilizer for blends of PVC with many polyolefins such as ethylene/propylene/diene terpolymer (Lee and Chen, 1987), high-density polyethylene (Xi et al., 1987), low-density polyethylene, and linear low-density polyethylene (Francis and George, 1992; He et al., 1997). The presence of chlorinated polyethylene in these blends resulted in improved impact resistance and finer phase morphology relative to the uncompatibilized blends. Each of the chlorinated polyethylenes used in these studies was synthesized by randomly chlorinating the polyethylene backbone, resulting in a random copolymer (Cinadr et al.,). A study on the tricomponent blend system involving PVC/chlorinated PE/ethylene propylene diene terpolymer revealed that the improvements of the impact strength of PVC were significant either at room temperature or at low temperature (– 12 °C). However, a loss of the yield tensile strength was observed (Lee and Chen, 1987). Chlorinated polyethylene (CPE) is a commonly used impact modifier of PVC, and its compatibility depends upon the chlorine content and the distribution of chlorine atoms on the polyethylene (PE) backbone. CPE with 36% chlorine is the optimum composition for obtaining the necessary features of impact, processing and strength (Klarić et al., 2000). Zhou et al. (Zhou et al., 2000) reported a recent study on the modification of polyvinyl chloride/chlorinated PE blends with ultrafine particles of polystyrene. The results showed that the tensile strength slightly decreased with the impact strength increasing significantly; the rheological and plasticizing properties of PVC/CPE blends also improved when a proper content of UPS was filled into PVC/CPE PVC/CPE (100/10) matrix. The measured impact strengths generally correlated well with SEM and TEM morphologies. This result was different from that of toughening plastics with elastomer. The effects of modifying blend of PVC with LLDPE by acrylic acid, maleic anhydride, phenolic resins, and p-phenylene diamine were investigated. Modification by acrylic acid and maleic anhdride in the presence of dicumyl peroxide was found to be the most useful procedure for improving the mechanical behavior and adhesion properties of the blend. The improvement was found to be mainly due to the grafting of the carboxylic acid to the polymer chains; grafting was found to be more effective in LLDPE/PVC blends than in pure LLDPE (Francis et al., 1992).

In a study on the natural weathering, photo and thermal degradation of the two-phase system like that of PVC and poly(iso butylene) (PIB), the mechanical properties improved considerably when the poly blends were degraded partially presumably because of the interactions between the polymers. The presence of PIB in PVC up to 20% gave optimum stabilization to PVC, whereas small amounts of PVC in PIB destabilized PIB considerably. The presence of PIB in a PVC-PIB blend leads to considerable suppression of dehydrochlorination at 100°. Thus, the blends became more stable than the homopolymers (Kolawole and Olugbemi, 1985). Hydrogenated segmented poly[butadiene-block-((styrene-co-acrylonitrile)-block-butadiene)$_n$] block copolymers, were tested for their compatibilizing capacities for (10/90) LDPE/PVC blends. The acrylonitrile content of the SAN blocks of the block copolymers was found to be an extremely important factor for their miscibility with the PVC phase. When the SAN blocks of the block copolymers have the proper composition, they are excellent agents for the dispersion of the polyethylene phase in the blend into smaller domains. The compatibilized blends showed toughness comparable to the virgin PVC. Scanning electron micrographs showed improved adhesion between the dispersed PE phase and the PVC matrix (Kroeze et al., 1997).

A ternary blend consisting of chlorinated polyethylene (CPE), PVC and ENR was investigated to define the miscibility regime. In this ternary blend, where ENR played the role of a compatibilizer, it was established that different amounts were required to cause miscibility for the different ratios of CPE/PVC; the highest at equal concentrations of the chlorinated components (Koklas et al., 1991). In these cases, SAN, ENR, and so on, are referred to as the compatibilizers of the PVC blends. They compatibilize the blends and account for the miscibility of the participating blend components. In a recent study by Ivanova et al. (Ivanova et al., 2009) rheological and diffusion properties (mass transfer) of binary PVC–CPE blends with wide composition range (PVC:CPE = 100/0, 90/10, 80/20, 60/40, 40/60, 20/80, 10/90, 0/100 wt./wt.%) were investigated. The rheological characteristics of the compositions were measured on a capillary viscometer. From the rheological parameter pseudoplasticity of the investigated systems is evaluated in terms of the flowing index, which is shown in Figure 14.4. Flow index is measured with variation of PVC content in the blend, keeping temperature constant. Four different temperatures of 433 (1), 443 (2), 453 (3), and 463 K (4) are chosen for this purpose. From the graph, it is evident that the pseudoplasticity of the PVC–CPE blends is considerably increased by raising the PVC content in the blends, especially at higher temperatures and PVC content above 20 wt%. They have concluded from the sorption analysis data from the low-molecular-weight compound sorption data and gas permeability analysis that due to their smaller diffusion coefficients, systems with higher PVC content should be regarded as more appropriate for coating and barrier material applications.

Important information about the structure of the blends can be obtained from the changes in the flowing process activation energy as well the relaxation proces E_R which are shown in Figure 14.5. Changes of activation energies associated with the mass transfer process are closely related to (1) the length of the penetrant diffusion path, (2) the amount of the microvoids, and (3) the relationship of the amounts of permeable and unpermeable structural arrangements in the composites. As the CPE content increases or the PVC content decreases, irregularly arranged chlorine atoms of the CPE macromolecules promote the flow of CPE-rich blends in comparison to PVC-rich blends, at defined processing temperatures, that is, E_R of the polymer systems decreases with increasing CPE content in the blend, and relaxation will also follow the same trend.

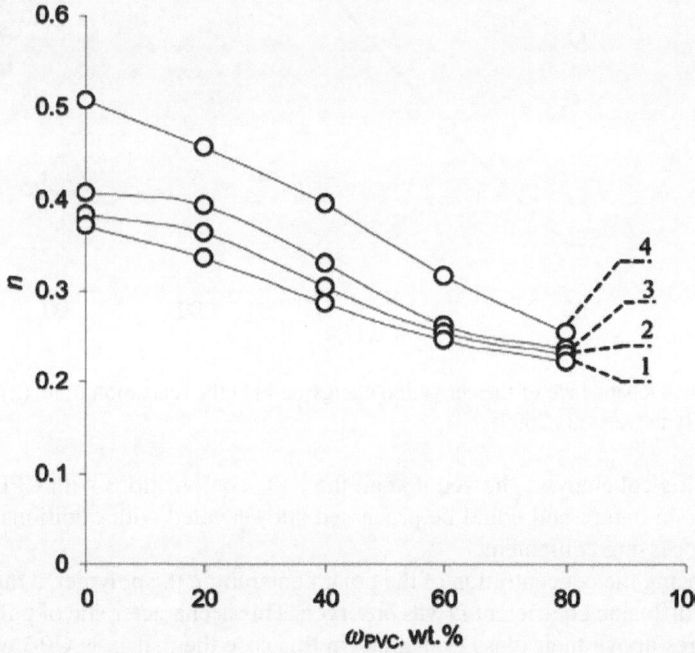

FIGURE 14.5 Dependence of the flowing index n on the PVC content at 433 K (1), 443 K (2), 453 K (3), and 463 K (4) (Ivanova et al., 2009)

Nonlinearity of the *R E* dependence on the blend content could be explained, to a certain extent, with structural changes in the polymer melt, as a result of which the intermolecular interaction is weakened, which undoubtedly also influence the relaxation processes in the polymer. The kinetics of the diffusion of low-molecular compounds and permeability of gases depends on the microheterogeneous structure of the investigated PVC–CPE blends as well as on the amount of the microvoids on the interfacial boundary area, locked there during thermoplastic mixing of PVC and CPE. Different mass transfer mechanisms can govern the diffusion process in various structural areas of the PVC–CPE blend.

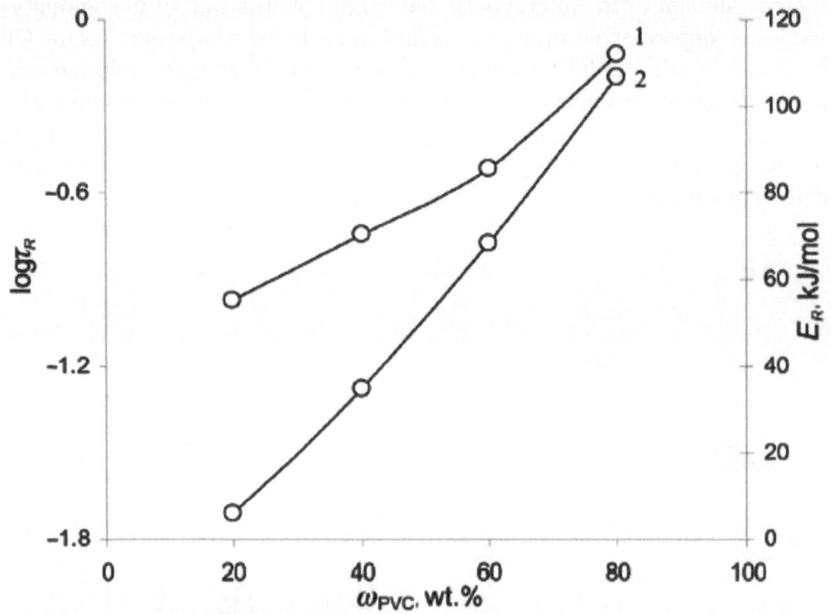

FIGURE 14.6 Dependence of the activation energy and (1) the relaxation time, (2) on the PVC content (Ivanova et al., 2009)

The rheological analysis showed that all the PVC compositions with CPE were thermoplastic in nature and could be processed and recycled with traditional thermoplastic processing equipment.

By increasing the concentration of the polar penetrant in the polymer, a rapid increase of the diffusion coefficient *D* was observed. This is characteristic of polymers at temperatures above their glass transition (in this case the matrix is CPE with its glass transition at about 255 K).

Asymmetric double cantilever beam and peel test experiments were completed by Eastwood et al. (Eastwood and Dadmun, 2002) to evaluate the ability of multiblock or blocky distributed chlorinated polyethylenes (bCPEs) to strengthen the PVC/POE interface compared to that of randomly distributed chlorinated polyethylene (rCPE). Additionally, the dependence of molecular weight and chlorine content of the bCPE (composition) will be evaluated to ascertain the influence of these parameters on the compatibilization process. Chlorinated polyethylenes to compatibilize poly(vinyl chloride) (PVC) and polyolefin elastomer (POE) blends. A series of chlorinated polyethylenes that are blocky in nature (bCPEs) with varying composition (% chlorine), and molecular weight (melt index) were used for this experiment.

The bare PVC/POE interface (i.e., samples with no copolymer at the interface) exhibit very weak interfaces, that is, the interfacial fracture toughness of the PVC/POE interface is 1.8 J/m^2 (Eastwood and Dadmun, 2002). The addition of chlorinated polyethylenes at the interface, to create a trilayer, gives much stronger interfaces. The effect of molecular weight on the ability of the blocky chlorinated polyethylene to strengthen the PVC/POE interface is shown in Figure14.7. Results indicate that improvement in the interfacial adhesion between the PVC and the POE is dramatically more pronounced with the bCPEs than with the rCPE. In addition, the optimum bCPE composition was determined to be 20% chlorine, and the interfacial adhesion force was found to increase with increasing molecular weight.

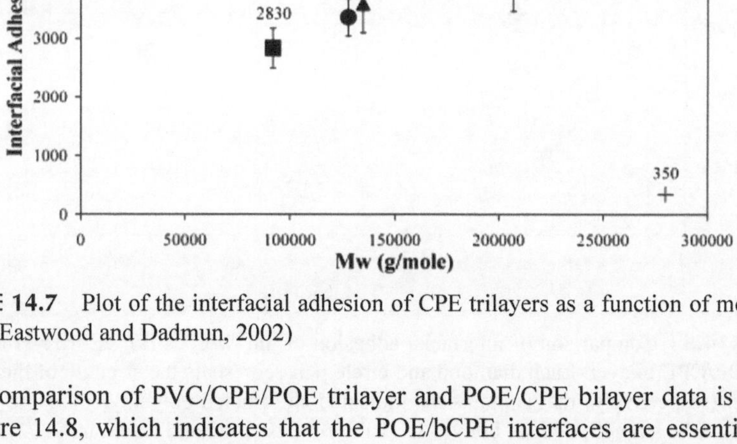

FIGURE 14.7 Plot of the interfacial adhesion of CPE trilayers as a function of molecular weight (Eastwood and Dadmun, 2002)

A comparison of PVC/CPE/POE trilayer and POE/CPE bilayer data is shown in Figure 14.8, which indicates that the POE/bCPE interfaces are essentially as strong as the compatibilized PVC/POE interfaces. This phenomenon implies that the strength of the compatibilized PVC/POE interfaces is governed by the POE/CPE interaction.

To evaluate the extent of compatibility of bCPE DSC analysis and XRD, studies made of the blend samples and the pure components have indicated the presence of cocrystallization which is evident by both the test. The melting behavior of a 50:50 blend is then compared to the DSC curves of the pure components, as shown in Figure14.9, for a blend of POE and b0.3-15. In addition, a DSC curve of a physically separated blend of POE, and b0.3-15 is included to provide a reference thermogram for a blend where cocrystallization cannot occur. All samples were annealed for 16 days at 95 °C before these thermograms were obtained. Large melting peaks in between the bCPE and POEs melting, peaks and slight shoulders as is shown in Figure14.9 DSC analysis of 50:50 POE/bCPE blends revealed melting peaks that were in between melting peaks of the individual polymers, and melting peaks found for the physically separated blends were noticeably absent from the 50:50 blends. This combination of data, therefore, strongly suggests that the POE will cocrystallize with each of the bCPE, providing a possible mechanism for the strengthening of the POE/PVC interface by the blocky copolymers.

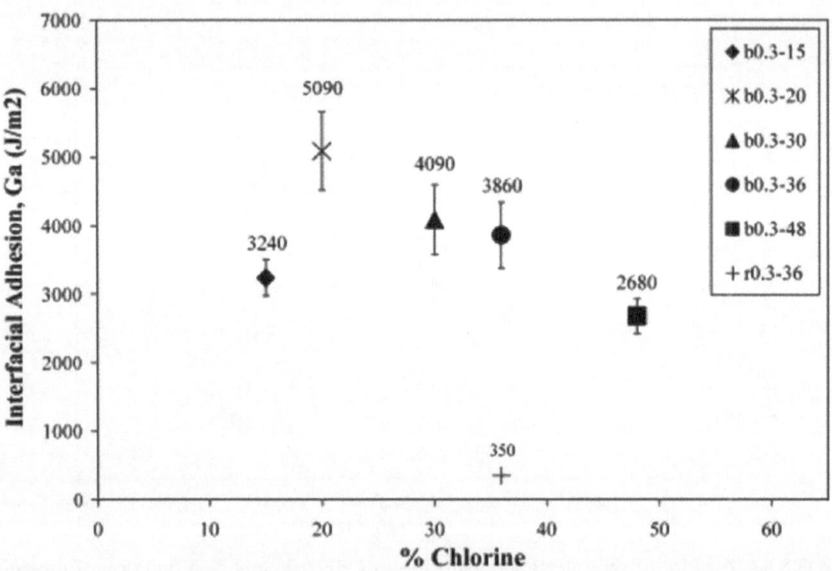

FIGURE 14.8 Comparison of interfacial adhesion of the PVC/CPE/POE trilayer and that of the POE/CPE bilayer. Each diamond and circle pair represents the strength of the bilayer or trilayer (respectively) that contains the indicated chlorinated polyethylene. Note that the similarity of the strength of each bilayer and trilayer (Eastwood and Dadmun, 2002).

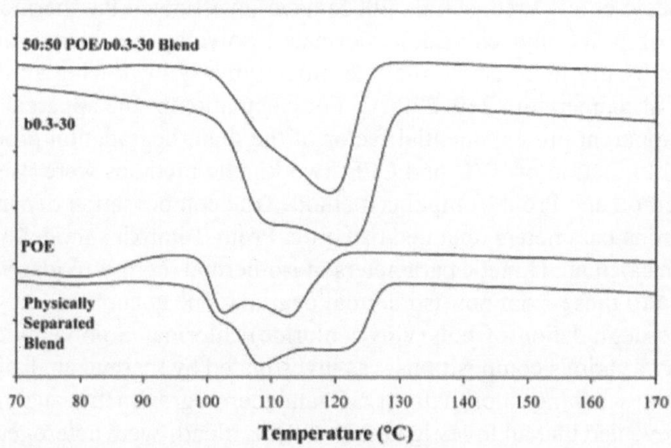

FIGURE 14.9 DSC curves of the b0.3-30, the POE, a 50:50 POE/b0.3-30 blend, and the physically separated 50:50 POE/b0.3-30 blend (Eastwood and Dadmun, 2002)

Qualitatively, Figure14.10 shows that the (110) and (200) peaks of the blend are not broader than those of the neat polymers, suggesting cocrystallization. More quantitatively, all blends display Bragg peaks in between that of the pure bCPE and POE, suggesting the formation of new cocrystallites. The interplanar distance for the pure b0.3-15, d110, is equal to 4.096 A°, while upon blending with POE, the interplanar distance shifted to 4.119 A°. Therefore each analysis of the crystalline structure of the POE/bCPE blends strongly indicates that cocrystallization occurs between the bCPE and POE.

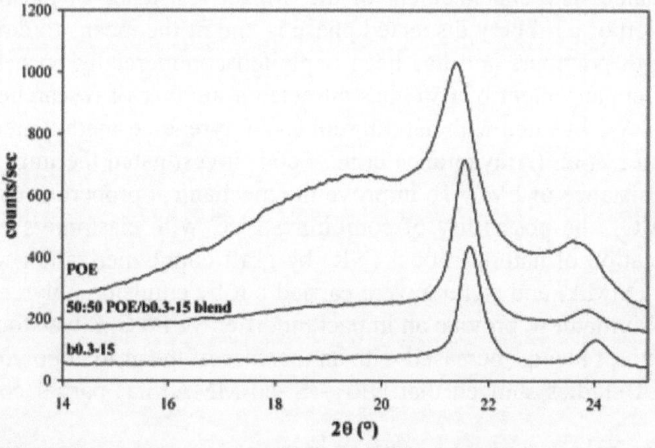

FIGURE 14.10 X-ray diffraction curves of the b0.3-15 copolymer, the POE, and a 50:50 POE/b0.3-15 blend (Eastwood and Dadmun, 2002)

N.S. Vraecic et al. (Vrecic et al., 2005) have investigated the thermo-oxidative degradation of poly(vinyl chloride)/chlorinated polyethylene blends of different compositions by means of isothermal thermogravimetry in flowing atmosphere of synthetic air at temperature 240–270 °C. For calculation of the apparent activation energy and apparent pre-exponential factor of the main degradation processes,that is, dehydrochlorination of PVC and CPE, two kinetic methods were used: isoconversional method and Prout-Tompkins method. True compensation dependency between Arrhenius parameters obtained using the Prout-Tompkins model, was found. However, the calculated kinetic parameters of isothermal thermo-oxidative degradation are close to those from non-isothermal degradation. In another study, the thermo-oxidative degradation of poly(vinyl chloride)/chlorinated polyethylene (PVC/CPE) blends of various compositions was investigated by thermal analysis methods like differential scanning calorimetry (DSC) and thermogravimetric analysis (TGA). DSC study revealed that all investigated PVC/CPE blends were heterogeneous. The main reaction of PVC degradation at moderate temperatures is dehydrochlorination and also a dominant reaction in CPE degradation, but the thermo-oxidative stability and degradation mechanism are different. It was found that CPE had a stabilizing effect on thermo-oxidative degradation of PVC and interactions of blends components with their degradation products occurred (Vrecic et al., 2004).

14.3 BLENDS OF PVC WITH ELASTOMERS AND THEIR COPOLYMERS

Unmodified unplasticized poly(vinyl chloride) (PVC-U) has the disadvantage of being prone to occasional brittleness and is notch sensitive. Extensive research and development work has therefore been carried out to formulate polymers with high-impact resistance. The enhancement of the impact resistance of plastics through the introduction of a rubbery dispersed phase is one of the means to develop high-impact strength polymers and has been exploited commercially on a large scale. The study of enhancement of PVC has attracted a number of researchers. In a recent study of PVC blended with natural rubber-g-(styrene-co-methyl methacrylate), W. Arayapranee et al. (Arayapranee et al., 2004) investigated the improvement of the impact resistance of PVC. To improve the mechanical properties of poly(vinyl chloride) (PVC), the possibility of combining PVC with elastomers was considered. Modification of natural rubber (NR) by graft copolymerization with methyl methacrylate (MMA) and styrene were carried out by emulsion polymerization by using a redox initiator to provide an impact modifier for PVC. It was found that the impact strength of blends increased with an increase of the graft copolymer product content. DMA studies showed that NR-g-(St-co-MMA) has partial compatibility with PVC.

The EPDM-PVC and MMA-g-EPDM-PVC blend as developed by D Singh, V.P. Malhotra and J.L. Vats (Singh et al., 1999) showed two glass transition temperatures indicating the incompatibility of these blends, and the Tg increased with increasing graft content. The tan δ curve from DMA studies indicated three dispersion regions for all blends arising from the α, β, and γ transitions of the molecules. The SEM micrographs of EPDM-PVC showed less interaction between the phases in comparison to MMA-g-EPDM-PVC blends.

In another related study, core-shell structured grafted copolymer particles of polybutadiene grafted PMMA (PB-g-PMMA, MB) prepared by emulsion polymerization by Wu et al. (Wu et al., 2004) were used to modify PVC by melt blending. The results indicated that the samples with the best impact strength could be obtained when the core shell weight ratio of PB to PMMA is lower than 93:7 and the addition of modifier with the ratio core to shell of 93:7 could reduce the domain size of the dispersed phase. Furthermore, the compatibility and properties of the blends were greatly enhanced and improved. The modifier particles could be well dispersed in the PVC matrix. An allied study has been made by Sahaet al. (Saha, 2001) on the miscibility of PVC with polychloroprene blends. Compatibility of PVC/PCP blend system was determined by Zelinger-Heidingsfeld criterion at 30 °C. The Stockmayer Fixman equation related intrinsic viscosity and $\overline{M}_w$ of polymers and was used to calculate polymer-solvent interaction. It was found that although PVC and PCP were incompatible in general, they exhibit limited compatibility at high PCP concentration. A rheological study of an amorphous polychloroprene (PCP) with PVC blends during fusion processes has been carried out with respect to their compatibility over the complete composition range. Experimental results show that PCP can promote the processability and fusion of PVC. Torque-rpm behavior exhibits the "entanglement plateau" effect and the polymer-solvent interaction lying beyond the tolerated limit. Temperature and rpm promote additional gelation of PVC particles. It has also been found that the mixing sequence has a profound influence on the processability and compatibility of the melt blend system studied (Saha, 2001).

A widely studied PVC blend is epoxidized natural rubber-50 (ENR50)/ poly(vinyl) chloride (PVC), which is found to form miscible blends at any compositions ratio (Ishiaku et al., 1999). The excellent miscibility between ENR50 and PVC is believed to be induced by the highly polar epoxide groups within the ENR50 molecules (Perera et al., 2000). PVC is also anticipated to provide high-tensile strength and good chemical resistance, while ENR can act as a plasticizer for PVC. In fact, the mechanical properties of ENR50/PVC have been widely studied and reported in the literatures (Ishiaku et al., 1999). The compatibility of ENR with PVC containing 25 and 50 mol % oxirane groups (ENR 25 and ENR 50) indicated that the latter was miscible at the segmental level throughout the composition range while the former was only partial. The conclusion was that the introduction of the oxirane groups gave a stiffer polymeric plasticizer compared to other macromolecules plasticizing PVC with pendant active groups (Perera et al., 2000).

C. T Ratnam et al. (Ratnam and Zaman, 1999) has fabricated blends with 50/50 PVC/ENR composition and 1–5 phr tribasic lead sulfate (TBLS) at 6 and 30 min mixing time in Brabender. These blends were irradiated using a 3.0 MeV electron accelerator with doses ranging from 0 to 200 kGy. The suitable dose of TBLS was selected as 2 phr which rendered optimum tensile properties of the blend upon irradiation and 30 min Brabender mixing time produced highest tensile strength indicating homogeneous mixing; tan δ curves also supported the observation of increasing homogeneity with increasing mixing time. In another attempt by Ratnam et al. (Ratnam and Zaman, 1999) the influence of multifunctional acrylates (MFA) such as TMPTA, HDDA and EHA as a crosslinker on the 70/30 PVC/ENR blend along with irradiation was investigated. Study of mechanical properties along with gel content and glass transition temperature revealed that except elongation, all the properties are improved with increasing irradiation dose and further improved by addition of MFA. Among the MFA, TMPTA can be useful as an efficient crosslink enhancer to PVC/ENR blends. Fourier transform infrared spectroscopy (FTIR) study indicated that radiation-induced crosslinks formed in PVC/ENR blends to have influenced by TMPTA and increased the cross-link density. The single glass transition temperature obtained confirms that the blend remains miscible upon irradiation with the presence of TMPTA.

The stability of 70/30 poly(vinyl chloride)/epoxidized natural rubber (PVC/ENR) blends was studied by C.T. Ratnam et al. (Ratnam and Zaman, 1999), in a different work. The result revealed that tribasic lead sulfate is efficient in stabilizing PVC/ENR blends as well as in enhancing the blend properties upon irradiation and addition of zinc octoate proved to play a major role in providing good color retention and radiation stability to the blends with relatively low improvement in blend properties upon irradiation. They have also studied the effects of three different types of antioxidants, a hindered phenol, a phosphite, and a hindered amine light stabilizer (HALS) on the radiation-induced cross linking and oxidative degradation of the same blend with a cross-linking agent. Among all the antioxidants, the hindered phenol is the most effective one in this case and induced post irradiation reaction as evident from the Fourier transform infrared spectroscopy study. The analysis indicated the radiolysis products of hindered phenol in PVC/ENR blends.

It was evident from Monsanto rheometry, solvent swelling and infrared spectroscopic studies that miscible blend of PVC and epoxidized natural rubber (ENR) could be cross linked during high temperature moulding for prolonged periods (Ramesh and Dey, 1993). The cross linking had no effect on blend miscibility. The physical properties, however, depended on the blend composition. As the PVC content increased, the blend behavior changed from elastomer to be glassy.

In a recent work by Nawi et al. (Nawi et al., 2012) commercially acquired TiO_2 photocatalyst (99% anatase) powder was mixed with epoxidized natural rubber-50 (ENR50)/polyvinyl chloride (PVC) blend by ultrasonication and immobilized onto glass plates as TiO_2-ENR50-PVC composite via a dip-coating method. Photo etching of the immobilized TiO_2-ENR50-PVC composite was investigated under the irradiation of a 45W compact fluorescent lamp. The composite samples are characterized by chemical oxygen demand (COD) analysis, scanning electron microscopy-energy dispersive X-ray (SEM-EDX), spectrometry, thermogravimetry analysis (TGA), and fourier transforminfrared (FTIR) spectroscopy. The BET surface area of the photo etched TiO_2 composite was observed to be larger than the original TiO_2 powder due to the systematic removal of ENR50, while PVC was retained within the composite. This is consistent with the results of SEM whereby most of the non-etched immobilized TiO_2-ENR50-PVC surface was covered with the polymer blend. However, compared with pure TiO_2, the surface areas of the photo etched immobilized TiO_2-ENR50-PVC composite were increased due to the decreased pore diameter. This gives the impetus for using ENR50/PVC blend as the adhesive for immobilizing the TiO_2 powder onto solid supports such as glass plates, since their subsequent etching process enhances the surface properties of the immobilized system.

TABLE 14.1 Microstructures of TiO_2 powder, nonetched TiO_2-ENR50-PVC composite, and photo etched TiO_2-ENR50-PVC composite (Nawi et al., 2012)

Samples	BET surface area ($m^2\ g^{-1}$)	Total pore volume ($cm^3\ g^{-1}$)	Average pore diameter (nm)
TiO_2 powder	10.17	0.0237	9.348
TiO_2 composite	5.44	0.0161	11.846
Etched TiO_2 composite	15.24	0.0244	6.418

The SEM micrograph of the nonetched TiO_2-ENR50-PVC composite showed in Figure14.11 (b) exhibits extensive aggregations of TiO_2 particles with irregular emergence of pores with visible depths. The catalyst particles were heavily agglomerated and are seen to be covered by whitish layers which are believed to be the polymer blends. As seen in Figure 14.11 (c), the photo etched TiO_2-ENR50-PVC composite via irradiation for 9 h clearly shows the elimination of the sticky whitish layers that enveloped the TiO_2 particles even though aggregations of TiO_2 particles and porous depths are still observed.

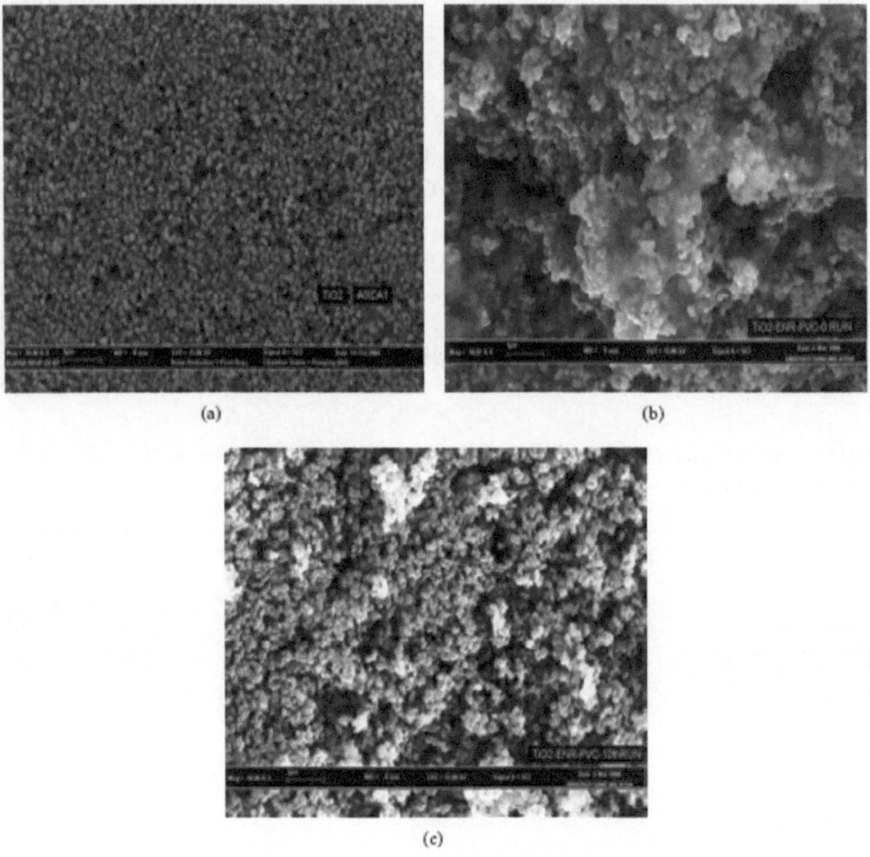

FIGURE 14.11 SEM micrographs of (a) TiO_2 powder, (b) nonetched TiO_2-ENR50-PVC composite, and (c) photo-etched TiO_2-ENR50-PVC composite under 10,000 × magnification (Nawi et al., 2012)

Nawi et al. (Nawi et al., 2012) also evaluated the photomineralization capability of the prepared immobilized TiO_2 composite by using methylene blue (MB) as the model pollutant. The reusability and reproducibility of the TiO_2 composite in the degradation of MB were also tested. It also exhibited better photocatalytic efficiency than the TiO_2 powder in a slurry mode and was highly reproducible and reusable.

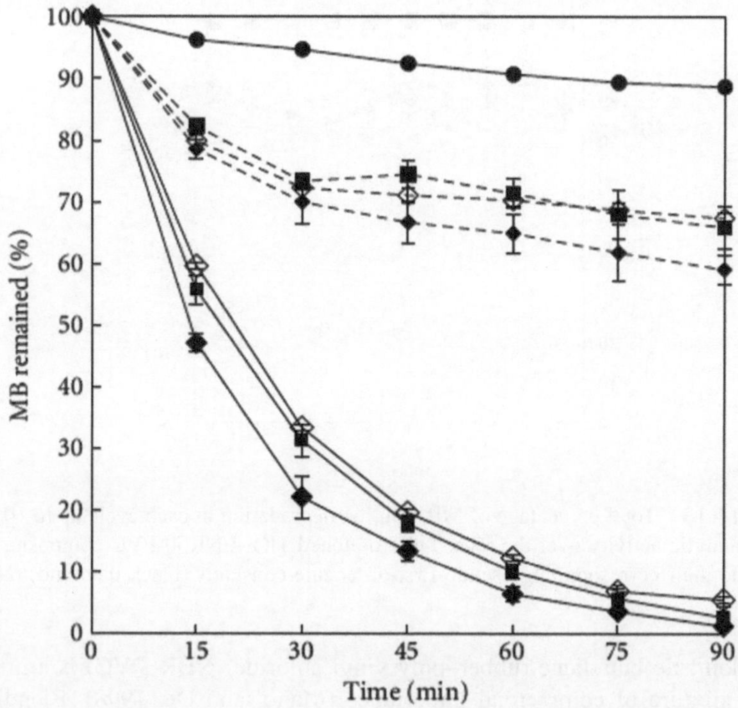

FIGURE 14.12 Degradation of MB by photolysis (black circle), TiO_2 powder (black square), nonetched TiO_2-ENR50-PVC (white diamond), and photo etched TiO_2-ENR50-PVC composite (black diamond) via irradiation under 45W compact fluorescent lamp with UV leakage of 4.35Wm^{-2} (continuous line) and adsorption in the dark (dashed line). (Nawi et al., 2012)

As shown in Figure14.12, only 10.53 ± 1.15% of MB was photolyzed within and sustainability of the photo-etched immobilized TiO_2-ENR50-PVC composite in the photocatalytic degradation of MB is shown in Figure14.13. The photocatalytic efficiency of TiO_2 composite was consistent throughout the 10 subjected cycles whereby more than 98% of the MB was consistently degraded from the first cycle until the tenth cycle of applications. About 93% of the 20mg L^{-1} MB was mineralized over a period of 480 min. The presence of SO4^{2-}, NO^{3-}, and Cl^{-} anions was detected in the mineralized solution where the solution pH was reduced from 7 to 4. The photo-etched immobilized TiO_2-ENR50-PVC composite system therefore possesses good sustainability and reusability upon its recycled applications. It was also noted that the fabricated photocatalyst plate was essentially intact after 10 cycles of applications and can certainly be reused for many more applications. More than 98% of MB removal was consistently achieved for 10 repeated runs of the photo-etched photocatalyst system.

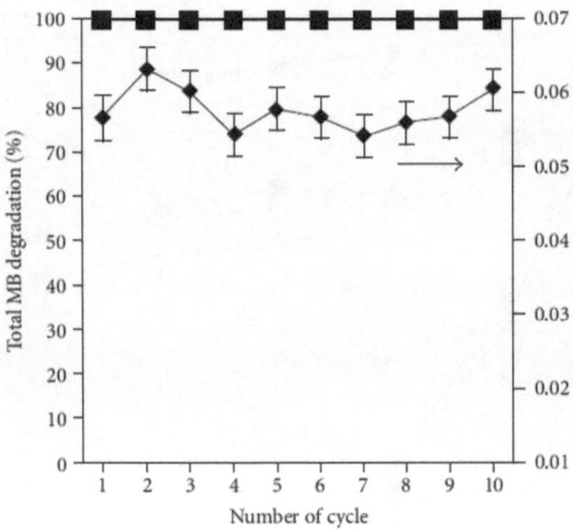

FIGURE 14.13 Total percentage of MB solution degradation in each cycle up to 10 cycles of photocatalytic activity over the reused photo-etched TiO_2-ENR50-PVC composite (black square) and their corresponding pseudo-first-order rate constants (black diamond) (Nawi et al., 2012)

Acrylonitrile butadiene rubber–poly vinyl chloride (NBR–PVC) is a miscible physical mixture of commercial importance (Manoj and De, 1998). Blending of polymers is made to improve the physical, thermal, and mechanical properties as well as to modify the processing characteristics and cost reduction of the final product. Efficiency of blending depends upon the nature of both the polymers. As both NBR and PVC are polar in nature they form a compatible blend (Lapa et al., 2002; Thormer et al., 1984). The NBR can act as a permanent plasticizer for PVC in applications like wire and cable insulation, food containers, pond liners used for oil containment, and so on. The presence of PVC improves aging and chemical resistance of NBR in applications like feed hose covers, gaskets, conveyor belt covers, printing roll covers, and so on. PVC also vastly improves abrasion resistance, tear resistance, and tensile properties. It also adds gloss and improves finish of the extruded stock and imparts flame-retardant character. NBR/PVC blends can be conveniently milled, extruded, and compression-molded using traditional processing equipments for natural and synthetic rubbers. One difficulty in forming successful blends of NBR and PVC is the lack of suitable stabilizers for PVC which do not affect NBR. Work based on fabrication and also modification of this conventional blend has been developed already starting from three decades ago. Recent works are based on successful crosslinking of NBR in the blend by different vulcanization technique (Mark and Bilaes, 1986; Supri and Yusof, 2004) and also blend composite formation (Supri and Yusof, 2004) Vulcanization occurs when two radicals produced on neighboring polymer units recombine. These radicals are produced by a chemical agent, such as peroxide or sulfur or alternatively by radiation, such as electron beam or gamma radiation (Hoffman et al., 1989; Mark and Bilaes, 1986). The cross-linking reaction involves the allylic chlorine sites in poly (vinyl chloride) and the CN group in the nitrile rubber. The dynamic mechanical analysis reveals that such a self-crosslinkable plastic-rubber blend is miscible in different blend ratios. The structure and dynamics of the binary blends of PVC/NBR ternary blend were investigated by DMA analysis and solid state NMR. Samples were prepared by mechanical blending at 150 °C. The effect of the change in the composition of the polymer blends, on the tan δ peak width, the tan δ max and the area under the tan δ curve were used to understand the miscibility and damping properties. Solid state ^{13}C NMR measurements were carried out to determine several relaxation time parameters. Cross-polarization times and carbon relaxation times were interpreted based on the heterogeneity of the matrix. Results confirmed phase separation between 1 and 6 nm length scale.

It has been reported by Manoj et al. (Manoj et al., 1993) that PVC /NBR rubber blends undergo a self cross-linking reaction (crosslinking without the aid of any external curing agents) during processing at elevated temperatures, resulting in an increase in torque as studied by means of a Brabender Plasticorder. The extent of reaction depended on the blend composition, processing temperature, shear rates, fill factor, and the presence of a PVC stabilizer. The chemical reaction as analyzed by IR spectroscopy showed that the reaction involved the hydrolysis of the nitrile group by HCl liberated during the degradation of PVC.

A study by A.K. Sen and G.S. Mukherjee (Sen and Mukherjee, 1993) reported the use of inverse gas chromatography to investigate the thermodynamic compatibility of blends of PVC and nitrile rubber (NBR) as a function of blend composition and acrylonitrile content of NBR. The values of the polymer-polymer thermodynamic interaction parameters and the solubility parameter of the polymers and their blends were determined with the help of the measured retention data for various polar and nonpolar probes in the pure and mixed stationery phases of these polymers. The two polymers exhibited fair compatibility which increased with increasing content of acrylonitrile in NBR.

The structure and dynamics of the binary blends of PVC/NBR ternary blend were investigated by DMA analysis and solid state NMR. Samples were prepared by mechanical blending at 150 °C. The effects of morphological parameter, especially rubber particle size distribution on the brittle-ductile transition of PVC/NBR blends with well dispersed NBR particles have been studied by Z.H. Lin and coworkers. PVC had been super toughened with well dispersed NBR particles. It had been argued that the rubber particle size distribution was one of the morphological parameters dominating the toughness and the toughening mechanisms of PVC/NBR blends, in addition to the average rubber particle size, rubber particle volume fraction, and average matrix ligament thickness. A narrow rubber particle size distribution favors the rubber toughening of PVC (Liu et al., 1998). Another related study by the same workers (Liu et al., 1998) reported the influence of rubber particle spatial distribution on the toughening and stiffening efficiency of PVC/NBR blends quantitatively. Blends with the pseudo network morphology or the morphology of well dispersed rubber particles were used to elucidate the influence. The spatial distribution and matrix ligament thickness considerably affected the impact strength of PVC/NBR blends. On the other hand, the tensile modulus of the blends was a function of rubber content but not of particle spatial distribution. The dependence of toughening and stiffening efficiency on the spatial distribution was stronger than on the matrix ligament thickness. The psuedo network morphology was thus demonstrated to be very efficient in generating PVC/NBR blends simultaneously with high toughness and high stiffness.

The influence of interfacial adhesion on the tensile properties including toughness of PVC-NBR blends with the morphology of well dispersed rubber particles had been investigated using two types of blends containing NBR 26 (26 wt% acrylonitrile) having higher interfacial adhesion strength than the second type that contains NBR 18. The secant modulus and yield stress of the blends were found to be independent of interfacial adhesion. On the other hand, the elongation at break and toughness depended strongly on the interfacial strength (Liu et al., 2001).

Blends of PVC with varying contents of plasticizer and finely ground powder of waste nitrile rubber rollers were prepared over a wide range of rubber content through high temperature blending. The percentage elongation, flexural crack resistance, and impact strength of blends increased considerably over those of PVC. The waste rubber had a plasticizing effect. Blends of waste plasticized PVC and waste nitrile rubber showed promising properties. The electrical properties and limiting oxygen index decreased with increasing rubber and plasticizer content (Ghaisas et al., 2004).

The compatibility of hydrogenated poly(butadiene-co-acrylonitrile) (HNBR) with PVC was investigated (Sotiropoulou et al., 1993). DMA, DSC, and phase contrast microscopy indicated that the blends were miscible at the segmental level.

In a related study (Li and Chan, 2001), acrylonitrile-butadiene copolymers (NBR) with different acrylonitrile contents were melt mixed with PVC. TEM results indicated that all samples blended with different NBR, including NBR-33 and NBR-41, were heterogeneous, but the dispersion of NBR-33 was the finest. The results of chlorine loss study showed that the radiation sensitivity of PVC was improved by the addition of NBR and the extent of improvement was strongly related to the miscibility.

K. J. Kim (Kim, 2012) has studied the effect of addition of organo bifunctional silane (TESPD) into silica contained NBR/PVC blend. The addition of the TESPD into silica filled NBR/PVC compound leads to an increase in the degree of cross-linking by formation of a strong three-dimensional network structure between silica surface and rubber matrix via coupling reaction, which results in improved mechanical properties and processability.

The effect of acrylic acid (AAc) on the torque, stabilization torque, mechanical energy, swelling behavior, mechanical properties, thermal stability, and morphological characteristics of recycled poly(vinyl chloride)/acrylonitrile–butadiene rubber (PVCr/NBR) blends was studied by Ismail et al. (Ismail et al., 2005). The blends were prepared by melt mixing at a temperature of 150 °C and rotor speed of 50 rpm. It was found that PVCr/NBR + AAc blends exhibit higher stabilization torque, mechanical energy, stress at peak, and stress at 100% elongation, but lower elongation at break and swelling index than those of PVCr/NBR and PVCv/NBR blends due to increased compatibility of the incorporation of AAc in the blend. SEM study of the tensile fracture surfaces of the blends also supports the observation. However, thermal gravimetry analysis of the blends showed that the presence of AAc decreased the thermal stability of PVCr/NBR blends.

Gheno et al. (Gheno et al., 2010) have applied dynamic vulcanization process to fabricate employed in the melt state of elastomers with thermoplastics. In this study, a vulcanized thermoplastic e system based on sulfur (S)/tetramethylthiuram disulfide (TMTD) and mercaptobenzothiazyl disulfide (MBTS) was obtained with the dynamic vulcanization process by Gheno et al. The formation of crosslinks was characterized by differential scanning calorimetry (DSC) and Fourier transform infrared (FTIR) spectroscopy. The blend was obtained with properties such as mechanical strength, Young's modulus, hardness, and abrasion fatigue. The phase morphology was investigated using atomic force microscopy (AFM) operating in the tapping mode-AFM. The phase images of the dynamically vulcanized blends showed an elongated morphology, which can be associated to the formation of crosslinks that give the material its excellent mechanical properties.

In the work by Hafiz et al. (Hafezi et al., 2007) physicomechanical properties including tensile strength, elongation at break, hardness, resilience, and aging of acrylonitrile butadiene rubber–poly vinyl chloride (NBR–PVC) blend cured by sulfur and electron beam has been investigated. Results show that the physicomechanical properties, except resilience, of NBR–PVC blend cured by the electron beam improved by 15% compared with blends cured by the sulfur system. The samples were further aged by keeping in an oven at 100 °C for 48 h, for example, the NBR–PVC blend cured by electron beam exhibited higher tensile properties and hardness before aging and after aging than the sulfur curing system. Figures 14.14 and 14.15 show an increase in TS, 7% before aging and 6% after aging; in EB, 30% before aging and 23% after aging. It is because of higher cross-linking density and –C–C– bond existed in cross-linked NBR–PVC blend cured by electron beam.

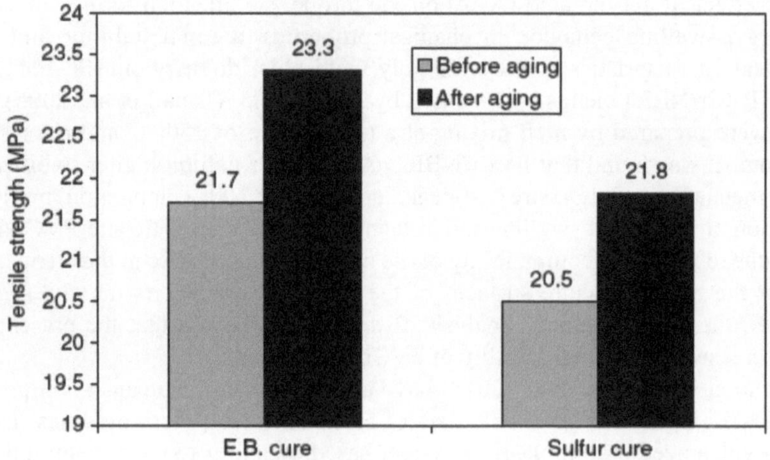

FIGURE 14.14 Comparison of tensile strength of sulfur and electron beam cured NBR–PVC blend (Mark and Bilaes, 1986)

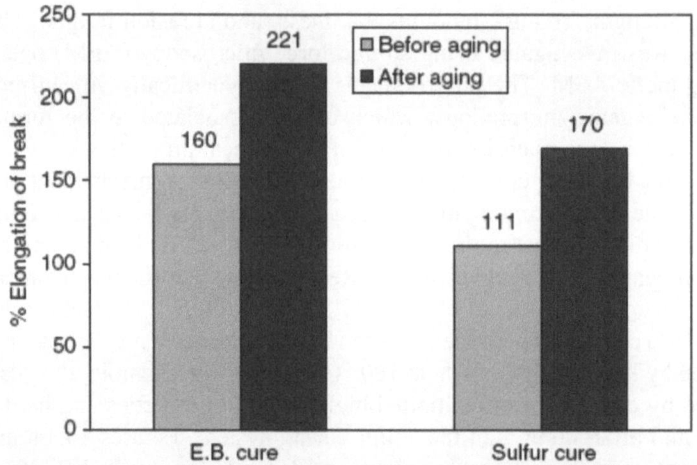

FIGURE 14.15 Comparison of elongation at break of sulfur and radiation cured NBR/PVC blend (Mark and Bilaes, 1986)

Hafezi et al. (Hafezi et al., 2007) have concluded about the advantages of using the electron beam curing system over conventional techniques. The thermal or chemical processing vary with the application, but advantages include one or more of the following: reduced cure time and energy investment; higher process throughput; operation at ambient temperatures with no effluent; more uniform product; unique product properties; and better control on curing process. Therefore high energy electron beam is a suitable method for cross-linking of NBR–PVC blend.

14.4 BLENDS OF PVC WITH VARIOUS METHACRYLATES/ ACRYLATES

The most important type of polymeric modifiers are the methacrylates and the acrylates, the incorporation of which is expected to act both as processing aid and impact modifier. The incorporation of acrylates in PVC is analogous to the rubber toughening of glassy polymers as these are well known to be rubbery in character having glass transition temperatures close to or below room temperature. The rubbery acrylates when incorporated within PVC exert their modifying influences in improving the toughness of the blended systems. The increment in ductility is associated with a simultaneous decrease in load bearing characteristics as manifested by modulus and ultimate tensile strength. This decrease is however, far compensated against the reduction observed when the rigid PVC was converted to flexible PVC by use of suitable plasticizers (Chakrabarti et al., 2004).

14.4.1 PHYSICO-MECHANICAL MODIFICATION

In an early study on the polyblends of poly(vinyl chloride) with various methacrylate copolymers, the effect of blending on the physical and mechanical properties has been investigated. Copolymers of methylmethacrylate with methyl acrylate, ethyl acrylate, butyl acrylate and 2 ethylhexyl acrylate in 80:20 and 50:50 proportions with respect to methylmethacrylate have been prepared and characterized by nuclear magnetic resonance spectroscopy. Polyblends of PVC and such polyacrylates have been prepared in 80:20 ratio by melt blending technique and characterized by thermomechanical analysis to study the glass transition behavior vis-a-vis the compatibility of these blends. Mechanical properties of these blends revealed a substantial increase in impact strength particularly when long chain acrylate polymers like butyl acrylate and 2 ethyl hexyl acrylates are used; however, there is a decrease in the yield stress and initial modulus. A shift from brittle failure to ductility has been observed in blends of PVC on incorporation of these acrylate copolymers (Higashida et al., 1992).

Similar trends were noted on polyblend formation of PVC with various methacrylates and acrylates when PVC was blended with poly(methyl methacrylate) (PMMA), poly(ethyl methacrylate) (PEMA), poly(butyl methacrylate) (PBMA), poly(ethyl acrylate) (PEA) and poly(butyl acrylate) (PBA) (Chakrabarti and Chakraborty, 2006; Chakrabarti and Chakraborty, 2007). The blends were characterized with respect to their physicomechanical, thermomechanical, thermal and morphological properties. The blends were prepared with PVC and the different methacrylates and acrylates in varying composition ratios of each polymer. The mechanical, thermal and morphological parameters under consideration varied according to the percentage incorporation of the modifying polymer with the major polymer matrix PVC. The resultant mechanical properties of the PVC-acrylate series conform to the fact that the α-hydrogen of vinyl chloride unit of chain segments can interact with the ester carboxylate group (H-bond acceptor of the acrylic polymer) to form a hydrogen bond, which is assumed to play a key factor in achieving miscibility and enhanced mechanicals (Olabisi, 1975: 63; Prudhome, 1982). Considering the specific case of PVC-PMMA blends, a short shift in frequency characteristic of C-Cl dipole can be observed in the IR spectra of the base PVC compound and its blends with PMMA. The typical absorbance due to C-Cl in the base compound at 704 cm^{-1} undergoes a decrease in frequency (693 cm^{-1}) in the blend in presence of methacrylate esters which may possibly be attributed to a feeble interaction between the C-Cl dipole and the β hydrogen of the methacrylate esters.

Both the modulus and the ultimate tensile strength exhibit a decreasing trend with increasing proportion of PMMA to the extent of 10% of PMMA incorporation beyond which, there is a steady increase in both of these parameters within the ranges of concentration studied. The somewhat plasticized PVC as has been used in various blends undergoes further plasticization initially with the incorporation of stiff and rigid PMMA which, by virtue of its bulky molecular structure throws apart the stiff and rigid chains of PVC and thus synergizes the function of the already present conventional plasticizer DOP which in turn reduces the dipole-induced dipole interactions exerted by polar C-Cl bonds on the neighbouring chains. The plasticizing influence of DOP is thus enhanced. Beyond 10% of PMMA incorporation within PVC, both the toughness and elongation at break decreases. From the optical micrographs, it can be observed that as the number of dispersed PMMA particles increases there is also a rise in the coalescing tendency of the particles with progressive increases in PMMA content. Thus, the pathlength for cracks to propagate decreases and hence the reduction in toughness and EB% is observed. The rigid PMMA particles may be assumed to exert their influences as its content increases in the matrix. The decrease in the number of cracks attributed to the increasing sizes of domains coupled with the intrinsic rigidity of the dispersed PMMA phase may be assumed to offset the large increase in toughness of PVC observed at the earlier stages of PMMA incorporation. However, the tough PVC blends with strength almost equivalent to unmodified plasticized PVC results. The results have been elucidated in Figures 14.16 and 14.17.

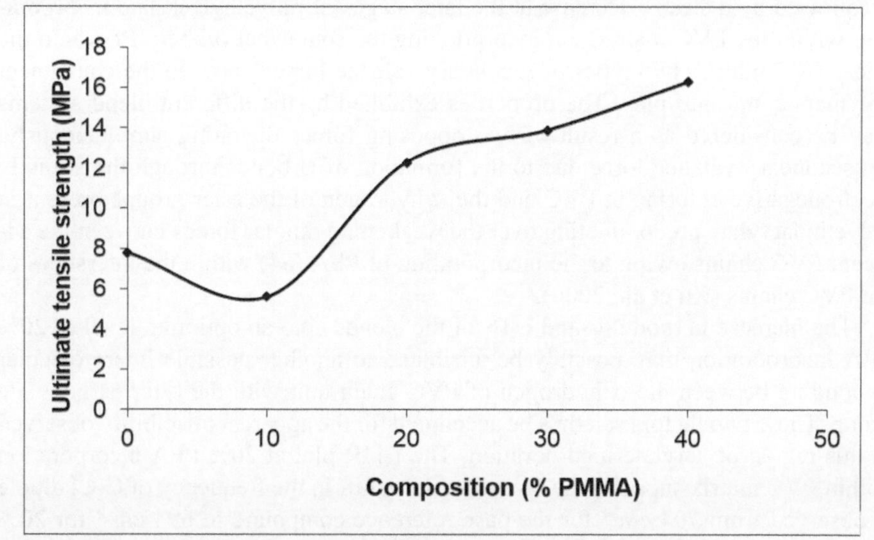

FIGURE 14.16 Variation of ultimate tensile strength of PVC-PMMA blends with variation of PVC-PMMA blend ratio (w/w) (Higashida et al., 1992)

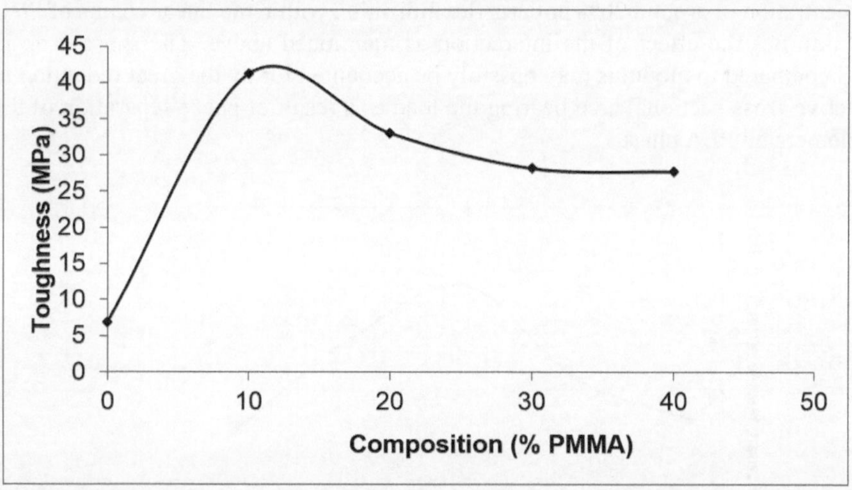

FIGURE 14.17 Variation of toughness of PVC-PMMA blends with variation of PVC-PMMA blend ratio (w/w) (Higashida et al., 1992)

The modulus and ultimate tensile strength of PVC-poly(ethyl acrylate) systems when considered as a function of composition display an initial upward swing and is followed by a steady decrease at the later stages of polyethylacrylate incorporation within the PVC matrix. On incorporating the somewhat rubbery PEA into the linear PVC matrix, two types of secondary valence forces between the constituent polymers come into play. The properties exhibited by the different blend systems may be considered as a result of two opposing forces operating simultaneously; the secondary valance force due to the formation of H bonds through the strongly electronegative chlorine in PVC and the α hydrogen of the ester groups present in polyethylacrylate predominating over the weakening dipolar forces between the adjacent PVC chains owing to the incorporation of PEA (64) within the interstices of the PVC chains (Xu et al., 2000).

The increase in modulus and UTS of the blends upto an optimum level of 20% PEA incorporation may possibly be attributed to another possible intermolecular H bonding between the α hydrogen of PVC chain unit with the ester carboxylate group. These two factors can thus be accounted for the apparent miscibility observed in this region of acrylate incorporation. The FTIR plot at 20% PEA incorporation within PVC matrix supports the fact wherein a shift in the frequency of C-Cl dipole is observed from 704 cm^{-1} for the base reference compound to 694 cm^{-1} for 20% PEA incorporation. At the later stages of PEA incorporation, the plasticizing influence of the rubbery polyethylacrylate moieties appear to play the major role because of the proportionate decrease in PVC content. The possibility of H-bond formation reduces because of the relatively less abundance of either the C-Cl or of α H from the PVC. An increase in the percentage of PEA beyond an approximate threshold concentration of around 20% imparts flexibilization within the linear chains of PVC and nullifies the effect of the interaction as mentioned above. The sharp drop in UTS compared to modulus may possibly be accounted for by the great reduction in effective cross-sectional area bearing the load as a result of phase separation of the agglomerated PEA phase.

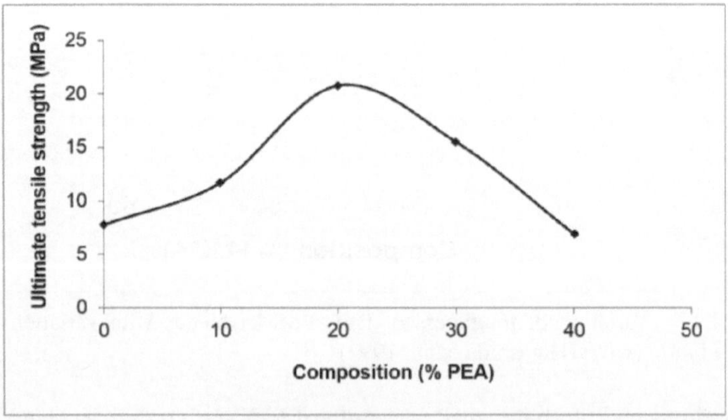

FIGURE 14.18 Variation of ultimate tensile strength of PVC-PEA blends with variation of PVC-PEA blend ratio (w/w) (Chakrabarti and Chakraborty, 2006)

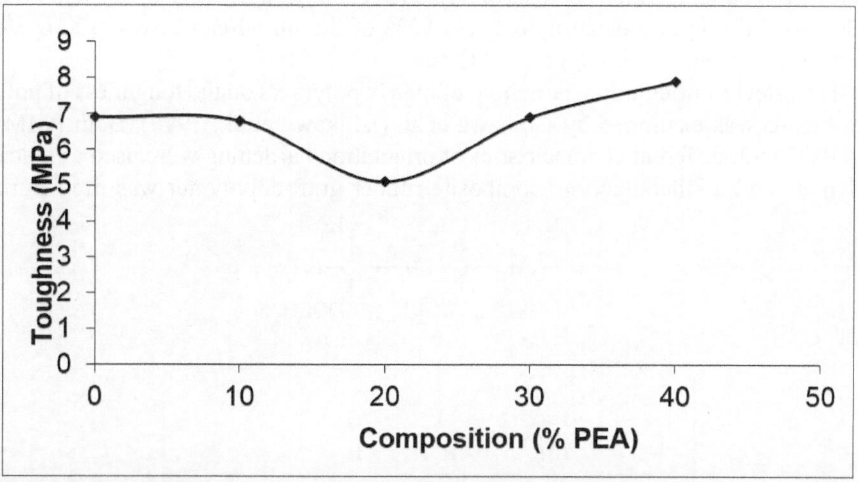

FIGURE 14.19 Variation of toughness of PVC-PEA blends with variation of PVC-PEA blend ratio (w/w) (Chakrabarti and Chakraborty, 2007)

The toughness values are obtained from the stress-strain curves and not from the conventional Charpy or Izod impact tests. Hence their units are obtained in MPa as observed. The variation in ultimate tensile strength and toughness of PVC-PEA blends have been shown in Figures 14.18 and 14.19.

The effect of polymer-polymer interaction on the miscibility and macroscopic properties of PVC/PMMA, PVC/PS and PMMA/PS blends revealed that the miscibility of the components was characterized by the Flory-Huggins interaction parameter or by quantities related to it. All three polymer pairs form heterogeneous blends but the strength of molecular interactions is different in them, the highest being observed in PVC/PMMA system resulting in partial miscibility of the components and beneficial mechanical properties. The structure of these blends depends strongly on composition. A significant change in structure and properties are observed. The PVC/PS and PMMA/PS pairs are immiscible, though the results indicate the partial solubility of the components (Fekete et al., 2005). The characteristics of polymer electrolytes based on a poly(vinyl chloride) (PVC) /poly(ethyl methacrylate) (PEMA) blend was reported by H.S. Han et al. (Han et al., 2002). The PVC/PEMA-based polymer electrolyte consists of an electrolyte-rich phase that acts as a conducting channel and a polymer-rich phase that provides mechanical strength. The mechanical strength of the PVC/PEMA-based polymer electrolyte was found to be much higher than that of PVC/PMMA-based polymer electrolyte (poly(methyl methacrylate), PMMA) at the same PVC content, and even comparable with that of the PVC-based polymer electrolyte. The blended polymer electrolytes showed ionic conductivity higher than 10^{3}S cm^{1} and electrochemical stability up to at least 4.3 V.

A prototype battery, which consists of a $LiCoO_2$ cathode, a MCMB anode, and PVC/PEMA-based polymer electrolyte, gives 92% of the initial capacity at 100 cycles upon repeated charge-discharge at 1 °C rate.

The effect of orientation hardening of matrix polymers on the toughness of polymer blends was examined by Ishikawa et.al. (Ishikawa et al., 1996). Both PMMA and PVC with different characteristics of orientation hardening were used as matrix polymers and a silicon/acrylic composite rubber graft copolymer was used as the

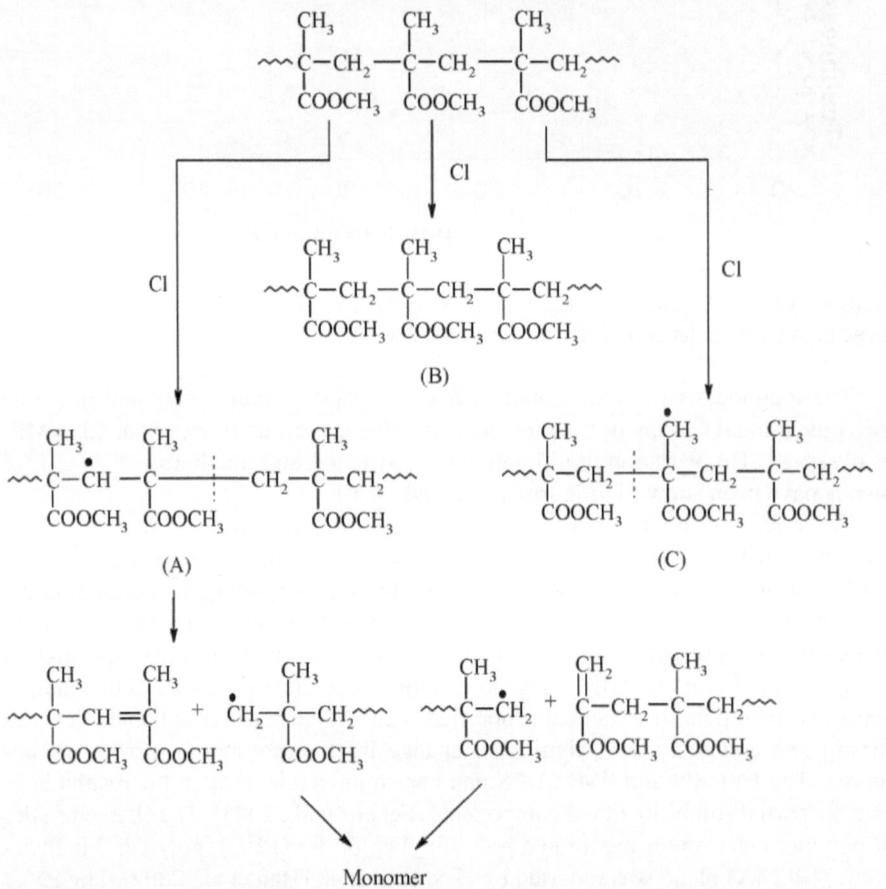

SCHEME 14.1 Scavenging of chlorine radicals by PMMA to minimize dehydrochlorination

Secondly, even if the HCl is produced, it fails to catalyze the process of subsequent degradation as it reacts with the ester group of PMMA as follows:

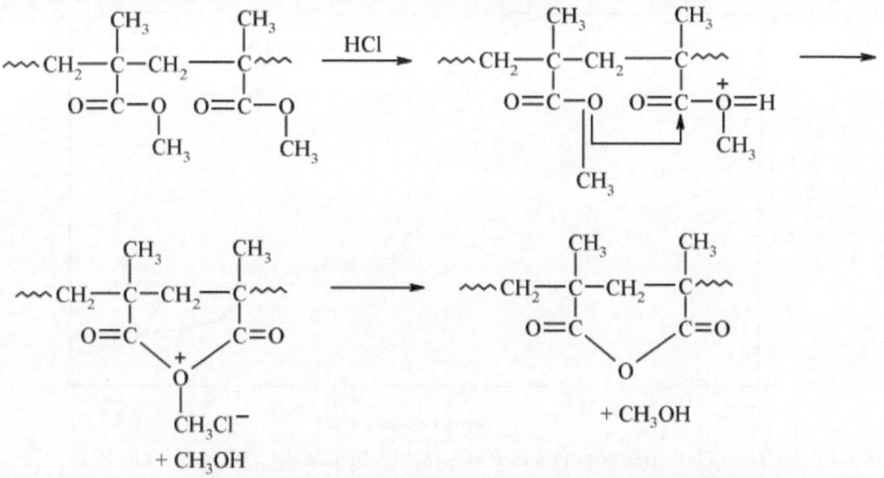

SCHEME 14.2 Scavenging of HCl by PMMAto prevent subsequent degradation

The chloride radical Cl. causes the PMMA chain to break and subsequently unzip quite readily at low temperatures. Reaction with HCl on the otherhand, presumably a slower process, gives rise to anhydride rings which act as locking groups reducing the zip length of depolymerization and thus stabilizing the chain. Moreover, the unzipped monomers (MMA) are considered to be very good radical scavengers and thus scavenge the chloride radicals immediately as they are formed.

In this way there is a mutual stabilizing effect whereby PVC is stabilized by PMMA in blend by not allowing the Cl. radical to form hydrogen chloride and subsequently not permitting the HCl even if formed by forming an anhydride type of structure.

Thus, the methacrylates and acrylates provide an effective means of modifying the base polymer PVC depending on its percentage incorporation. The thermal behavior of PVC – PMMA blends have been given in Figures 14.20 and 14.21 where the thermogravimetric and thermomechanical behavior have been displayed, respectively.

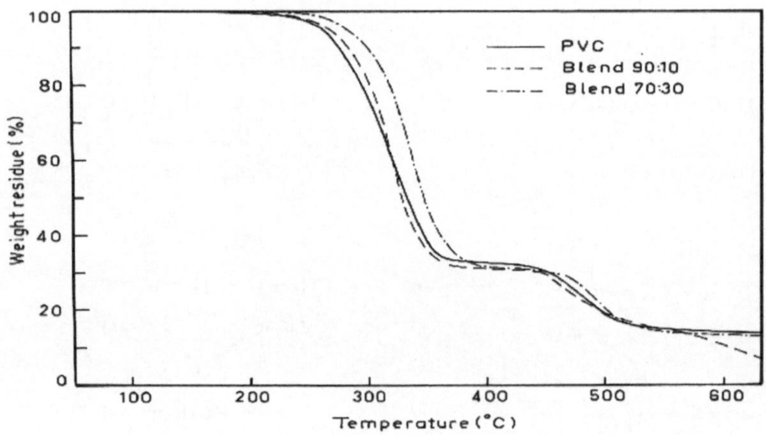

FIGURE 14.20 TGA thermograms of PVC-PMMA blends (Mcneill and Nell, 1970)

In another study based on PVC blended with PMMA or polyoxymethylene (POM) with lithium acetate as a stabilizing agent, the significant effect of lithium acetate on the thermal properties has been investigated. It causes the initial decomposition temperatures to increase by about 60–150 °C for PVC-POM blends, a substantial suppression of the volatile products evolution for PVC/PMMA blends and an improvement in the surface morphology for both polymer systems by lowering the degree of roughness. The origin of these effects was discussed by analysis of the intermolecular complexation between metal salt and PVC structural arrangements in the blends. Such interactions may lead to the formation of long range, directional-specific structural regularities (Vorenkamp et al., 1985). The miscibility behavior of poly(ethyl methacrylate) (PEMA), poly(n-propyl-methacrylate) (PPMA), poly(n-butyl methacrylate) (PBMA) and poly(n-amyl methacrylate) n(PAMA) with PVC was investigated using differential scanning calorimetry. All of the blends were miscible, that is, showed a single Tg when subjected to an appropriate thermal history. A lower critical solution temperature was detected in PVC/PBMA and PVC/PAMA blends but not in PVC/PEMA and PVC/PPMA blends (although it must exist at temperatures higher than 200 °C which was not accessible experimentally due to severe degradation of PVC). The observations led to the conclusion that there was a miscibility window in terms of the polymethacrylate structure and the greater degrees of miscibility were found in PEMA and PPMA blends (Perrin and Prudhomme, 1991). The improvement of the miscibility of PMMA and PVC with moderate chlorination of PVC was demonstrated by determination of cloud-point curves and glass transitions of the mixtures. Improvement of miscibility was observed, caused by the specific interaction between the carbonyl group of PMMA and predominantly the CHCl group of CPVC. The decrease in miscibility observed when more strongly chlorinated PVC was used could be ascribed to the relatively high concentration of

CCl2 groups which had a less favorable interaction with the carbonyl group (Varenkamp and Challa, 1988).

The in situ polymerization of vinyl chloride with various polyacrylates and polymethacrylates (PMA) had been studied by Walsh et al. (Walsh and Cheng, 1984). Poly (methyl acrylate) and poly(ethyl acrylate) (PEA) had previously produced two phase blends with PVC by solvent casting but PEA was shown to be miscible with PVC when blends were produced by in situ polymerization. PMA and polyoctyl acrylate were found to be immiscible with PVC, whereas other polyacrylates and polymethacrylates with intermediate ester group concentrations were found to be miscible. This phenomenon was reflected in the glass transition temperature of the polymers. It was found that the polymers having intermediate ester group concentrations showed the strongest interactions.

Rajulu et al. (Rajulu et al., 1999) studied the ultrasonic velocity and refractive index of PVC/PMMA blend solutions in cyclohexanone at 30 °C. The variation of these parameters with blend composition has been found to be linear, again indicating the miscibility of PVC with PMMA. A series of blends of PVC with (i) PMMA or (ii) polyoxymethylene (POM) with polyethylene glycol (PEG) as a thermal energy storage material has been investigated by DSC, TGA, SEM and FTIR. It was found that PEG has a significant effect on their thermal properties of the blends which results in the initial decomposition temperatures to increase by about 70 °C for PVC/POM blends, a substantial suppression of the volatile products evolution for PVC/PMMA blends and for both polymer systems, an improvement in the surface morphology in terms of uniformity (Pielichowski et al., 1999).

The thermomechanical curves of the base reference compound PVC and the PVC-PMMA blend samples indicate that in all cases , the probe is slightly pushed up by the expansion of the samples up to a temperature around 50 °C. Once the sample starts softening, the loaded probe penetrates the sample at a rate inversely related to the moduli of the various samples. The extent of penetration indicates that at the initial stages of PMMA incorporation, the lowering of rigidity due to breakdown in regularity of chain structure leads to softness whereas at later stages, the slow and gradual association of the stiff and rigid PMMA particles probably accounts for the decreased penetration within the range of PMMA incorporation studied (Chakrabarti and Chakraborty, 2006). Once the blend has softened completely, the molten sample undergoes expansion till the upper test temperature is reached. At this stage, while PVC shows breakdown and sharp fall, the other blend samples remain stable and exhibit stabilization over PVC.

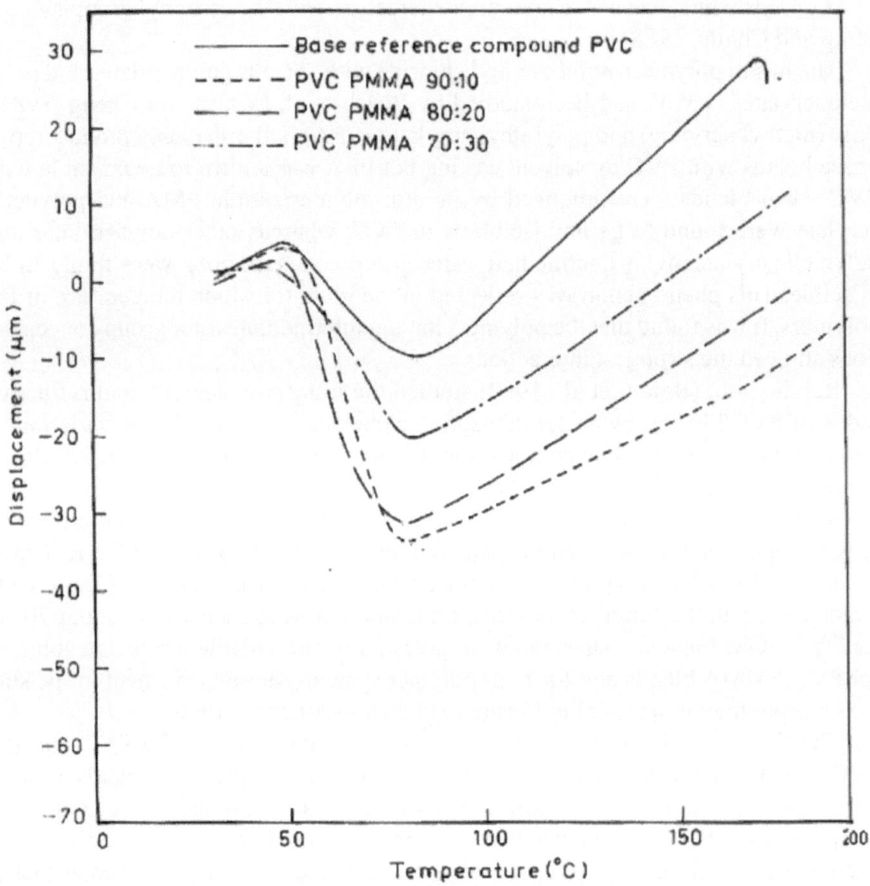

FIGURE 14.21 Thermomechanical curves of PVC-PMMA blends (Mcneill and Nell, 1970)

From a different study, it was seen that the transition behavior of blends of PVC with PMMA isotactic (i-PMMA) and syndiotactic (s-PMMA) was determined in the temperature range −150 °C to 130 °C by the thermally stimulated depolarization currents method (TSDC) and was found that the iPMMA and PVC formed an incompatible system over the entire concentration range, while in sPMMA/PVC blends, some compatibility probably existed but only for concentrations in sPMMA not higher than 10 wt% (Schurer et al., 1975; Verschueren et al., 1982). Hence it could be concluded that PVC was more miscible with syndiotactic PMMA than with isotactic PMMA.

A study on chain transfer grafting of butyl acrylate onto PVC in emulsion following the synthesis of graft copolymers showed that for the chain transfer method, the grafting efficiency does not increase as Flory's chain transfer theory predicted, but decreases with increasing copolymerization temperature. Thermogravimetry results showed that the thermal stability of the graft copolymer was better than those of PVC and PVC/poly (butyl acrylate) blend (Zhou et al., 1994). From the dynamic mechanical study of PVC-PMMA blends, it was revealed that in the glassy state, a very large secondary relaxation in the range of 100–325 K results from the combination of secondary relaxations of PVC and PMMA and only one main relaxation at 364 K associated to the glass rubber transition (Flores et al., 1995).

14.4.3 PVC-METHACRYLATE/ACRYLATE BLEND MORPHOLOGY

The optical micrographs of PVC-PMMA system display the typical structure of a heterogeneous polymer blend. There exists a continuous phase of PVC and the dispersed phase of PMMA (Figures 14.22 and 14.23). In conformity to our observation, in SEM studies, the PVC sample shows white specks and microvoids as structural defects which are expected with rigid glassy chain structures. The formation of PVC resin matrix is preceded by the in situ formation of PMMA particles (Yang et al., 1998). Thus, while the PVC resin particles undergo fusion and subsequent crystallization on cooling and the neighbouring chains exert dipole-dipole interaction through the C-Cl bonds, the already formed PMMA particles inhibit such interaction, and plasticize the resin matrix beyond the extent to which it was plasticized by the DOP already present in the master formulation. However, the extent of plasticization that has been achieved at the initial stages of PMMA incorporation appears to be missing at later stages possibly because (i) the co-continuous formation of matrices and (ii) the increasing number of PMMA particles generated at higher concentration undergo aggregation and intend to exert their own rigid and brittle behavior. However, there seems to be a combined effect of the brittle behavior of PMMA phase coupled with possible interaction between the ester group of PMMA and the C-Cl of PVC. The expected increment in toughness is attributed to breakdown in regular chain structure of PVC brought about by PMMA particles and their aggregates.

The blend properties of methacrylates and acrylates with PVC can further be enhanced by cross linking either of the polymers or both the polymers giving rise to interpenetrating polymer networks or IPNs. The IPNs provide a different mode of modifying PVC based on interaction of the cross linked polymers.

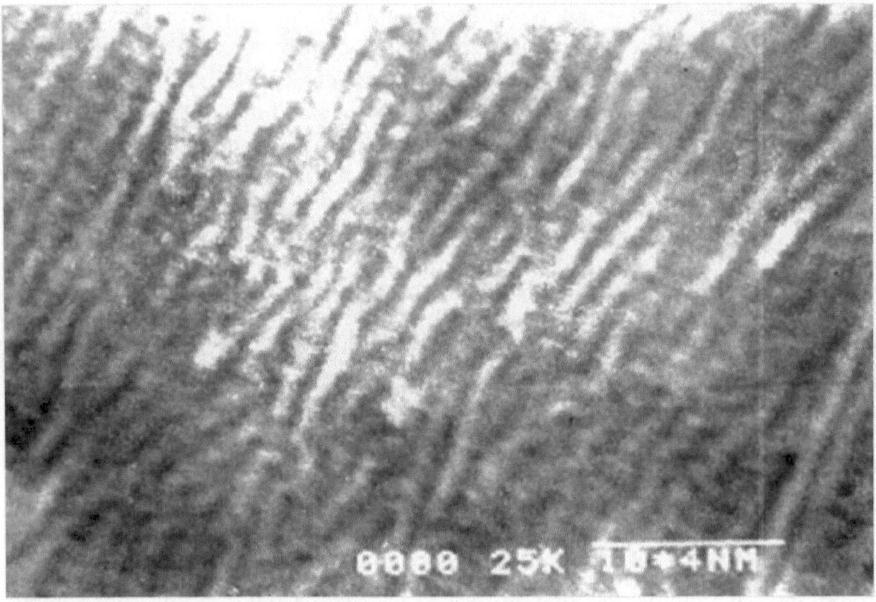

FIGURE 14.22 SEM of PVC:PMMA 90:10 (Yang et al., 1998)

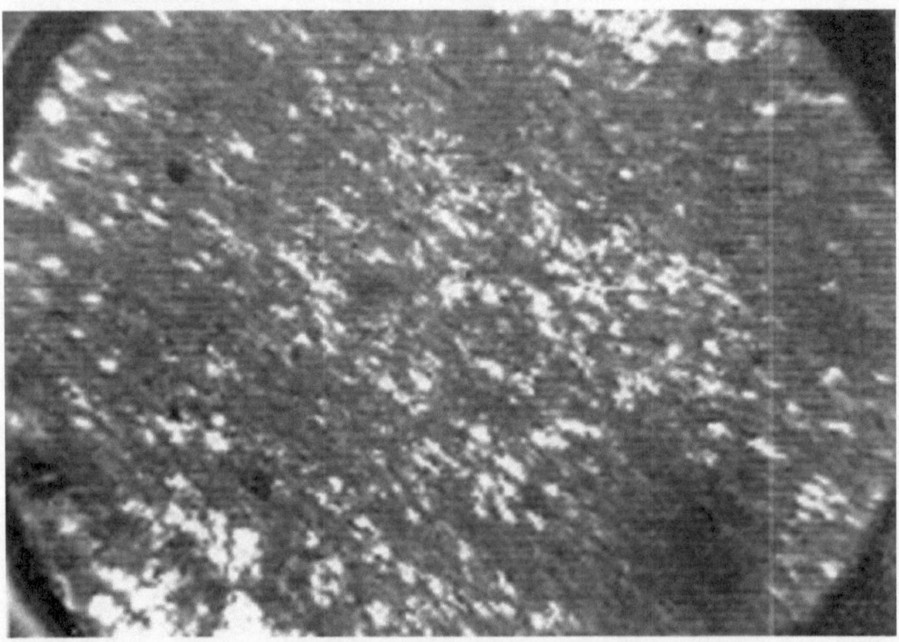

FIGURE 14.23 Optical micrograph of PVC:PMMA 80:20 (Yang et al.,1998)

14.5 OTHER BLENDS OF PVC

The effects of nanoscale calcium carbonate (nano-$CaCO_3$) particles on the mechanical properties of different ductile polymer matrices were investigated by N. Chen and his coworkers (Chen et al., 2004). Polyvinyl chloride (PVC) and PVC/Blended (Blendex ® 338) blend were used as the matrix in this study. The nano-$CaCO_3$ particles were observed to be dispersed uniformly on the nanoscale in both PVC and PVC/Blendex blend by means of transmission electron microscopy. The impact strength, flexural modulus, and Vicat softening temperature of PVC and PVC/Blendex blend were significantly enhanced after addition of 0–15 phr nano-$CaCO_3$, but the tensile properties of the two matrices showed different changes in the presence of nano-$CaCO_3$. The yield strength and elongation at break of PVC could be increased by the addition of nano-$CaCO_3$, while those of PVC/Blendex were decreased. Dynamic mechanical thermal analysis showed that the addition of nano-$CaCO_3$ led to an increase in storage modulus and glass transition temperature of both PVC and PVC/Blendex blend.

Blendex®338 is an ultra high rubber ABS impact modifier resin based on polybutadiene rubber. It offers superior impact efficiency especially at low temperatures, reduces notch sensitivity when alloyed with engineering polymers, disperses easily, and provides abrasion and chemical resistance to polymers.

The miscibility of PVC with PEO, in PVC/PEO blends was investigated by viscometric, microscopic, and thermal analyses. Viscometric and thermal results showed that the PVC/PEO blends were miscible. Polymer polymer interaction parameters, determined by depression of the melting temperature are negative and showed to be dependent on the molecular weight of PVC. The miscibility of the PVC/PEO blend was explained as a result of donor-acceptor interactions between chlorine atoms of PVC, as a weak acceptor species, and oxygen atoms of the PEO as a donor species (Neiro et al., 2000).

Yiyeon Kim and his co-workers (Kim et al., 1999) investigated the compatibility of blends of plasticized PVC (p-PVC) and thermoplastic polyurethane (TPU) using a DMA analyzer and SEM. Two kinds of TPU with different ratios of hard to soft segments, that is, TPU 90 and TPU 70 were compared. The p-PVC, TPU 90, and p-PVC TPU 70 blends with variable weight ratios were prepared by melt blending. It was found that TPU with a lower hard segment (i.e. TPU 70) was more compatible with plasticized PVC than TPU with a higher hard segment (i.e. TPU 90) was over the composition ranges examined. It was concluded that the compatibility of plasticized PVC and TPU were dependent on the ratio of hard to soft segments in TPU.

Mechanical and processing properties of blends of poly(vinyl chloride) (PVC) with di-isodecyl phthalate (DIDP) and thermoplastic polyurethane (TPU) were compared with di-2-ethylhexyl phthalate (DOP). The influence of processing conditions on the mechanical properties was studied by changing the content of the additives and using different twin screw speeds. The mechanical properties (tensile strength, modulus, and energy) of PVC/DIDP presented the same behavior as PVC/DOP at all concentrations. The mechanical characteristics of PVC/DOP and PVC/DIDP show that these systems are probably affected by the speed of the twin screw. Polyurethane blends exhibit better structural properties compared to the other plasticizers. The twin screw speed exhibited no influence on the stress-strain property profile of the PVC/TPU blends. The results indicate a minor influence on the decrease of PVC properties when the plasticizer used is TPU (Pita et al., 2002).

The DSC results of binary blends of PVC and poly(vinyl butyral) (PNB) prepared by solution blending revealed a high degree of molecular mixing of the two polymers exhibiting one major glass transition temperature (Tg) whose position on the temperature scale is lowered with increasing level of PVB. The thermal stability of the blends was found to increase with the increase in the PVB content in the blend. (Mohamed and Sabaa 1999).

Polymer blends consisting of unplasticized poly(vinyl chloride) (UPVC), and indulin lignin (IL) were prepared by compression molding. The original UPVC formulation was free from titanium dioxide TiO_2. The durability of the blends was evaluated by comparing the mechanical and the thermal properties of the blended and unblended PVC containing 10 per hundred parts (phr) of TiO_2 before and after artificial weathering. The thermal testing results showed a high degree of mixing of the two polymers. The glass transition temperature (T_g) of the blends was exhibited by one major transition between the two T_g's of PVC and lignin. There was an increase in tensile strength and decrease in elongation at yield of PVC by lignin addition. Lignin addition did not affect the tensile strain properties of PVC after weathering (Raghi et al., 2000).

V. Mano and his coworkers studied the conductivity, thermal, mechanical, and electrochemical properties of PVC/polypyrrole blends. The blends were prepared by oxidative chemical polymerization of pyrrole in the vapor phase in PVC films impregnated with $FeCl_3$. I.R. reflectance spectra suggested that the polymerization occurred preferentially on the matrix surface producing sandwich type structures. DMA studies suggested a certain degree of miscibility among the polymeric components of the blends (88. Mano et al., 1996).

A method was presented by Nguyen et al. (Nguyen et al., 1992) for studying the anomalies in the diffusion of gases and vapors in dense polymer films in the transient regime. It was applied to the diffusion of CO_2 and water vapor in plasticized PVC-potato-starch blends. The apparent diffusion coefficient of CO_2 in plasticized PVC remained constant in the transient regime, while that of water vapor first increased, passed through a maximum and finally decreased to a constant value.

Solution cast specimens of PVC/PU blends were studied by means of infra red, DSC, and DMA measurements. The results indicated that it was possible to change the morphology of the blends significantly by proper selection of the structure of the soft segments. The polyester soft segments were more compatible with PVC than were the polyether ones. Information about the distribution of hydrogen bonding between the different acceptors was used to discuss the segregation and mixing of the hard and soft segments at different PVC contents of the blends (Xiao et al., 1987).

Blends of PVC with polyesters poly(butylene adipate), poly(hexamethylene sebacate), poly(2,2-dimethyl,1,3-propylene succinate), and poly(1,4-cocylohexanedimethanol succinate) were found to exhibit a single, composition dependent glass transition. However, mixtures of poly(ethylene succinate), poly(ethylene adipate), and poly(ethylene orthorphthallate) with PVC were found not to be miscible. Thus there was an optimum density of ester groups in the polymer chain for achieving maximum interaction with PVC. Too few or too many ester groups resulted in immiscibility with PVC (Ziska et al., 1981).

Four saturated polyesters poly(hexamethylene adipate), poly(ethylene adipate), poly(hexamethylene terephthalate) and poly(ethylene terepathalate) were prepared. The resulting materials were characterized by IR and ^{1}H NMR, end group analysis and gel permeation chromatography by S.Y Tawfik and others. The effect of blending these polyesters (5–10%) with poly(vinyl chloride) (PVC) in the melt was investigated in terms of changes in the thermal behavior of PVC by studying the weight loss after 50 min at 180 °C, color changes of the blend before and after aging for one week at 90 °C, the variation in glass transition temperature and the initial decomposition temperature. The results gave proof for the stabilizing role played by the investigated polyesters against the thermal degradation of PVC. The best results are obtained when PVC is mixed with 5% aliphatic polyesters rather than with aromatic ones. This is well illustrated not only from the increase in the initial decomposition temperature (IDT), but also from the decrease of % weight loss and from the lower extent of discoloration of PVC, which is a demand for the application of the polymer. It was also found that blending PVC with 5% of the four investigated polyesters before and after aging for one week at 90 °C gave better mechanical properties than that of the unaged PVC blank (Tawfik et al., 2006).

The surfaces of polymeric blends of poly vinyl chloride and poly ethylene terephthalate have been treated by Kureshi et al. (Qureshia et al., 2007) with reactive (N_2^+) gas plasma to understand the effects of low energy ions on the surface modification of polymeric blends. These effects were determined by microhardness tester, TGA/DSC analysis, and morphology study by atomic force microscope (AFM). Figure 14.24 shows the plot of Vickers' microhardness (Hv) versus applied load (P) for pristine and plasma treated samples.

It is evident from the graph that the hardness becomes independent of load for loads more than 400 mN. Though hardness is a surface phenomenon at lower load at higher loads, beyond 400 mN, the hardness value represents the true hardness value of bulk and it is consequently independent of the load. The reason is attributed to the fact that the plasma generated excited species interacts with the surface of the polymeric films and leads to cross-linking which improve the hardness of the samples which is also corroborated with TGA thermograms (Figure 14.25).

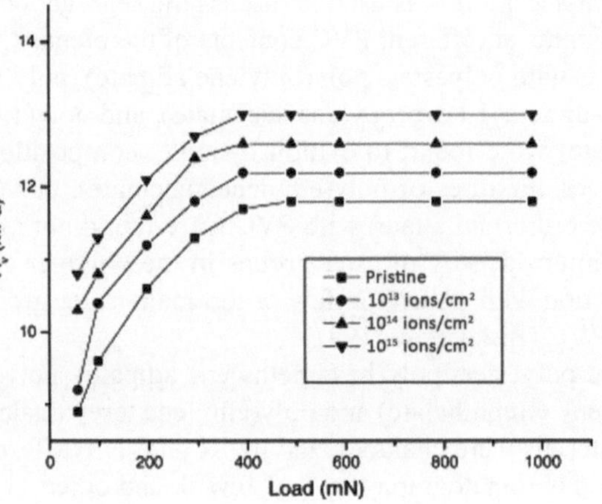

FIGURE 14.24 Hardness (Hv) versus applied load (P) for pristine and plasma treated PVC and PET blend films (Qureshia et al., 2007)

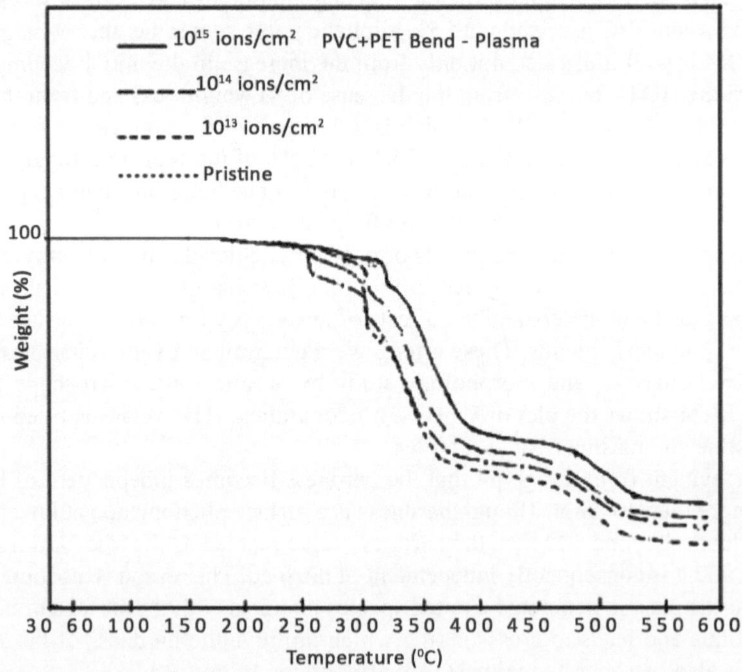

FIGURE 14.25 TGA thermograms of pristine and plasma treated polymeric blends (Qureshia et al., 2007)

The TGA thermograms reveal that the cross-linking dominants or system became organized increasing thermal stability of the polymeric blend system.

The atomic force microscopy (AFM) shows that the surface average roughness (Ra) of the film surface after nitrogen plasma implantation increases from 21.08 nm to 104.89 nm as fluence increases (Figure 14.26a–c) and also it is clear that the sputtering effect is not homogeneous throughout the surface.

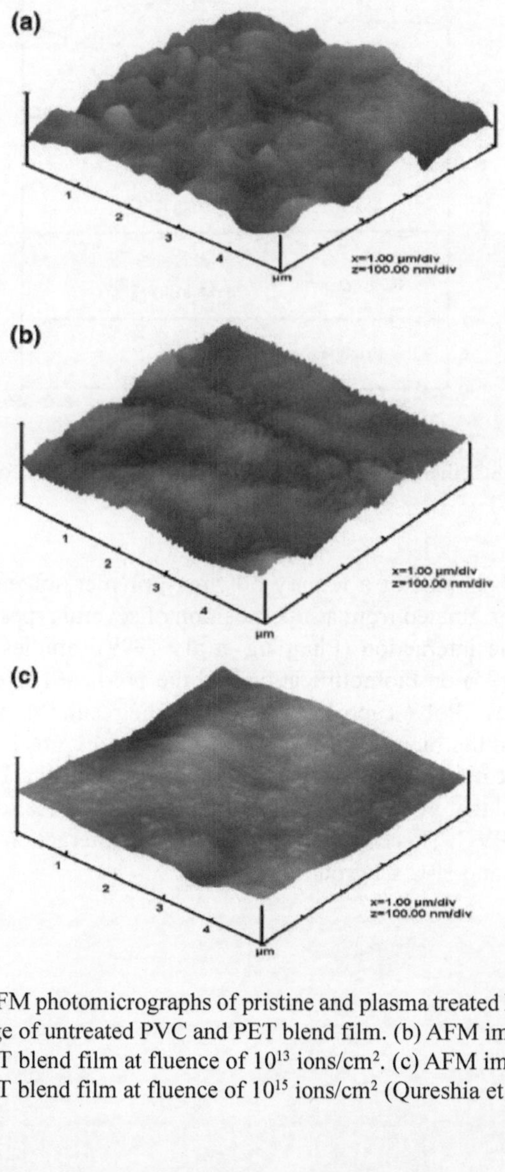

FIGURE 14.26 AFM photomicrographs of pristine and plasma treated PVC and PET blend films. (a) AFM image of untreated PVC and PET blend film. (b) AFM image of plasma (N_2^+) treated PVC and PET blend film at fluence of 10^{13} ions/cm^2. (c) AFM image of plasma (N_2^+) treated PVC and PET blend film at fluence of 10^{15} ions/cm^2 (Qureshia et al., 2007).

The DSC thermograms (Figure14.27) revealed a quite complex behavior in the temperature range 250–350 °C, where it is seen that with increase in fluency the DSC exotherm changes into DSC endotherm.

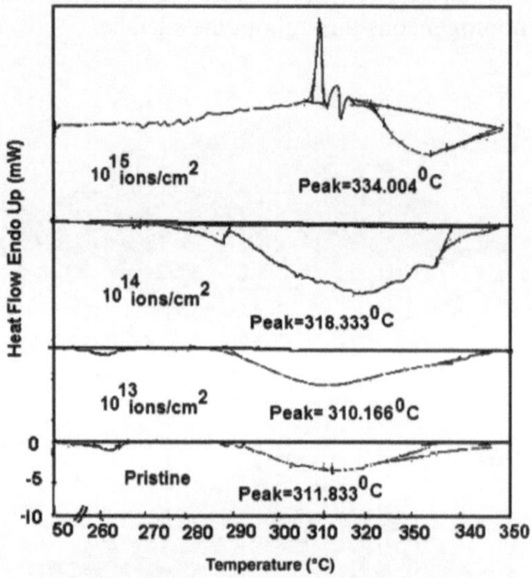

FIGURE 14.27 DSC thermograms of pristine and plasma treated polymeric blends (Qureshia et al., 2007)

The viscosity behavior of a ternary polymer-polymer solvent system of PVC with polystyrene, originated from a superposition of several types of hydrodynamic and thermodynamic interaction (Pingping et al., 1999). Studies were carried out by Campos et al. (95) on biomodification and the pre-heat (130 °C) influence on Poly(vinyl chloride) / Poly(-caprolactone) films. The results showed that heat pre-treatment improved the biomodification of PVC films (Figure 14.28), but it inhibited the biomodification of the PVC/PCL film. Regarding the PVC/PCL blend it could be concluded that PCL avoided both the thermal degradation and the thermal biodegradation of PVC. This could be attributed to the interaction between carbonyl groups of the PCL and H-C-Cl groups of the PVC.

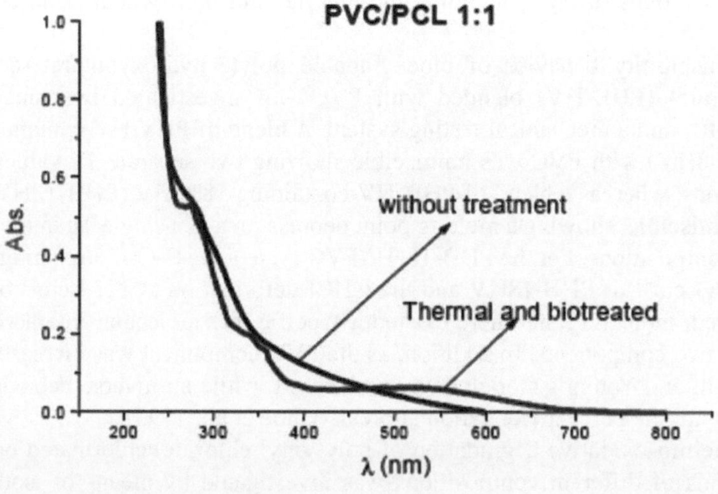

FIGURE 14.28 UV-Vis Absorption spectra of PVC/PCL 1:1 films without treatment, thermal and biotreated with *P. chrysosporium/A. fumigatusref* (Campos et al., 2005)

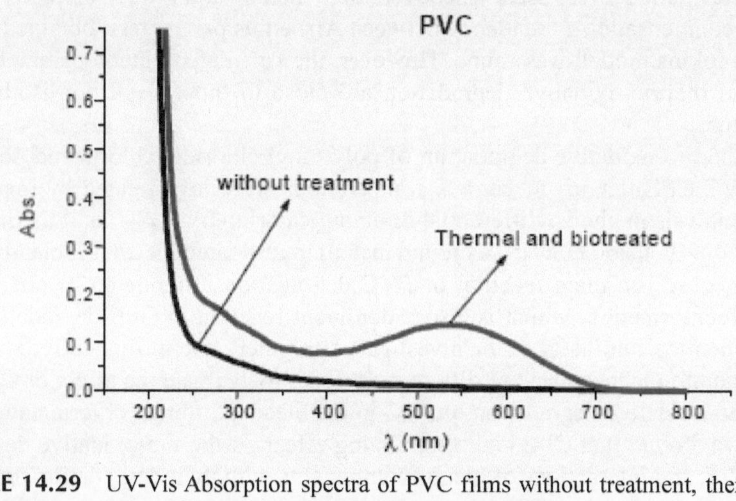

FIGURE 14.29 UV-Vis Absorption spectra of PVC films without treatment, thermal and biotreated with *P. chrysosporium/A. fumigatus* (Choe et al., 1995)

The UV-Vis. spectrum of the thermal biotreated blend (Figure14.29) showed that the bands of carbonyl groups and polyenes decreased (260–340 nm and 450–650 nm, respectively) in comparison to the PVC spectrum (Figure14.29). It seemed that the PCL protected the thermal degradation of the PVC and the PVC protected the biodegradation of the PCL, that is, the –C = O groups of PCL interacted with the H-C-Cl groups of PVC, avoiding the thermal and biodegradation in both the polymers.

The miscibility behavior of biodegradable poly(3-hydroxybutyrate-co-3-hydroxyvalerate) (PHB-HV) blended with PVC was investigated by using DSC, DMA, FTIR, and a mechanical testing system. A blend of PHV-HV containing 8% HV (PHV-8HV) with PVC was immiscible showing two separate Tg values in all compositions, whereas a blend of PHB-HV containing 18% HV (PHB-18HV) with PVC was miscible, showing a melting point depression and a single Tg in the whole range of compositions. For the PHB-18HV/PVC system the C-O-C stretching vibration at 1183 cm^{-1} of PHB-18HV and the CHCl deformation at 1254 cm^{-1} of PVC were shifted, indicating that there existed a specific intermolecular interaction between the two components. In addition, as the PVC component was increased, tensile strength and Young's modulus were increased, while an inverse behavior was observed with the percent elongation at break (Choe et al., 1995).

The thermo-oxidative degradation of poly(vinyl chloride)/chlorinated polyethylene blends of different compositions was investigated by means of isothermal thermogravimetry in flowing atmosphere of synthetic air at temperature 240–270 °C. The main degradation processes are dehydrochlorination of PVC and CPE. For calculation of the apparent activation energy and apparent pre-exponential factor two kinetic methods were used: isoconversional method and Prout-Tompkins method. True compensation dependency between Arrhenius parameters, obtained using Prout-Tompkins model, was found. However, the calculated kinetic parameters of isothermal thermo-oxidative degradation are close to those from non-isothermal degradation.

The thermo-oxidative degradation of poly(vinyl chloride)/chlorinated polyethylene (PVC/CPE) blends of various compositions were investigated by means of thermal analysis methods: differential scanning calorimetry (DSC) and thermogravimetry (T_g). By using DSC it was found that all investigated PVC/CPE blends were heterogeneous. The main reaction of PVC degradation at moderate temperatures is dehydrochlorination, which is also a dominant reaction in CPE degradation. In spite of chemical similarity of the investigated polymers, thermo-oxidative stability and degradation mechanism are different. In order to evaluate the effect of CPE on the thermo-oxidative degradation of PVC in the blends, different criteria have been used. It was found that CPE had a stabilizing effect on thermo-oxidative degradation of PVC and interactions of blends components with their degradation products occurred.

Epoxidation of conjugated polyenes formed by thermal degradation of poly(vinyl chloride) (PVC) was carried out in cyclohexanone and tetrahydrofuran solution with *m*-chloroperoxybenzoic acid (mCPBA). PVC was thermally degraded in the solid state under continuous nitrogen flow at 200 °C for 30 min leading to 0.6 mol% double bonds in the polymer chain as determined from the UV-visible spectrum of the degraded polymer. Cyclohexanone proved to be an inappropriate solvent for the epoxidation reaction probably due to its competitive reaction with the peroxide. UV-visible and FTIR spectroscopic analyses indicate that the epoxidation of polyenes by mCPBA in THF occurs in high yields leading to polyepoxy sequences in the PVC chain, and there are no signs of side reactions, such as oxidation of double bonds in polyenes to carbonyl groups. To our knowledge this is the first report on epoxidation of polymers containing conjugated double bond sequences. The resulting new epoxidized PVC might be useful in a variety of applications, such as fully miscible secondary epoxy stabilizer for PVC itself without plasticizing effect, starting material for epoxy curing and co-curing for reactive blends, and so on (Szakacs and Ivan, 2004).

The thermal behavior of a series of solution cast blends of PVC/chlorinated 2,4-toluene diisocyanate based polyurethane (PVC/CPU) polymers have been examined. It has been found that the decomposition proceeds through a two-step route; the main decisive degradation stage occurring in the 200–320 °C temperature range was found to be a result of parallel reactions of PVC and PU decomposition. This was also confirmed by Ozawa-Flynn Wall kinetic analysis – the activation energy remained constant for degrees of conversion >0.3. The reasons for better thermal stability of some PVC/CPU blends can be explained by analysis of specific interactions between the C=O groups of the urethane segments and the α hydrogen of the chlorinated polymer or a dipole–dipole interaction. On the other hand, the rate of diffusion of volatile products through microphase domain structure may differ due to changes in morphology and the spatial arrangement thus considerably affecting the overall decomposition route (Pielichowski and Hamerton, 2000).

Another related study on the thermal behavior of PVC blends with two novel dyes 3-(4-bomophenylozo)-9-(2,3 epoxypropane) carbazole (I) and 3-(4-nitrophenylozo)-9-(2,3 epoxypropane) carbazole II revealed that these additives can be used as effective dyes while having little effect on the thermal stability of PVC upto 200 °C. These data were also confirmed by analysis of hydrogen chloride evolution by GC and FTIR (Hamerton et al., 1998).

Thermal degradation of blends of PVC with polydiphenylsiloxane (PDPS) and polydimethyldiphenylsiloxane (PDMDPS) has been studied by thermogravimetry (TG) and differential thermal analysis (DTA) over the whole composition range (McNeill and Neil, 1970). From the TG and DTA data, curves have been constructed to enable the experimental weight loss behavior to be compared with the component polymers. Activation energy and order of reactions were calculated for the blends and component polymers using TG data. The results show slight destabilization at low loading of PDPS and PDMDPS but for compositions with 50% or more PDMDPS and PDPS, both polymers are stabilized, particularly the PDMDPS blends which show much slower weight loss.

A thermal degradation study by I.C. McNeill and D. Neil (McNeill and Neil, 1970) involving PVC/PMMA mixture yielded only hydrogen chloride and methyl methacrylate as major products of degradation. Minor products included carbon dioxide, methyl chloride, and benzene. The changes which occurred in PMMA in the course of the degradation were studied by extracting the polymer with toluene from the partly degraded blend. It was found that chain scission had occurred and that anhydride structures had been formed and the polymer showed differences in degradation behavior compared with the original PMMA sample. The first was the attack on PMMA by chlorine radicals produced during the dehydrochlorination of PVC; the second was the reaction between methacrylate ester groups and the hydrogen chloride.

It has been found that 3-(2,4 dichlorophenylazo)-9-(2,3-epoxypropane) carbazole can act as an efficient colourant and thermal stabilizer of PVC. It hinders evolution of hydrogen chloride, thus contributing to the stabilization of the system, probably via an aromatic alkylation reaction. The thermal stability has been estimated by TGA and FTIR studies on the blend of PVC with the colorant (Pielichowski and Hamerton, 1998).

A very recent publication by Rinaldi et al. (Rinaldi, 2006) showed that the solid phase photopolymerization of pyrrole in PVC matrix results in the formation of electrically conductive films. The blend obtained has low conductivity and rather poor electroactivity due to the loss of conjugation length of polypyrrole (PPy) provoked by hologenation. Micrographs of cryofractured surfaces suggested two distinct phases and thermogravimetric analysis revealed a low thermal stability of the blend.

A conductivity study on PVC-PMMA-$LiAsF_6$-DBP polymer blend electrolyte by S. Rajendran, T. Uma, and T. Mahalingam (Rajendran et al., 2000) revealed the effect of PVC-PMMA blend ratio on ionic conductivity. The temperature dependence of the conductivity of the polymer films obeys the Vogel-Tamman-Fulcher (VTF) relationship.

In the work by Lakshmi et al. (Lakshmi, 2011), Poly (o-toluidine) (PoT), a derivative of polyaniline, is prepared by chemical oxidation polymerization and is blended with polyvinylchloride (PVC) to achieve self supported films. These PoT-PVC blend films were irradiated by 60 MeV Si^{5+} ions at different fluences and evolved gases were monitored online by Residual Gas Analyzer (RGA).

FTIR transmission spectra of pristine and irradiated PoT-PVC blend films are shown in Figure14.30. After irradiation the overall transmission intensity of the polymer blend decreases which indicates some disturbance in the structure of polymer blend films due to the chain scissoring and bond breaking in the polymer chains. The C-Cl stretch intensity at 625 cm^{-1}, the intensity of substituted benzene ring vibration at 680 cm^{-1}, and the C-N stretch intensity at 971 cm^{-1} decreases. The intensity of ring vibrations of benzenoid and quinoid groups at 1602 and 1427 cm^{-1} decreases, which may be due to the deformation in ring structure of polymer blend after irradiation. The C-H stretch vibration at 2850–3000 cm^{-1} deforms after irradiation.

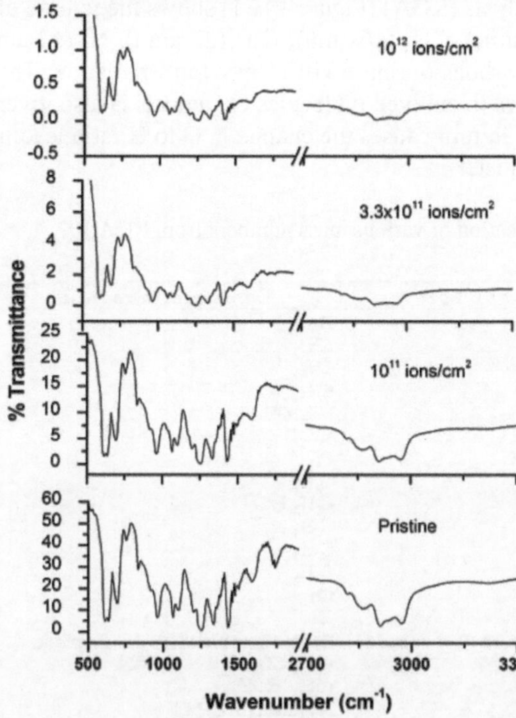

FIGURE 14.30 FTIR of pristine and irradiated PoT-PVC blends (Lakshmi, 2011)

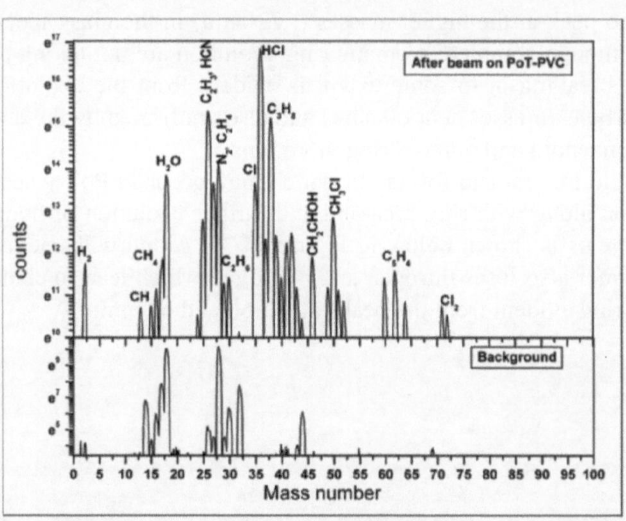

FIGURE 14.31 RGA spectrum showing mass number versus counts (Lakshmi, 2011)

Residual gas analysis (RGA) (Figure 14.31) shows the release of gases like H(2 amu [atomic mass unit]), CH_4 (16 amu), C_2H_6(27 amu), N_2 (17 amu) and various fragments of hydrocarbons during Swift Heavy Ion irradiation. The assignment of various mass numbers to evolved molecules or radicals is also given in Table 14.2. The release of gases in turn causes the change in C to H ratio leading to the formation of carbon rich clusters.

TABLE 14.2 Identification of various mass numbers from RGA 107

Mass number	Molecules or radicals
2	H_2
13	CH
15	CH_3
16	CH_4
17	NH_3, OH
18	H_2O
24,25	C_2
26	C_2H_2
27	C_2H_3, HCN
28	C_2H_4, N_2
29	C_2H_5
30	C_2H_6
35	Cl
36	HCl
37,38,39,40,41,42 and 43	Evolution of various (C_3H_8) fragments
44	C_3H_8, CO_2
46	(CH_3CHOH)
49, 50, 51 and 52	Fragments of CH_3Cl
60, 62 and 64	Ring breaking (C_5H_4) fragments
71 and 72	Cl_2

There is no peak at the higher masses (>72 amu) in the mass spectrum, which indicates that though complete aromatic ring evolution are not taking place but the aromatic ring breaking in to some extent is evident from the evolution of higher mass hydrocarbons (masses near 60 amu) and also confirmed by the decrease in IR intensity of benzenoid and quinoid ring vibrations.

The possible free radical formation which may occur in PoT when exposed to swift heavy ion along with ring breaking and further evolution of hydrogens from the ring fragments is shown below in Figure 14.32. Along with methyl groups (–CH_3) the polymer also loses nitrogen and hydrogen which lead to chain scissoring and ring breaking evident from the peaks at 64, 60 and 62 amu.

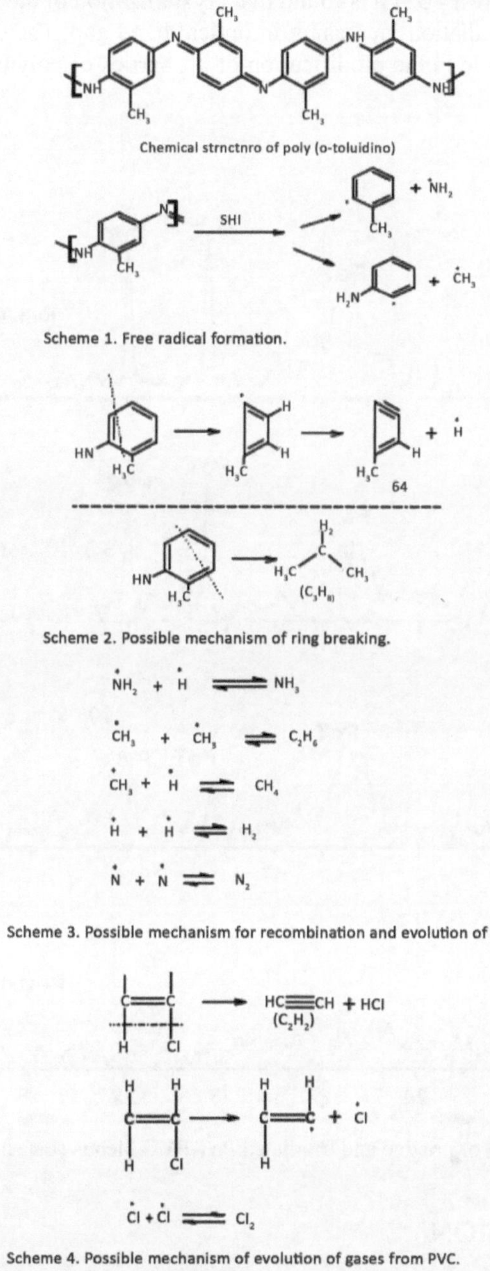

FIGURE 14.32 Free radical formation which may occur in PoT when exposed to swift heavy ion (Lakshmi, 2011)

From XRD (Figure 14.33) it is found that crystallization of the polymer blend is taking place after irradiation increasing in optical band gap. The electronic energy loss plays a crucial role in the modification of properties of polymers after irradiation.

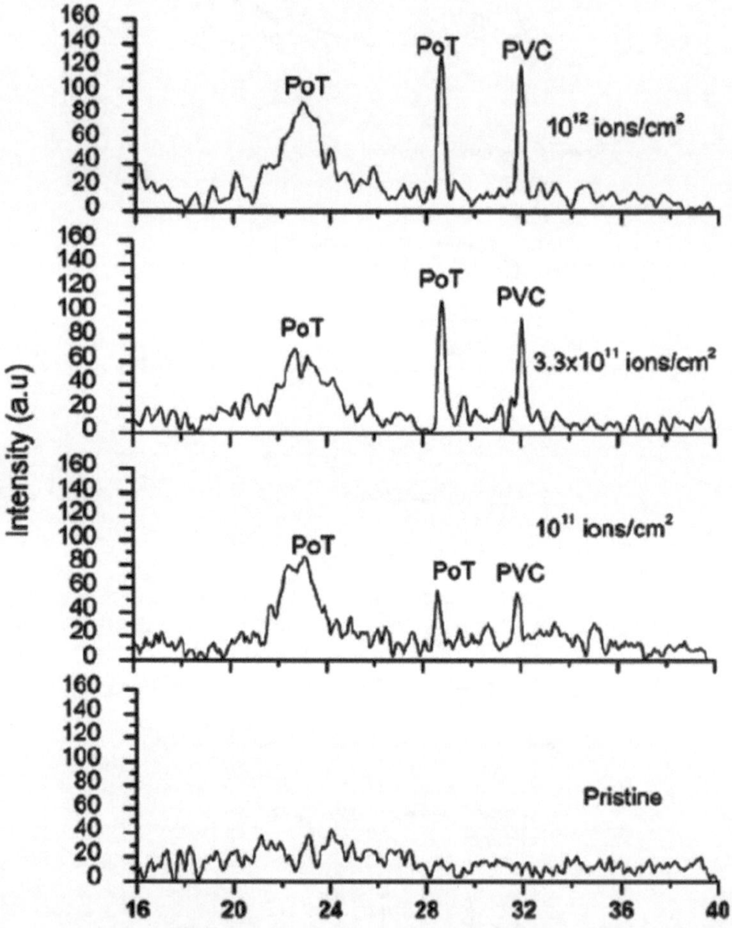

FIGURE 14.33 XRD of pristine and irradiated PoT-PVC blends (Lakshmi, 2011)

14.6 CONCLUSION

This review deals with the effects which are brought by blending various polymers and copolymers with PVC. The blend components involve a variety of ranges including the polyolefins and their copolymers, methacrylates and acrylates, as well as different types of elastomers. The incorporation of these polymers within PVC exhibit modified results which ultimately influences the end use properties. However, the modification of the various characteristics and morphology evolution cannot be arrived at by blending PVC with a single kind of polymer but the selection of the polymer for blending is highly oriented to the specific application area. So, there is a wide scope for more investigation to select the role of individual polymers, different co-polymers and compatibilizers to bring modification in different properties of PVC. Hence, further research in this area will yield a more comprehensive picture about the dependencies of polymer architectures and characteristics on the detailed mechanisms involved when processing a compatibilized blend. The technology of blending to enhance the properties of PVC is however, an ongoing continuous trend. In this context, it is worthy to mention that the process of forming blend composite with PVC offers an emerging idea for future scope of research and upgradation of PVC properties.

KEYWORDS

- **Poly(vinyl chloride)**
- **Blend**
- **Elastomers**
- **Plastomers**

REFERENCES

Akovali, G., Zorun, T. T., Bayramli, E., Erinc, N. K. *Polymer*, *1998*, **39**, 1363–1368.

Arayapranee, W., Prasassarakich, P., Rempel, G. L. *J. Appl. Polym. Sci.*, 2004, **93**, 1666–1672.

Campos, A., Franchetti, S. M. M. *Brazilian Archives of Bio. Technol.*, 2005, **48**(2), 235–243.

Chakrabarti, R., Chakraborty, D. J. *Appl. Polym. Sci.*, 2006a, **99**, 2033–2038.

Chakrabarti, R., Chakraborty, D. J. *Appl. Polym. Sci.*, 2006b, **102**, 3698–3703.

Chakrabarti, R., Chakraborty, D. J. *Appl. Polym. Sci.*, 2007, **105**, 1377–1384.

Chakrabarti, R., Das, M., Chakraborty, D. J. *Appl. Polym. Sci.*, 2004, **93**, 2721–2730.

Chen, N., Wan, C., Zhang Y., Zhang, Y. *Polym. Test.*, 2004, **23**, 2, 169–174.

Choe, S., Cha Y-J., Lee, H-S., Yoon, J. S., Choi, H. J., *Polymer*, 1995, **36**, 26, 4977–4982.

Cinadr, B. F., Lepilleur, C. A., Backman, A. L., Detterman, R. E., Schmitz, T. J. United States Patent no. 6,124,406.

Eastwood, E. A., Dadmun, M. D. *Polymer*, 2002, **43**, 6707–6717.

Encyclopedia of Polym. Sci & Engg., 1988, **12**, 426.

Fang, Z., Xu, C., Bao, S., Zhao, Y. *Polymer*, 1997, **38**, 131–133.

Fang, Z. P., Ma, G. W., Shentu, B. Q., Cai, G. P., Xu, C. W. *Eur. Polym. J.*, 2000, **36**, 2309–2311.

Fekete, E., Földes, E., Pukanszky, B. *Eur. Polym. J.*, 2005, **41**, 727–736.

Flores, R., Perez, J., Cassagnau, P., Michel, A., Cavaille, J. Y. *J. Appl. Polym. Sci.*, 1995, **60**, 1439–1453.

Francis, J., George, K. E. *J. Elast. Plast.* 1992, **24**, 151.

Francis, J., George, K. E., Joseph, R. *Eur. Polym. J.*, 1992, **28**, 1289–1293.

Ghaisas, S. S., Kale, D. D., Kim, J. G., Jo, B. W. *J. Appl. Polym. Sci.*, 2004, **91**, 1552–1558.

Gheno, S. M., Passador, F. R., Pessan, L. A. *J. Appl. Polym. Sci.*, 2010, **117**, 3211–3219.

Hafezi, M., Khorasani, S. N., Ziaei, F., Azim, H. R. *J. Elast. Plast.*, 2007, **39**, 151–164.

Hamerton, I., Pielichowski, K., Pielichowski, J, Stanczyk, P. *Eur. Polym. J.*, 1998, **34**, 5–6, 653–657.

Han, H. S., Kang, H. R., Kim, S. W., Kim, H. T. *J. Pow. Sour.*, 2002, 112, 461–468.

He, P., Huang, H., Xiao, W., Huang, S., Cheng, S. *J. Appl. Polym. Sci.*, 1997, **64**, 2535.

Higashida, N., Kressler, J., Yukioka, S., Inone, T. *Macromolecules*, 1992, **25**, 5259.

Hoffman, W. Rubber Technology H Book, Hanser, New York, 1989.

Ishiaku, U. S., Ishak, Z. A. M., Shonaike, G. O., Simon, G. P. Polymer Alloys and Blends, Eds. Marcel Dekker, New York, USA, pp. 663, 1999.

Ishikawa, M., Yanagase, Y. S. A., Ito, M., Yamamoto, N. *Polymer*, 1996, **37**, 5583–5588.

Ismail, H., Supri, A., Yusof, M. M. *J. Appl. Polym. Sci.*, 2005, **96**, 2181–2191.

Ivanova, T., Elksnite, I., Zicans, J., Meri, R. M., Kalnins, M., Kalkis, V. *Proc. Eston. Academ. Sci.*, 2009, **58**, 1, 29–34.

Jarusa, D., Summersb, J. W., Hiltnera, A., Baer, E. *Polymer*, 2000, **41**, 3057–3068.

Kim, K.J. *J. Appl. Polym. Sci.*, 2012, 124, 2937–2944.

Kim. Y., Cho W-J., Ha. C-S. J. *Appl. Polym. Sci.*, 1999, **71**, 3, 415–422.

Klarić, I., Vrecic, N. S., Roje, U. J. *Appl. Polym. Sci.*, 2000, **78**, 166–172.

Koklas, S. N., Sotiropoulou, D. D., Kallitsis, J. K., Kalfoglou, N. K. *Polymer*, 1991, **32**, 66–72.

Kolawole, E. G., Olugbemi, P. O. *Eur. Polym. J.*, 1985, **21**, 187–193.

Kroeze, E., ten Brinke, G., Hadzüoannou, G. *Polymer*, 1997, **38**, 379–389.

Lakshmi, G. B. V. S., Avasthi, D. K., Prakash, J., Siddiqui, A. M., Ali, V., Khan, S. A., Zulfequar, M. *Adv. Mat. Lett.*, 2011, **2**, 2, 125–130.

Lapa, V. L. C., Visconte, L. L. Y., Affonso, J. E. S. *Polym. Testing.*, 2002, **21**, 443–447.

Lee, Y. D. Chen, C. M. *J. Appl. Polym. Sci.*, 1987, **33**, 1231–1240.

Li, J. X., Chan, C. M. *Polymer*, 2001, **42**, 6833–6839.

Liu, Z. H., Wu, L. X., Kwok, K. W., Zhu, X. G., Qi, Z. N., Choy, C. L., Wang, F. S. *Polymer*, 2001, **42**, 1719–1724.

Liu, Z. H., Zhang, X. D., Zhu, X. G., Li, R. K. Y., Qi, Z. N., Wang, F. S., Choy, C. L. *Polymer*, 1998, **39**, 5019–5025.

Mano, V., Felisberti, M. I., Matencio, T., Paoli, M-A D. *Polymer*, 1996, **37**, 23, 5165–5170.

Manoj, N. R., De, P. P., De, S. K. *J. Appl. Polym. Sci.*, 1993, **49**, 133–142.

Manoj, N. R., De, P. P. *Polymer*, 1998, **39**, 733–741.

Mark, H. F., Bilaes, N. M. *Encyclo. Polym. Sci. Eng.*, 1986, **4**, 422–423.

Mathew, A., Deb, P. C. *J. Appl. Polym. Sci.*, 1992, **45**, 2145–2151.

,McNeill. I.C., Neil D. *Eur. Polym. J.*, 1970, **6**, 4, 569–583.

Mikkonen, R., Savolainen, A. *J. Appl. Polym. Sci.*, 1990, **39**, 1709–1725.

Mohamed, N. A., Sabaa, M. W. *Eur. Polym. J.*, 1999, **35**, 9, 1731–1737.

Nawi, M. A., Ngoh, Y. S., Zain, S. M. *Int. J. Photoenergy*, 2012.

Neiro, S. M., da S., Dragunski, D. C., Rubira, A. F., Muniz, E. C. *Eur. Polym. J.*, 2000, **36**, 3, 583–589.

Nguyen, X. Q., Sipek, M., Nguyen, Q. T. *Polymer*, 1992, **33**, 17, 3698–3705.

Olabisi, O. *Macromolecules*, 1975, **9**, 316.

Perera, M. C. S., Ishiaku, U. S., Ishak, Z. A. M. *Polym. Degradation Stability*, 2000, **68**, 393–40.

Perrin, P., Prudhomme, R. E. *Polymer*, 1991, **32**, 146–1473.

Pielichowski, K. *Eur. Polym. J.*, 1999, **35**, 27–34.

Pielichowski, K., Hamerton, I. *Polymer*, 1998, **39**, 1, 241–244.

Pielichowski, K., Hamerton, I. *Eur. Polym. J.*, 2000, **36**, 1, 171–181.

,Pingping, Z., Haiyang, Y., Yiming, Z. *Eur. Polym. J.*, 1999, **35**, 5, 915–921.

Prudhome, R. E. *Polym. Engg. Sci.*, 1982, **22**, 90.

Qureshia, A., Singha, N. L., Sharmaa, A., Mukherjee, S., Rakshitc, A. K., Tripathi, A., Avasthi, D. K. *Surface Coating and Technology*, 2007, **201**, 19–20, 8278–8281.

Raghi, S. E., Zahran, R. R., Gebril, B. E. *Mater. Lett.*, 2000, **46**, 6, 332–342.

Rajendran, S, Prabhu, M. R, Rani, M. U. *Int. J Electro. Sci.*, 2007, **3**, 282–290,7

Rajendran, S., Uma, T., Mahabingan, T. *Eur. Polym. J.*, 2000, **36**, 12, 2617–2620.

Rajulu, A. V., Reddy, R. L., Raghavendra, S. M., Ahmed, S. A. *Euro. Polym. J.*, 1999, **35**, 1183–1186.

Ramesh, P., Dey, S. K. *Polymer*, 1993, **34**, 4893–4897.

Ratnam, C.T. *Polym. Testing*, 2002, **21**, 93–100.

Ratnam, C.T., Zaman, K. *Polym. Degr. Stab.*, 1999a, **65**, 99–105.

Ratnam, C. T., Zaman, K. *Nuc. Ins. Meth. Phys. Res.*, 1999b, **152**, 335–342.

Rey, L., Duchet, J., Galy, J., Sautereau, H., Vouagner, D., Carrion, L. *Polymer*, 2002, **43**, 4375–4384.

Rinaldi, A. W., Kunita, M. H., Santos, M. J. L., Radovanovic, E., Rubira, A. F., Girotto, E. M. *Eur. Polym. J.*, In press, 2006.

Saha, S. *Eur. Polym. J.*, 2001a, **37**, 2513–2519.

Saha, S. *Eur. Polym. J.*, 2001b, **37**, 399–410.

Schurer, J. W., de Boer, A., Challa, G. *Polymer*, 1975, **16**, 201–204.

Sen, A. K., Mukherjee, G. S. *Polymer*, 1993, **34**, 2386–2391.

Singh, D., Malhotra, V. P., Vats, J. L. *J. Appl. Polym. Sci.*, 1999, **71**, 1959–1968.

Sotiropoulou, D. D., Avramidou, O. E., Kalfoglou, N. K. *Polymer*, 1993, **34**, 2297–2301.

Supri, H. I., Yusof, A. M. M. *Polym. Testing*, 2004, **23**, 675–683.

Szakacs. T., Ivan. B. *Polym. Degrad. Stab.*, 2004, **85**, 3, 1035–1039.

Tawfik, S. Y., Asaad, J. N., Sabaa, M. W. *Polym. Degrad. Stab.*, 2006, **91**, 2, 385–392.

Thormer, J., Bertram, H., Benn, O., Hurrnik, H. U. S. Patent No. 4, 1984, 435–532.

Pita, V.J.R.R., Sampaio, E. E. M., Monteiro, E. E. C. *Polym. Testing*, 2002, **21**, 5, 545–550.

Varenkamp, E. J., Challa, G. *Polymer*, 1988, **29**, 86–92.

Verschueren, J., Janssens, A., Ladang, K., Niezette, *J. Polymer*, 1982, **23**, 395–400.

Vorenkamp, E. J., Brinke, G., Meijer, J. G., Jager, H., Challa, G. *Polymer*, 1985, **26**, 1725–1732.

Vrecic, N. S., Klaric, I., Kovacic, T. *Polym. Degrad Stab.*, 2004, **84**, 1, 23–30.

Vrecic, N. S., Ricic, B., Klar, I., Kovacic, T. *Polym. Degrad Stab.*, 2005, **90**, 3, 455–460.
Walsh, D. J., Cheng, G. L. *Polymer*, 1984, **25**, 495–498.
Wu, G., Zhao, J., Shi, H., Zang, H. *Eur. Polym. J.*, 2004, **40**, 2451–2456.
Xi, X., Xie, M., Keqiang, C. *Polym. Engg. Sci.*, 1987, **27**, 391.
Xiao, F., Shen, D., Zhang, X., Hu, S., Xu, M. *Polymer*, 1987, **28**, 13, 2335–2345.
Xu, X., Toghiani, H., Jr. Pittman, C.U. *J. Polym. Sci. Eng.*, 2000, **40**, 2027.
Yang, Y., Fujiwara, H., Chiba, T., Inoue, T. *Polymer*, 1998, **39**, 2745.
Yen, C. S., Hassan, A., Hasim, *S. J. Teknol.* 2003, **39**(A), 107–116.
Zhou, Q., Yang, W., Wu, Q., Yang, B., Huang, J., Shen. *J. Eur. Polym. J.*, 2000, **36**, 1735–1740.
Zhou, Z. M., Zhu, M., Wu, X. D., Quan, D. *J. Polym.*, 1994, **35**, 2888–2892.
Ziska. J. J., Barlow, J. W., Paul. D. R., *Polymer*, 1981, **22**, 7, 918–923.
Zulfiqar, S., Ahmad, S., *Polym. Degrad. Stab.*, 1999, **65**, 2, 243–247.

CHAPTER 15

RADIATION PROTECTION PROPERTIES OF NATURAL RUBBER COMPOSITES

EDYTA KUSIAK and MARIAN ZABORSKI

Department of Chemistry, Institute of Polymer & Dye Technology, Technical University of Lodz, Poland 116 Zeromskiego Str., 90-924 Lodz, Poland

E-mail: edyta.kusiak@p.lodz.pl; marian.zaborski@p.lodz.pl

CONTENTS

ABSTRACT

The purpose of the study was to determine the composition of the elastomers characterized by absorption properties of the X-ray radiation. Natural rubber (RSS I) was applied as the elastomer and matrix and bismuth (III) oxide (Bi_2O_3), bismuth oxochloride (BiOCl), gadolinium (III) oxide (Gd_2O_3), bismuth (III) gallate ($C_7H_5BiO_6{\cdot}xH_2O$), and barium sulfate ($BaSO_4$) were added as fillers. The vulcanizates morphology was assessed by scanning electron microscope (SEM). X-ray absorption measurements were carried out using isotopic source ^{241}Am (60 keV) with 40 kBq activity. The mechanical properties and hardness (Shore method) of the composites were examined. The cross-link density was calculated using the Flory–Rehner equation. The samples were subjected to UV aging. The natural rubber composites containing Gd_2O_3, $BaSO_4$ and Bi_2O_3 provided a good degree of radiation comparison to other samples. The mechanical properties of vulcanizates were above the 16 MPa, with the observation result that UV aging coefficient of composite from gadolinium oxide was at the level 0.8. This means that the degradation degree of sample was not high.

15.1 INTRODUCTION

Radiation is widely used in medicine, diagnostic X-ray or computer tomography (CT) to diagnose illness, but the longer a person is exposed to radiation, the more energy the body will absorb from the radiation. The increasing use of CT has sparked concern over the effects of radiation dose on patients, particularly for those who had repeated CT scans (Lee and Chhem, 2010). The typical radiation dose for one CT of the abdomen is 10 mSv or 500 X-ray and equal to 3.3 year equivalent effective dose from natural background radiation (Frasher and Altizer, 2006). Due to the stochastic biologic effects of ionizing radiation low levels in diagnostic imaging (Brenner et al., 2001) is expected that imaging centers should strive to achieve a radiation dose as low as possible for medical diagnosis. In recent times, much attention was devoted to the potential additional lifetime risk of cancer incidence and death from cancer related to computed tomography (Amis et al., 2007; Brenner and Hall, 2007; Muhogora et al., 2010). It is obvious that for children, the exposure risks are higher than for adults.

The organ and tissue absorbed doses were measured for three different pediatric CTs, and was observed that the highest average absorbed dose is to the brain and eye lens for the brain examination (Brady et al., 2011). Radiation is attenuated by different materials. Putting shields between the person and the radiation source will reduce the amount of radiations.

Lead is a good shielding material for gamma rays and X rays; it has high atomic mass number and high density, but this element has its influences on health and environment (Kaushik at al., 2010; Mheemeed at al., 2012). Radiopaque polymers offer similar radiation-shielding properties with low density as compared to that of lead and fewer toxic (Nisha and Joseph, 2006).Blends have been prepared by incorporating radio pacifying agents such as heavy-metal powders cadmium (Cd), indium (In), tin (Sn), antimony (Sb), cesium (Cs), barium (Ba), gadolinium (Gd) embedded in natural rubber or various polymers (McCaffrey et al., 2007). Radioopaque polymer–salt complexes are produced by the incorporation of a radioopaque heavy-metal salt such as bismuth bromide or uranyl nitrate into an appropriate polymer ligand via chelation (Nirmala et al., 2006). Inorganic salts of a heavy element or organic compounds containing a heavy atom substituent as physical mixtures with an appropriate polymer can also impart radiopacity (Galperin et al., 2007).

The parameter μ/ρ, effective electron density ($N_{e,eff}$) and an effective atomic number (Z_{eff}) is a basic quantity for determining the penetration of gamma-ray photons in matter (Chanthima et al., 2012). The total mass attenuation coefficients (μ/ρ) are important in fundamental physics and many applied fields, the accurate μ/ρ values for X- and γ-rays in several materials are essential for some fields such as, nuclear and radiation physics, radiation dosimetry, biological, medical, agricultural, environmental and industrial (Sharanabasappa et al., 2010). In composite materials, a single number cannot represent the atomic number uniquely across the entire energy range as the partial interaction cross-sections have different atomic number, Z, dependence. This number is a very useful parameter for many fields of scientific, technological, and engineering applications. The Z_{eff} is a convenient parameter for representing the attenuation of X-rays in a complex medium and particularly for the calculation of the dose in radiation therapy (Demir and Han, 2009). When an incident gamma beam with the intensity of I_0 collides perpendicularly with an absorber with a thickness of d, the intensity (I) passing through the absorber can be evaluated with the following the Beer–Lambert's equation (equation 15.1)

$$I = I_0 e^{-\mu d} \tag{15.1}$$

where μ is gamma–ray linear attenuation coefficients of the absorber material (Celiktas, 2011).

Mass attenuation coefficient may be written as (equation 15.2):

$$\mu/\rho = \ln (I_0/I)/\rho d \tag{15.2}$$

Where ρ is the density of material, I_0 and I the incident and transmitted intensities, and d is the thickness of the absorber (Singh et al., 2006). The purpose of the study was to determine the composition of the elastomers characterized by absorption properties of the X-ray radiation. Preparing radioopaque materials from natural rubber (NR) containing different fillers as active substances.

15.2 EXPERIMENTAL

15.2.1 MATERIALS AND METHODS

Natural rubber (RSS I) from Malaysia was applied. Bismuth (III) oxide (Bi_2O_3) (Aktyn company), bismuth (III) oxochloride (BiOCl) (BASF company), gadolinium (III) oxide (Gd_2O_3), bismuth (III) gallate ($C_7H_5BiO_6 \cdot \times H_2O$) and barium sulfate ($BaSO_4$) were purchased from Sigma Aldrich Chemie GmbH. The choice of these fillers was dictated by properties, and they are environmentally friendly. Bismuth compounds are used orally in medicine for antacid action, and they have good gamma ray shielding properties (Ferraz et al., 2012) Bismuth (III) oxochloride is used in medicine (radiopaque medium), in cosmetics and in chemical industry (Novokreshchenova et al., 2005). A barium sulfate (BaSO4) base (Kim et al., 2012) and gadolinium base (McCaffrey et al., 2007) composite has a proven shielding effect.

TABLE 15.1 Composition of elastomer mixtures

Compound	Amount [phr]
Natural rubber RSS gat. I	100
2-mercaptobenzothiazole (MBT)	2
Zinc oxide	5
Sulfur	2
Stearic acid	1
Filler	10

Rubber mixtures were prepared by mixing natural rubber, zinc oxide, stearic acid, MBT (2-Mercapto benzothiazole), sulfur and filler. Full composition formula of the composites is given in Table 15.1. Mixing was done on a laboratory size two roll mixing mill at a friction ratio of 1:1.13. Dimensions of mixing mill rolls were about 400 × 200 mm. After complete mixing, the samples were kept for 24 h for maturation. The samples were vulcanized at the temperature 150 °C under pressure of 15 MPa. The vulcanizates obtained from this procedure had a thickness of about 1 mm. Tests for measuring the tensile strength were carried out according to PN-ISO 37:2007. The samples were held by two grips in a Zwick type 1435 tensile testing machine. The stretching velocity of the sample was 500 mm/min. Cross-link densities were determined rubber sample. The cross-link density was calculated using the Flory–Rehner equation (Flory and Rehner, 1943).

Intensities of absorbed radiation were measured using by a typical Canberra spectometry system with well collimated 3`` NaI(Tl) detector. As radiation source

^{241}Am (60 keV) with 40 kBq activity was applied. Activity of area under the surface photo peak was counted using appropriate software.

The samples were subjected to UV aging with an Atlas UV-2000 instrument. The total duration of the study was 120 h and consisted of two alternately repeating segments with the following parameters:

- daily segment (radiation intensity UV of 0.7 W/m^2, duration 8 h, temperature 60 °C);
- night segment (no UV radiation, duration 4 h, temperature 50 °C).

In order to estimate the resistance of the vulcanizates to UV aging, their mechanical properties and cross-link density after aging were determined and compared with the values obtained for the composites before UV aging. An aging factor (S) was calculated as the numerical change in the mechanical properties of the samples after and before aging (equation 15.3.)

$$S = (TS \cdot EB)_{\text{after aging}} / (TS \cdot EB)_{\text{before aging}} \quad (15.3)$$

The hardness of composites was measured by Shore method, using the Shore A scale according to PN – EN ISO 868:2004. The vulcanizates morphology was assessed by ZEISS emission gun scanning electron microscope (SEM). Small pieces of the uncured rubber were placed in liquid nitrogen for 5 min. They were recovered and fractured into two pieces to create fresh surfaces. The samples, 25 mm^2 in area and 6 mm thick, were coated with gold, and then examined and photographed in the SEM.

15.3 RESULTS AND DISCUSSIONS

15.3.1 EFFECT OF UV AGEING ON MECHANICAL PROPERTIES AND THE CROSS-LINKING DENSITY OF COMPOSITES

The mechanical properties and the cross-linking density of natural rubber composites with different fillers are given in Table15.2. The vulcanizates are characterized by good mechanical properties such as tensile strength, elongation at break. Especially, addition of BiOCl promoted to very good tensile properties (TS c.a. 30 MPa). The sample containing bismuth oxide showed the highest value of elongation at break. Influence of UV ageing for mechanical properties and the cross-linking density of natural rubber composites was on the good level. The ageing under the influence of UV radiation has brought about an increase in the stress at a relative elongation of 100%. Effect of UV ageing on the concentration of network nodes was low. An insignificant decrease in the cross-linking density was observed in the case of natural rubber vulcanizate containing Bi_2O_3. In other cases, values of ν_T were at comparable levels both before and after aging. The highest cross-linking density, amounting to 29×10^5 mol/cm^3, was shown by the vulcanizate with bismuth oxochloride.

TABLE 15.2 The effect of UV ageing on mechanical properties and the cross-linking density of natural rubber vulcanizates containing different fillers

Composite		SE100 [MPa]	TS [MPa]	EB [%]	$\nu_T \times 10^{-5}$ [mol/cm³]
NR/BiOCl	before UV ageing	1.1	30	609	29
	after UV ageing	1.3	26	459	29
NR/Bi_2O_3	before UV ageing	0.9	22	685	27
	after UV ageing	1.0	8	595	25
NR/Gd_2O_3	before UV ageing	1.0	15	580	27
	after UV ageing	1.2	14	546	28
NR/$C_7H_5BiO_6$	before UV ageing	1.0	17	591	22
	after UV ageing	1.1	14	518	24
NR/$BaSO_4$	before UV ageing	1.6	17	523	28
	after UV ageing	1.7	11	480	28

SE100 – stress at elongation 100%, MPa, *TS* – tensile strength, MPa, *EB* – elongation at break, %,ν_T – curing density, mol/cm³

Figure 15.1 shows the results of the UV ageing coefficient (S) which oscillated between 0.8 for Gd_2O_3 and 0.6 for $BaSO_4$; means that sample NR/Gd_2O_3 underwent degradation in small degrees.

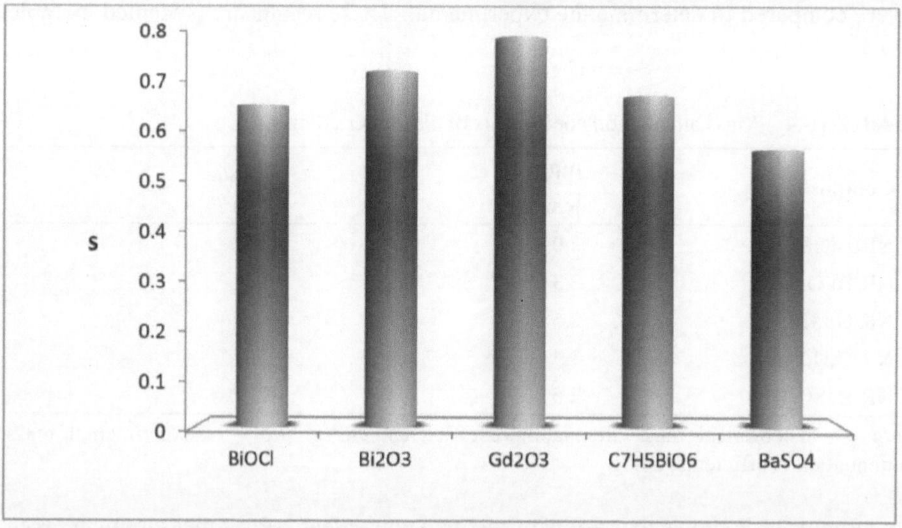

FIGURE 15.1 The UV ageing coefficient of NR vulcanizates containing different fillers

15.3.2 EFFECT OF DIFFERENT FILLERS ON HARDNESS OF NR COMPOSITES

Based on the hardness results, vulcanizate with gadolinium oxide were characterized by the lowest hardness (Table 15.3).

TABLE 15.3 The hardness of elastomeric blends

Composite	Hardness [Sh]
NR/BiOCl	55
NR/Bi_2O_3	46
NR/Gd_2O_3	42
NR/$C_7H_5BiO_6$	49
NR/$BaSO_4$	45

15.3.3 EFFECT OF FILLERS ON ABSORPTION PROPERTIES OF NR COMPOSITES

Based on the literature values of the mass attenuation coefficients were calculated theoretical coefficients of natural rubber composites containing different fillers and were compared to determine the experimental. These results are presented in Table 15.4.

TABLE 15.4 Mass attenuation coefficients of elastomer composites

Composite	$\mu/\rho_{teo.}$ [cm^2/g]	$\mu/\rho_{expe.}$ [cm^2/g]
NR/BiOCl	2.0	1.9
NR/Bi_2O_3	2.3	2.2
NR/Gd_2O_3	2.5	2.4
NR/$C_7H_5BiO_6$	1.4	1.2
NR/$BaSO_4$	2.5	2.3

$\mu/\rho_{teo.}$ – Theoretical mass attenuation coefficients, cm^2/g, μ/ρ_{expe} – experimental mass attenuation coefficients, cm^2/g

Figure 15.2 shows the results of the lead equivalents (Pn_{eq}) measurement taken for the energy commonly used in radiation medicine, 60 keV.

The lead equivalents for natural rubber composites containing Gd_2O_3 and $BaSO_4$ were 0.09; lower the degree of radiation absorption was for vulcanizate with Bi_2O_3.

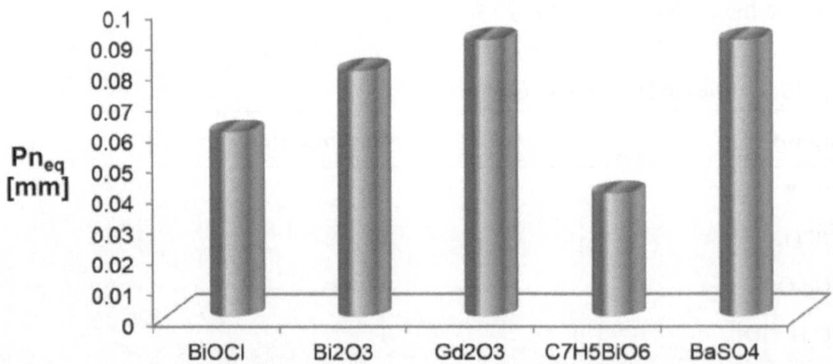

FIGURE 15.2 The lead equivalents of natural rubber composites (E_γ = 60keV)

15.3.4 MORPHOLOGY OF NATURAL RUBBER COMPOSITES

The dispersion degree of fillers in the elastomer matrix was determined on the base SEM photographs shown in Figure 15.3. SEM images of fractures of NR vulcanizates with different fillers did not show the formation of aggregates. This means that distribution of filler in the natural rubber was at a good level.

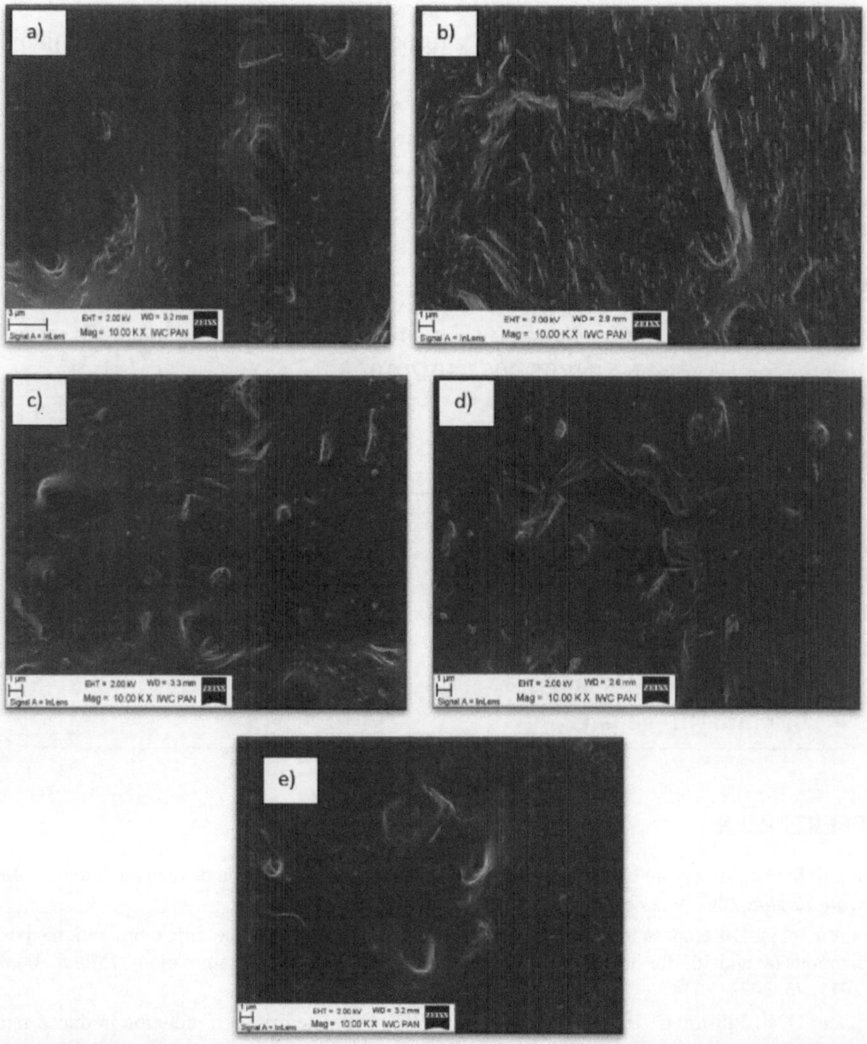

FIGURE 15.3 SEM image of NR composites containing (a) bismuth (III) oxide, (b) bismuth (III) oxychloride, (c) bismuth (III) gallate, (d) gadolinium (III) oxide, (e) barium sulfate (×10,000)

15.4 CONCLUSIONS

The vulcanizates of natural rubber containing different active substances are characterized by:

- high values of tensile strength (17–30 MPa)
- high resistance to aging due to UV (especially for composite from gadolinium (III) oxide)
- the best absorption of X-ray showed a sample containing Gd_2O_3.

Having regard to the absorption of radiation as well as the mechanical properties, natural rubber from gadolinium (III) oxide can be used as material protecting above X-ray.

ACKNOWLEDGMENTS

This work was financially supported by the Polish Ministry of Science and Higher Education under grant No. NR05-0087-10/2010.

KEYWORDS

- **Radiation protection**
- **Natural rubber composites**
- **Bismuth (III) oxide**
- **Bismuth oxochloride**
- **Gadolinium (III) oxide**
- **Bismuth (III) gallate**

REFERENCES

Amis E. S. et al. American College of Radiology white paper on radiation dose in medicine. *J. Am. Coll. Radiol.* **2007**, *4*, 272–84.

Brady Z. et al. Differences in using the international commission on radiological protection's publications 60 and 103 for determining effective dose in pediatric CT examinations. *Radiat. Meas.* **2011**, *46*, 2031–2034.

Brenner D. J.; Elliston C. D.; Hall E. J.; Berdon W. E. Estimated risks of radiation in duced fatal cancer from pediatric CT. *Am. J. Roentgenol.* **2001**, *176*, 289–96.

Brenner D. J.; Hall E. J. Computed tomography—an increasing source of radiation exposure. *New. Engl. J. Med.* **2007**, *357*, 277–2284.

Celiktas C. A method to determine the gamma-ray linear attention coefficient. *Ann. Nucl. Energy*. **2011**, *38*, 2096–2100.

Chanthima N. et al. Study of photon interactions and shielding properties of silicate glasses containing Bi_2O_3, BaO and PbO in the energy region of 1 keV to 100 GeV. *Ann. Nucl. Energy.* **2012**, *41*, 119–124.

Demir L.; Han I. Mass attenuation coefficients, effective atomic numbers and electron densities of undoped and differently doped GaAs and InP crystals. *Ann. Nucl. Energy.* **2009**, *36*, 869–873.

Ferraz K. S. O. et al. Investigation on the pharmacological profile of 2,6-diacetylpyridine bis(benzoylhydrazone) derivatives and their antimony(III) and bismuth(III) complexes. *European J. Med. Chem.*. **2012**, *53*, 98–106

Flory P. J.; Rehner J. Statistical mechanics of cross-linked polymer networks. I. Rubberlike elasticity. *J. Chem. Phys.* **1943**, *1*, 52.

Frasher D.; Altizer G. Consider the Dose. *J. Radiol. Nurs*. **2006**, *25*, 119–121.

Galperin A. et al. Radiopaque iodinated polymeric nanoparticles for X-ray imaging applications. *Biomaterials*. **2007**, *28*, 4461–4468.

Kaushik A.; Mondal A.; Dwarakanath B. S. Radiation protection manual: A publication of Institute of Nuclear Medicine and Allied Science (INMAS), DRDO, **2010**, Delhi, India.

Kim S.-C. et al. Medical radiation shielding effect by composition of barium compounds. *Ann. Nucl. Energy.* **2012**, *47*, 1–5.

Lee T.-Y.; Chhem R. K. Impact of new technologies on dose reduction in CT. *Eur. J. Radiol*. **2010**, *76*, 28–35.

McCaffrey J. P. et al. Radiation attenuation by lead and nonlead materials used in radiation shielding garments. *Med. Phys*. **2007**, 34, *2*, 530–537.

Mheemeed A. K. et al. Gamma-ray absorption using rubber—lead mixtures as radiation protection shields. *J. Radioanal. Nucl. Chem.* **2012**, *291*, 653–659.

Muhogora W. E. et al. Pediatric CT examinations in 19 developing countries: frequency and radiation dose. *Radiat. Prot. Dosim*. **2010**, *140*, 49–58.

Nirmala R. J.; Juby P.; Jayakrishnan A. Polyurethanes with radiopaque properties. *Biomaterials*. **2006**, *27*, 160–166.

Nisha V. S.; Joseph R. Preparation and characterization of radiopaque natural rubber. *Rubber Chem. Technol.* **2006**, 79, *5*, 870–880.

Novokreshchenova M. N. et al. Highly pure bismuth (III) oxochloride synthesis. *Chem. Sustain. Develop.* **2005**, *13*, 563–568.

Sharanabasappa et al. Determination of X-ray mass attenuation coefficients using HPGe detector. *Appl. Radiat. Isotopes*. **2010**, *68*, 76–83.

Singh N. et al. Gamma-ray attenuation studies of PbO-BaO-B_2O_3 glass system.

Radiat. Meas. **2006**, *41*, 84–88.

CHAPTER 16

BARIUM SODIUM NIOBATE FILLED POLYSTYRENE NANOCOMPOSITE AS A DIELECTRIC MATERIAL

ROSALIN ABRAHAM[1,4*], JAYAKUMARI ISAC[2], NANDAKUMAR K.[3,4], and SABU THOMAS[4]

[1]Department of Physics, St. Dominics College, Kanjirapally, Kottayam, Kerala, India – 686512

[2]CMS College, Kottayam, Kerala, India

[3]School of Pure & applied Physics, Mahatma Gandhi University, Kottayam-686 560, Kerala, India

[4]Center for Nanosciences and Nanotechnology, Mahatma Gandhi University, P.D Hills P.O Kottayam, Kerala, India

*School of Chemical Sciences, Mahatma Gandhi University, Kottayam, Kerala, India

*Corresponding author: Tel: 04828270257

E-mail: jollyrose@satyam.net.in

CONTENTS

ABSTRACT

Dielectric properties of BNN-PS (Barium Sodium Niobate-Polystyrene) nanocomposites were measured over a broad frequency range (100Hz to 13 MHz) and temperature ranges (28 °C–130 °C) to explore the possibility of their use as electronic materials, and characterize them on the basis of existing theories. The composites revealed marked departures from the law of physical mixtures for their dielectric properties. The dielectric constant and dielectric loss increase with increasing BNN content. However, the increase in losses is modest and the excellent dielectric properties of the composites are not adversely affected. At a constant temperature, the composites follow a linear relationship between logarithm of their dielectric constant and volume fraction of the ferroelectric filler. The system conforms to the Claussis–Mossotti equations. Dielectric permittivity values of the composites are modeled using different theoretical relationships. The presence of BNN nanofiller into polystyrene matrix is responsible for an increase of glass transition temperature, usually of about 9 °C, with respect to the neat polystyrene. This indicates high extent of polymer-filler interaction.

16.1 INTRODUCTION

Hard, brittle ferroelectric ceramics can be incorporated into polymer matrices (Abraham et al., 2012) to produce composite materials with superior mechanical properties, greater processability and flexibility. They can be easily molded into various useful shapes by versatile polymer processing techniques such as compression, extrusion, and injection molding. In particular, the composites, in which ceramic powders embedded in polymer matrices with (0–3) connectivity, are extremely flexible (Myounggu et al., 2008). The practical application of ferroelectric ceramics for electronic applications required the development of novel processes and devices. The development of capacitors that are compatible with superconducting and manufacturing materials capable of high frequency operations will be required for integrated devices and which can be prepared by the suitable inclusion of ferroelectric ceramics materials with polymers (Abraham et al., 2010). When polymeric materials serve practical uses, they are commonly mixed with other materials to achieve desired performances. Dielectric and conductive properties can be controlled over a broad range by ceramic inclusions (Abraham et al., 2011). It is important to study the electrical properties of these new materials, which would give information about their microstructure, composite dielectric behavior, frequency, and temperature dependence. The pace of research on thermal and dielectric properties of heterogeneous materials has accelerated in recent years. This is because electronics packaging has continuously provided the impetus pushing in the development of new materials in a fascinating and rich variety of applications (Yun et al., 2006). The principal applications for ceramics and ceramic composites are as capacitive elements in electronic circuits and as electrical insulation. For these applications, the properties of most concern are the dielectric constant and dielectric loss factor.

A dielectric material has interesting electrical properties because of the ability of an electric field to polarize the material to create electric dipoles. It is fundamental that the capacitance of a condenser is increased if the space between the conductors is filled with a dielectric material (Montedo et al., 2008). The dielectric constant is a measure of the extent to which the insulating material's surface interact with the electric field set up between the charged plates. The dielectric constant is dependent on two molecular level properties; the permanent "dipole moment" and the "polarizability" or the induced change in dipole moment due to the presence of an electric field (Michele et al., 2008). The permanent dipole moment is the average over the various dipole moments given rise by structural charge density differences over intramolecular distances (Kumar et al., 2007). Polarizability is the property which arises from changes in the molecular electron distribution induced by the applied electric field (Borcia et al., 2007). In the present work, we have synthesized a new class of ferroelectric ceramic material, barium sodium niobate, and prepared composites out of it by mixing it with polystyrene. A detailed discussion of the structural and mechanical properties of Barium Sodium Niobate (BNN)-Polystyrene (PS) (BNN-PS) composites has been published elsewhere (Abraham et al., 2009). In that paper, we reported a study on the structural and mechanical properties of the prepared composites. Here, in this paper, we discuss the dielectric properties of BNN-PS nanocomposites in detail. Dielectric mixing laws are examined for the composites. The values of dielectric constant and dielectric loss factor have been analyzed as a function of BNN filler loading on the basis of theoretical predictions for two-phase systems. SEM (Scanning Electron Microscope) and AFM (Atomic Force Microscope) pictures of selected composites have been made to characterize the morphology of the composites.

Dipole moment of each BNN particle is calculated using Claussis–Mossotti equations. Clausius–Mossotti approximation is one of the most commonly used equations for calculating the bulk dielectric properties of inhomogeneous materials (Ohad and David, 1997). It is useful when one of the components can be considered as a host in which, inclusions of the other components are embedded. It involves an exact calculation of the field induced in the uniform host by a single spherical or ellipsoidal inclusion and an approximate treatment of its distortion by the electrostatic interaction between the different inclusions. The Clausius–Mossotti equation itself does not consider any interaction between filler and matrix. This approach has been extensively used for studying the properties of two-component mixtures in which both the host and the inclusions possess different dielectric properties. In recent years, this approximation has been extensively applied to composites involving ceramics and polymers.

16.2 EXPERIMENTAL

16.2.1 MATERIALS

The filler, ferroelectric ceramic material (Barium Sodium Niobate (BNN)) was prepared by usual ceramic technique using reagent grade sodium carbonate, barium carbonate, and niobium pentoxide. Solid-state reaction technique was adopted for the preparation of BNN. The starting materials were $BaCO_3$ (Merck Ltd., Mumbai, India), Na_2CO_3 and Nb_2O_5 (CDH, New Delhi, India). The powders were taken by satisfying the stoichiometric ratios (equation 16.1), mixed, milled and calcined at a temperature of 1000 °C for 3 h. Sintering was carried out at a temperature of 1200 °C for 5 hrs.

$$3BaCO_3 + 2Na_2CO_3 + 5Nb_2O_5 \text{ ---- } Ba_3Na_4Nb_{10}O_{30} + 5CO_2 \quad (16.1)$$

Then these sintered samples were named as (Barium Sodium Niobate ($Ba_3Na_4Nb_{10}O_{30}$) or (BNN)). The specific properties of the Polystyrene (Merck Ltd., Mumbai, India) needed for this study are dielectric constant, molecular weight, and glass transition temperature and the values are 2.5, 208000, and 108 °C, respectively.

16.2.2 METHODS OF COMPOSITE PREPARATION

The melt mixing technique was chosen for preparing the composites because it allowed solvent free mixing for the ceramic filler. By melting at high temperature, molten polystyrene can easily penetrate between filler particles, which facilitate suitable mixing and allow avoiding air trapping into the composites. Consequently void free composites were obtained. Polystyrene-Ceramic composites were prepared in a Brabender plasticoder at a rotor speed of 60 rpm for about 6 min. The compositions of the composites were 10, 20, 30, and 40% by the volume of the filler. The mixed samples were compression molded into sheets of desired thickness by hydraulic press at a temperature of 180 °C and were used for different studies. The composites were named as BNN10, BNN20, BNN30 and BNN40, where 10, 20 30 and 40 represent the volume % of filler in the composite.

16.2.3 MEASUREMENTS

The morphology and microstructure of the composites were analyzed by means of high resolution scanning electron microscopy using a JEOL JSM 840–microscope. The average size of the dispersed particle is calculated using nanoparticle analyzer software attached to Atomic Force Microscope.

Dielectric measurements were performed with Hewlett-Packard-Japan HP 4192A impedance analyzer at a temperature range of 28 °C to 120 °C and at a frequency range of 100 kHz to 13MHz. Well-polished pellets of samples were given silver metallization for the measurement. The dielectric constant of the composites were calculated from the sample geometry and sample capacitance using the equation,

$$\varepsilon_c = \frac{Ct}{\varepsilon_0 A} \tag{16.2}$$

where ε_c is the dielectric constant of the sample, ε_0, is the absolute permittivity of vacuum (8.85×10^{-12} F/m), "A" the area, and "t" the thickness of the sample. C is the measured value of the capacitance of the sample.

16.3 RESULTS AND DISCUSSIONS

16.3.1 SEM AND AFM STUDIES OF THE COMPOSITES

Figures 16.1 (a, b), 16.2 (a–b) show the SEM and 16.2(c) shows the AFM image of some of the selected composites, which reveal that nanofillers are well dispersed and embedded rather uniformly through the PS matrix. The ceramic particles appear to be well dispersed both in low- and high-concentration composites. The filler particles are uniformly distributed in all composites and the particles are almost spherical in shape with irregular boundaries. In all composites filler particles are clearly embedded in the polymer matrix. It gives clear evidence to the (0–3) connectivity of the composites. The average particle diameter is found to be less than 100 nm in all BNN-PS composites. The average diameter of the nanoceramic is calculated by the software (Nanoscope particle analyzer V531r1) attached to AFM and reported in Table 16.1 and it is found that diameter is of the order of 58 nm for BNN.

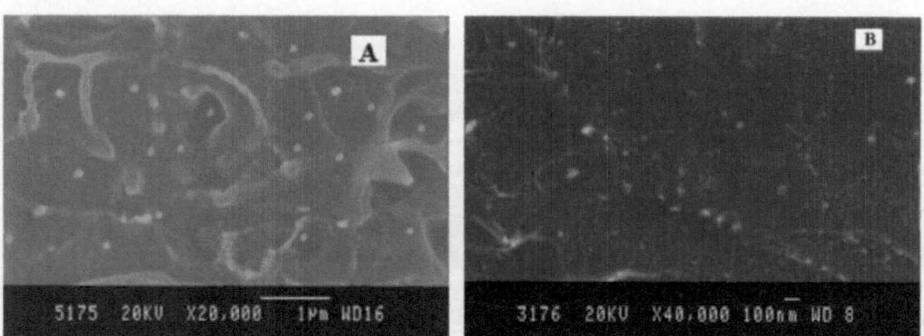

FIGURE 16.1 (a) and (b) SEM image of BNN20 at different magnifications

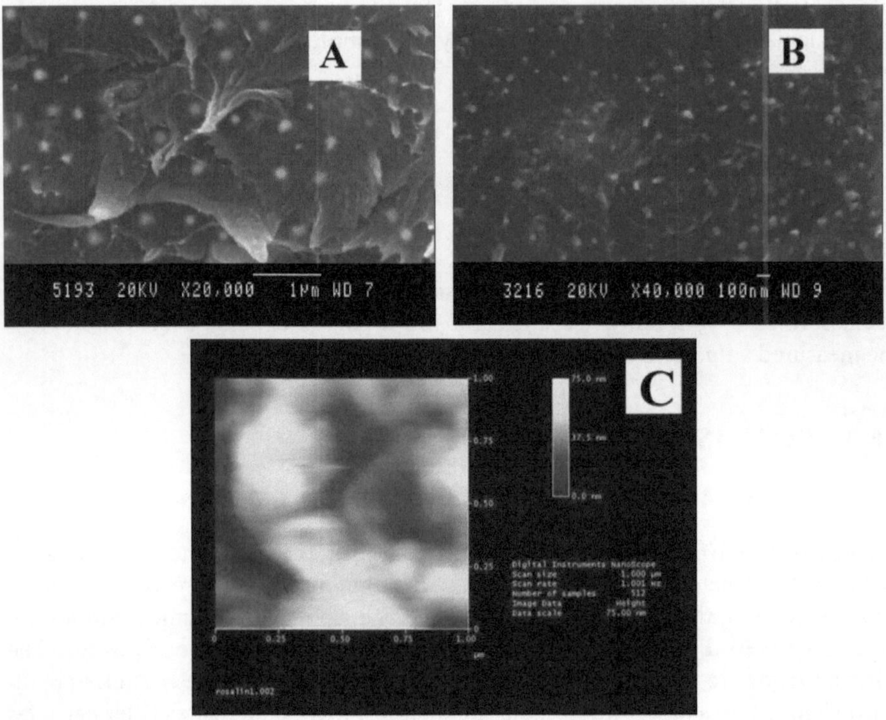

FIGURE 16.2 (a) and (b) SEM image of BNN40 at different magnifications, (c) AFM image of BNN10

TABLE 16.1 Particle diameters calculated using software

Dimension	Mean	Minimum
Height	2.58 nm	−0.91 nm
Area	10059nm^2	15.259 nm^2
Diameter	58.14 nm	4.40 nm
Length	119.38 nm	5.52 nm
Width	38.29 nm	5.52nm

16.3.2 DIELECTRIC CONSTANT OF THE COMPOSITES

The dielectric constant of BNN is greater than that of polystyrene, so the addition of BNN to the polymer matrix will result an increase in dielectric constant. Since the values of dielectric constant "ε"of the two ingredients, polystyrene and BNN are

2.55 and 430 respectively, it is clear from (Figure 16.3) that these composites do not obey the law of physical mixtures, as stated by

$$\varepsilon_c = \varepsilon_f v_f + \varepsilon_p (1 - v_f) \tag{16.3}$$

where v_f is the volume fraction of the filler, $\varepsilon_c, \varepsilon_p, \varepsilon_f$ are the dielectric constant of composites, polymer, and filler respectively. At room temperature and for a frequency of 1 MHz, the dielectric constant of the composites is found out. The calculated values of dielectric constant of the composites using physical law of mixing, its experimental values, and the inverse of experimental values are plotted in (Figure 16.3).

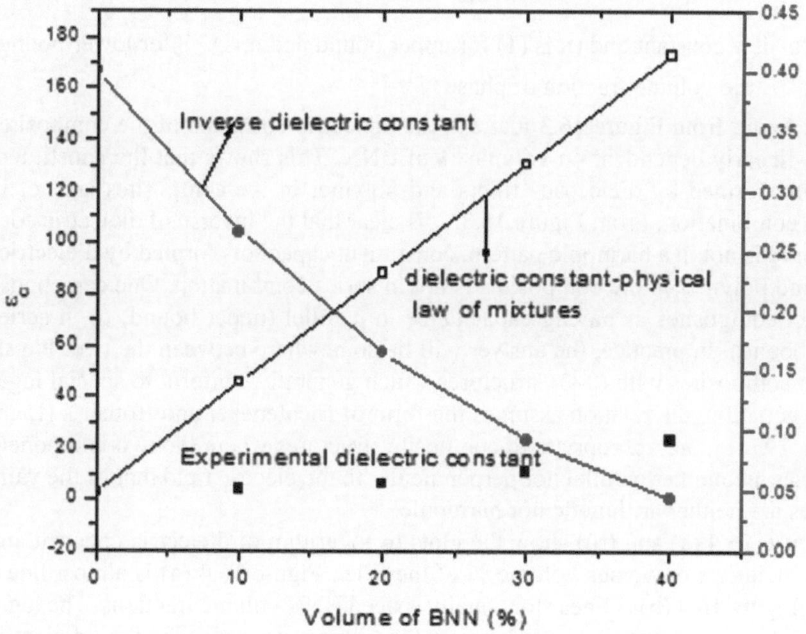

FIGURE 16.3 Dielectric constant (Theoretical and experimental and inverse) versus volume fraction of BNN

The present system of composites is a binary phase mixture of two dielectrically different materials, where BNN is ionic and polycrystalline and polystyrene are amorphous atactic and rather non-polar. A great variety of formulae has been suggested for the calculation of permittivity of heterogeneous mixtures.

DIELECTRIC MIXTURE RULES

Mixtures of ideal dielectrics can be most simply considered on the basis of layer materials with the layers either parallel or normal to the applied field. When the layers are placed normal to the applied field, the capacitance are additive. When layers are parallel to the applied field, the structure corresponds to capacitive elements in series, and the inverse capacitance are additive. In practice, the answer will lie somewhere between the two. These two are considered as upper bound and lower bound conditions. Upper bound and lower bound equations are the special cases of a general empirical relationship

$$\varepsilon_c^{\,n} = \sum_i v_i \varepsilon_i^n \tag{16.4}$$

where "n" is a constant and (n is (1) for upper bound and n is (−1)for lower bound) and "v_i" is the volume fraction of phase "i".

It is found from Figure 16.3 that ε_c, the dielectric constants of the composites are non-linearly dependent on volume % of BNN. This shows that the constituent capacitors formed by dielectrics fillers and polymer in the composites are not in parallel combination. From Figure 16.3, it is clear that the inverse of dielectric constant curve is not in a harmonic pattern, constituent capacitors formed by dielectrics fillers and polymer in the composites is not in series combination. One can choose to model composites as having capacitance in parallel (upper bound) or in series (lower bound). In practice, the answer will lie somewhere between the two. Physically, in composites with (0–3) structures which generally conform to special logarithmic equation, the relation assumes the form of Lichteneker and Rother's (Lichteneker, 1956) more appropriate to composite structures where the two-component dielectrics are neither parallel nor perpendicular to the electric field that is, the valid averages are neither arithmetic nor harmonic.

(Figure 16.4 (a) and (b)) show the plots of logarithm of dielectric constant and specific polarization versus volume % of the filler. Figure 16.4 (a) is almost linear and the Figure 16.4 (b) is linear for equally spaced BNN volume fractions. The logarithmic law of mixtures firmly confirms a logarithmic dependence of the dielectric constant of the composite on the volume fraction of the filler. So it is better to apply Lichtenecker's rule for two-phase system to BNN-PS composites. Hence, it is believed that the composites follow the "log law" relationship, originally proposed by Lichtenecker in which the dielectric constant of clean two-component system can be represented by

$$\log \varepsilon_c = \log \varepsilon_p + v_f \log\left(\frac{\varepsilon_f}{\varepsilon_p}\right) \tag{16.5}$$

The plot of specific polarization, $\left(\frac{\varepsilon_c - 1}{\varepsilon_c + 2}\right)$ versus volume fraction is presented in (Figure 16.4 (b)). Linear fitting the Figure we will get the value of slope and intercept using Origin Graph Plotter software. The slope and intercepts are 1.4 and 0.35. The specific polarization of polystyrene is exactly matching to the intercept value (dielectric constant of PS lies between 2.5 and 2.6) as expected in accordance with Clausius-Mossotti equation modified by Lorentz and Lorentz applicable to the overall composite dielectrics (Vemulapally, 2004). The Clausius-Mossotti equation itself does not consider any interaction between filler and matrix (Blythe, 1974). The same equation is used for the calculation of dipole moment of BNN particle (Devan et al., 2007).

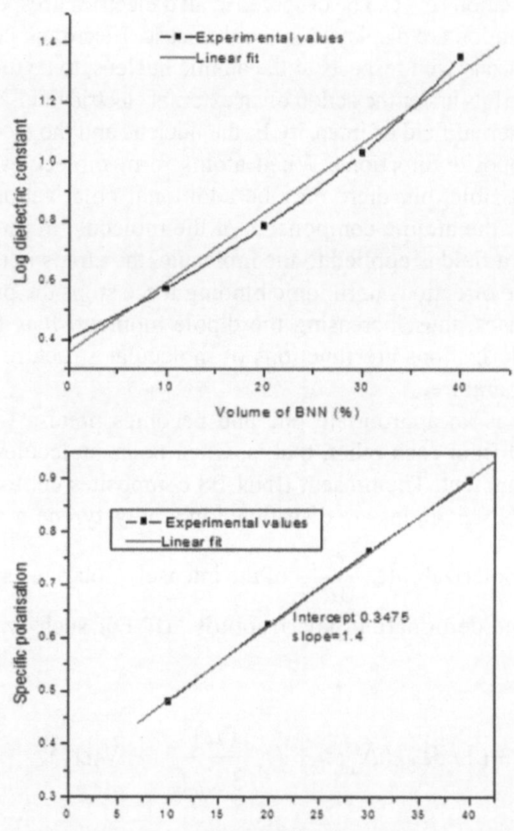

FIGURE 16.4 (a) Log ε versus volume% of filler in composites and (b) Specific polarization versus volume% of filler in composites

16.3.3 CALCULATION OF DIPOLE MOMENT OF BNN PARTICLE

The Clausius-Mossotti equation for a single–component system can be written as

$$\frac{\varepsilon_c - 1}{\varepsilon_c + 2} = \frac{N}{3\varepsilon_0}\left(\alpha + \frac{(\bar{\mu})^2}{3K_B T}\right) \tag{16.6}$$

where N is the number of molecules per unit volume, α is the deformational polarizability (both electronic and ionic polarization with polarizability factors "α_e" and "α_i"), $\frac{(\bar{\mu})^2}{3K_B T}$ is the dipolar polarizability, $(\bar{\mu})$ is the dipole moment, K_B is the Boltzman's constant, and T is absolute temperature.

Electronic polarization "α_e" can be observed in all dielectrics irrespective of whether other types of polarization are displayed in the dielectric. Electronic polarization is the displacement of electrons with respects to the atomic nucleus, to be more precise—the displacement of the orbits under the action of an external electric field. When the system is subjected to an external field of intensity E, the nucleus and the electron experience Lorentz forces in opposite directions. When atoms form molecules, electronic polarization is still possible, but there may be additional polarization due to a relative displacement of the atomic components of the molecule in the presence of an electric field. When a field is applied to the molecule, the atoms in the molecule are displaced in opposite directions until ionic binding force stops the process and ionic polarization "α_i" arises, thus increasing the dipole moment. It is found that electronic and ionic polarizations are functions of molecular structure and are largely independent of temperature.

Equation (16.6) is an appropriate one and becomes precise when polar molecules are separated from each other, that is, when polar molecules are distributed in a nonpolar environment. The present BNN-PS composites conform to this situation, where ionic BNN particles are distributed in a polystyrene matrix. For such a system, the dipolar polarizabilitiy, $\frac{(\bar{\mu})^2}{3K_B T}$ of the intensely polarized system is usually much higher than the deformation polarizabilitiy "α" For such systems the above equation becomes

$$\frac{\varepsilon_c - 1}{\varepsilon_c + 2} = (1/3\varepsilon_0)(N_1\alpha_1 + N_1\frac{(\bar{\mu})^2}{3K_B T} + N_2\alpha_2) \tag{16.7}$$

where "N_1" and "N_2"are respectively the number of molecules of BNN and polystyrene per unit volume of the composite. α_1 and α_2 are their corresponding deformational polarizability, and μ is the average dipole moment of BNN particles in the polystyrene matrix. The dipole polarizabilities of the intensely polarized system is usually much higher than the deformation polarizabilities α_1 and α_2. Neglecting the two terms involving subscripts 1 and 2 in (equation 16.7) and substituting in $N_1 = (d_1/M_1)\, N_A\, v_f$ where density of BNN d_1 = 5950 kgm^{-3}, molecular weight M_1 of BNN calculated from its stoichiometry is 1651 g, and N_A, Avogadro number, and v_f is the volume percentage of BNN. Figure 16.4 (b) is fitted by this relationship.

$$\frac{\varepsilon_c - 1}{\varepsilon_c + 2} = \frac{N_A d_1 (\bar{\mu})^2}{9\varepsilon_0 K_B T M_1} * v_f \tag{16.8}$$

Slope of the curve (Figure 16.4 (b)) we get,

$$\text{Slope} = \frac{N_A d_1 (\bar{\mu})^2}{9 K_B T M_1 \varepsilon_0} \tag{16.9}$$

$$\mu = \left(\frac{N_A d_1}{9 K_B T M_1 \varepsilon_0} \right)^{-1/2} \times (\text{Slope})^{1/2} \tag{16.10}$$

At room temperature, T = 300 K and the slope is 1.4. The calculated dipole moment of BNN in polystyrene matrix is 55.7 × 10^{-30} Cm. This dipole moment of BNN particles and the polarizability of the medium contribute for the high dielectric constant of the composites. Polystyrene, which is considered to be nonpolar, does in fact possess a very small dipole moment of the order of (0.3 D), due to the asymmetry of the phenyl side group in atactic polystyrene. Because of their anisotropic polarizablility, phenyl groups tend to orient with their greatest main axis of polarizabilitiy in the direction along which the E vector of an electric field and an induced dipole moment is produced in some phenyl groups. On the other hand, the induced dipole moments interact with other phenyl groups present in an ensemble of polystyrene. Both effects render it possible that the dielectric data of Polystyrene become ac field dependent if there is an internal degree of freedom concerning the phenyl-phenyl arrangement below the glass transition temperature (T_g).

16.3.4 MODELING

Several well-known dielectric equations are selected for this modeling. The two-phase mixtures are also represented by the Bottcher-Bruggeman formula (Bottcher, 1942) based on the spherical particle model where the filler is interacting with polymer. According to this formulae

$$\varepsilon_c = \frac{1}{4}\left(H + (H^2 + 8\varepsilon_f \varepsilon_p)^{1/2}\right) \tag{16.11}$$

where

$$H = (3v_f - 1)\varepsilon_f + (2 - 3v_f)\varepsilon_p$$

with $\varepsilon_f = 430$, the dielectric constant of the filler, and $\varepsilon_p = 2.5$, the dielectric constant of PS. The values of dielectric constant of the composites may be calculated from this equation and plotted against v_f in Figure 16.5.

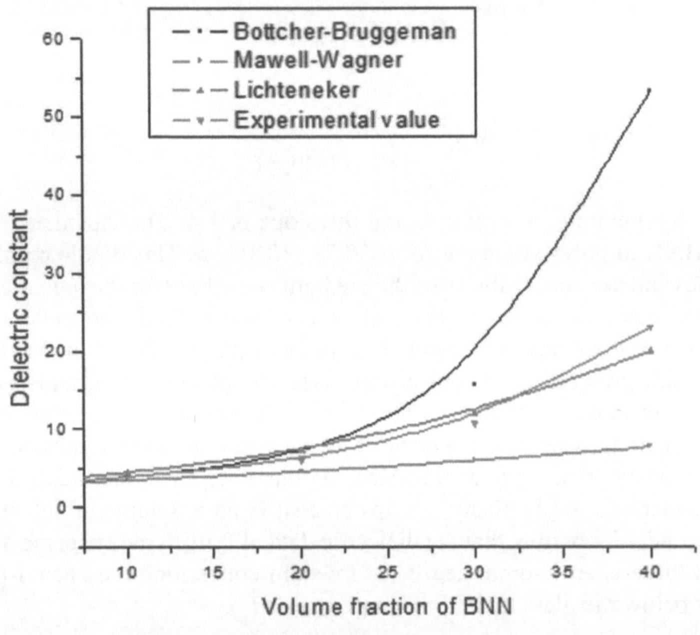

FIGURE 16.5 Comparison of the calculated dielectric constant by different laws

However, the Maxwell-Wagner-Sillars (Grossman and Isard, 1970) equation predicts as a complete solution of the Wagner-Raleigh theory for a system of one spherical particle uniformly distributed in another

$$\varepsilon_c = \varepsilon_p \frac{2\varepsilon_p + \varepsilon_f + 2v_f(\varepsilon_f - \varepsilon_p)}{2\varepsilon_p + \varepsilon_f - v_f(\varepsilon_f - \varepsilon_p)} \tag{16.12}$$

The behavior of the present system is in agreement with these equations up to 20% volume fraction as revealed by the plots of ε_c versus v_f as in (Figure 16.5). Beyond (20%), the experimentally observed values of "ε_c" lie in between the other plots coinciding with Lichtenecker's Logarithmic relations as in Figure 16.5.

This indicates that the shape, particle size distribution and concentration of the dispersed component do not permit a very high degree of chemical interaction as envisaged by Bottcher-Bruggeman. The composites under investigation consist of ionic BNN dispersed in polystyrene medium. Hence, it is likely that the magnitudes of both the short range and the long range interactions possible between the ions in the filled matrix are lessened by the imposition of a plastic environment on the ionic BNN.

16.3.5 DIELECTRIC CONSTANT, IMPEDANCE-FREQUENCY SPECTRA

When the dielectric material is subjected to an alternating field, the orientation of dipoles, and hence the polarization, will tend to reverse every time the polarity of the field changes (Nelson et al., 1991). At low frequencies the polarization follows the alterations of the field without any significant lag and the permittivity is independent of frequency and has the same value as in static field. When the frequency is increased the dipoles will no longer be able to rotate sufficiently rapidly so that their oscillations will begin to lag behind those of the field (Galasso, 1969). The above effect leads to a fall in dielectric constant of the material with frequency as in (Figure 16.6) (Vrejoiu et al., 2002).

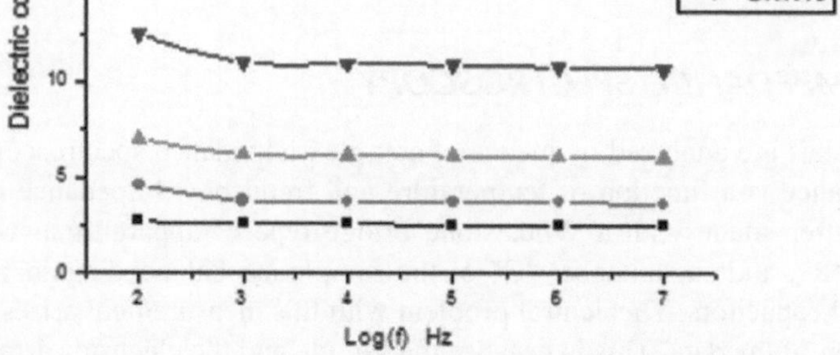

FIGURE 16.6 Dielectric constant versus frequency of PS and the composites

In nanomaterials, there is an additional chance of getting a high dielectric constant because of the large space charge polarization owing to the large surface area of a large number of individual grains. The interface contact area changes inversely as the radius of the particle (1/r). In a low-frequency regime electronic, ionic, dipolar, and space charge polarization play a dominant role in determining the dielectric properties of the materials (Newnham et al., 1971). In the BNN-PS composites, there is a finite contribution from the above mentioned polarization, which gives an initial high value for dielectric constant and decreases with frequency due to the changes in orientation polarization with frequency (Nakano et al., 2001).

It is observed that, up to one MHz, the permanent dipoles can follow the field quite closely and so dielectric constant is high. This observation is evidenced from (Figure 16.6) that the high dielectric constant values for the prepared samples fall with frequency for the composites (Yang et al., 2005). This indicates that at high frequency the mobility of polar groups in polymer chains is unable to contribute to the dielectric constant. At low frequency, the dielectric constant of the composite strongly depends on the dielectric properties of both polymer and ceramic contents, while at high frequency the dielectric constant becomes dependent primarily of the ceramic filler and its concentration. The frequency dispersion relation given by Habery and coworkers (Habery et al., 1968), and the recent work of Dilip et al. (Dilip et al., 2008) show that dielectric constant decreases with increasing frequency and reaches a constant value for each sample.

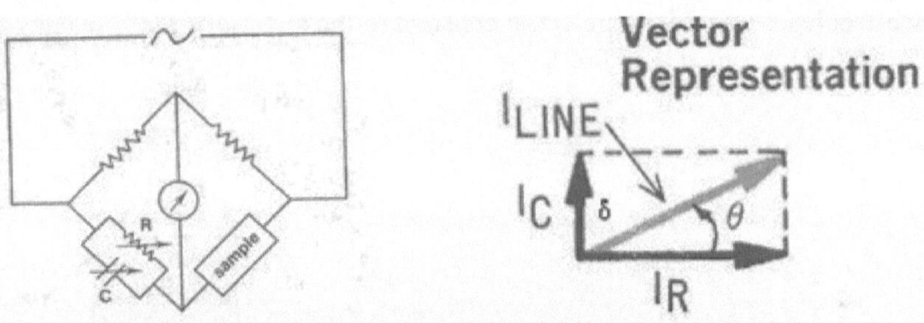

FIGURE 16.7 Vector representation of capacitive and resistive current

16.3.6 IMPEDANCE SPECTROSCOPY

The composites are analyzed by means of complex impedance spectroscopy as the total Impedance is a function of temperature and frequency. Impedance measurements are often made with a Wheatstone bridge type of apparatus in which the resistance, "R", and capacitance, "C" of the sample are balanced against variable resistors and capacitors. The central problem with this measurement arises over the interpretation of the data. This is because the sample and the electrode arrangement is electrically a "black box" whose equivalent circuit (i.e. its representation by some combination of R and C elements) is often unknown. The impedance spectra thus obtained are then processed through computer assisted electrochemical data analysis software that ideally fit to the experimental data. Figure 16.7 shows the impedance measurement terms used for analyzing the structure of BNN-PS composites at room temperature. The Impedance date we got for BNN-PS is fitted with a parallel RC circuit. Generally in parallel R C circuit, the total Impedance "Z" is,

$$Z = \frac{RX_c}{(R^2 + X_c^{\ 2})^{1/2}} \tag{16.13}$$

where $(R^2 + X_c^{\ 2})^{1/2}$ is the vector addition of the resistance and capacitive reactance. The impedance of a parallel RC circuit is always less than the resistance "R" or capacitive reactance "Xc" of the individual branches. The relative values of "Xc" and "R" determine how capacitive or resistive the circuit line current is. The one that is the smallest and therefore allows more branch current to flow is the determining factor. Thus if "Xc" is smaller than "R", the current in the capacitive branch is larger than the current in the resistive branch, and the line current tends to be more capacitive (Figure 16.8) (Jainwen et al., 2006).

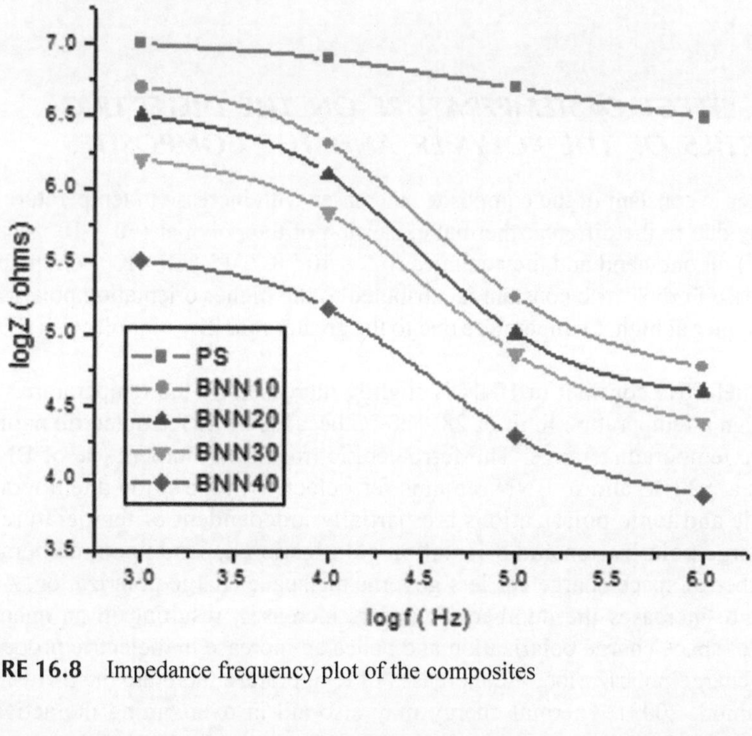

FIGURE 16.8 Impedance frequency plot of the composites

Frequency of the applied voltage determines many of the characteristics of a parallel RC circuit. Frequency affects the value of the capacitive reactance and so also affects the circuit impedance, line current, and phase angle, since they are determined by the value of "Xc". The higher the frequency of a parallel RC circuit, the lower is the value of "Xc". This means that for a given value of R, the impedance is also lower, making the line current larger and more capacitive (Parker and Elewell, 1966). The impedance measurement of the composites revealed that when the volume fraction increased from 10% to 40%, the composites remain in their capacitive characteristics. The a.c conductivity is a measure of resistive component and it depends on the value of "δ". No significant changes are observed for "δ" value from 10% to 40% volume fraction. The impedance Z-frequency curves were parallel curved lines and the phase angle θ about 90°, which are like ideal capacitors as in Figure 16.8 {Z depends on $\frac{1}{j2\pi fC}$} (Brisco, 1986). Dielectric measurements provide two fundamental electrical characteristics of materials: the capacitive (insulating) nature, which represents its ability to store electrical charge; and the conductive nature, which represents its ability to transfer electric charge (Sichel, 1981). However, in the composites the impedance values are still fairly high and showed a frequency dependence indicating that the particle-particle contacts are very weak and there are thin dielectric layers between the particles, which give strong capacitive effects (dominated by $\frac{1}{j2\pi fC}$).

16.3.7 EFFECT OF TEMPERATURE ON THE DIELECTRIC PROPERTIES OF THE POLYMER AND THE COMPOSITES

The dielectric constant of the composites increases with increase in temperature. It is essentially due to the different thermal expansion of the polymer ($50 \times 10^{-6}K^{-1}$ $300 \times 10^{-6}K^{-1}$) on one hand and the ceramics ($0.5 \times 10^{-6}K^{-1}$ $15 \times 10^{-6}K^{-1}$) on the other. The increase in dielectric constant is attributed to the higher orientation polarization of the polymer at higher temperature due to the greater mobility of molecules (Suzhu, 2000).

The dielectric constant of BNN is slightly increased by the temperature variations within a temperature limit of 28–120 °C because of its ferroelectric nature in the above temperature range. The ferroelectric transition temperature of BNN is observed at 530 °C and so BNN remains ferroelectric in the studied temperature. Electronic and ionic polarizations are partially independent of temperature, but space charge polarization and orientation polarization depend upon temperature. The number of space charge carriers governs the space charge polarization. As the temperature increases the number of carriers increases, resulting in an enhanced build up of space charge polarization and hence an increase in dielectric properties. In space charge polarization, the increases of temperature facilitate the diffusion of ions (Pramod, 2003). Thermal energy may also aid in overcoming the activation barrier for orientation of polar molecules in the direction of the field. In such cases relative dielectric constant increases when the temperature increases, the orientation of these dipoles is facilitated and this increases the dielectric polarization. As temperature approaches relaxation transition temperature, the dielectric constant increases dramatically because the dipoles gain sufficient mobility to contribute to the dielectric constant. But at very high temperatures the chaotic thermal oscillations of molecules are intensified and the degree of orderliness of their orientation is diminished, and thus the permittivity passes through a maximum (Mark et al., 2007). The glass transition temperature corresponds to the inflection point on the curves. In orientation polarization, the randomizing action of thermal energy decreases the tendency of the permanent dipoles to align themselves in the applied field at very high temperature. This levels off the dielectric constant with temperature at and above 100 °C that is nearly the glass transition temperature of PS (Jose and Jose, 1955).

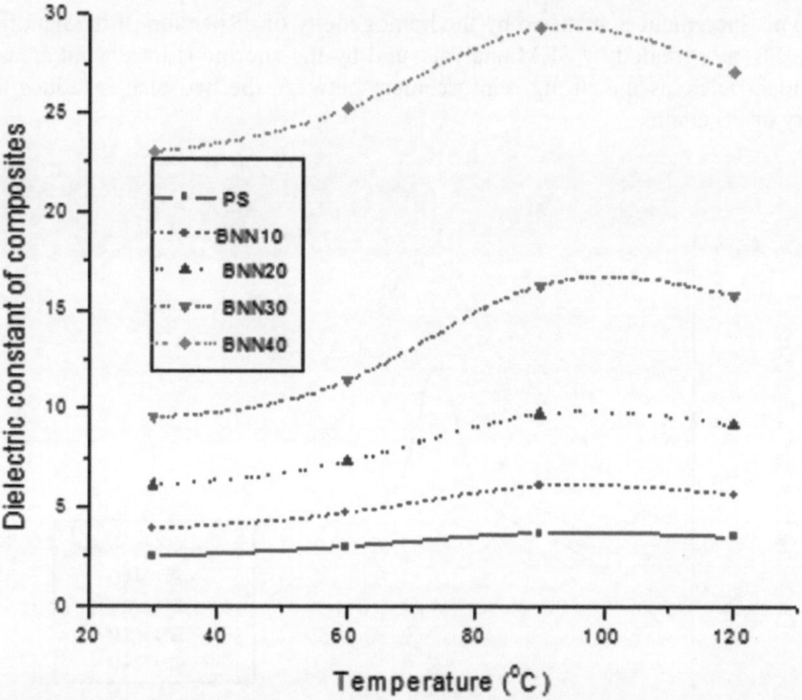

FIGURE 16.9 Temperature dependence of dielectric constant of PS and composites

16.3.8 CONDUCTIVITY AND GLASS TRANSITION TEMPERATURE

The ac conductivity (σ) of the prepared composites are calculated using the formulae

$$\sigma = 2\pi f \varepsilon_0 \varepsilon_c \tan(\delta) \tag{16.14}$$

where ε_0 is the dielectric constant of vacuum $\varepsilon_{c,}$ the relative dielectric constant of the composites, "f" the applied frequency that is, one MHz.

Although the glass-rubber transition of polymer structure, "Tg" as observed, depends largely on the chemical nature of the polymer chain that is, chain flexibility, molecular structure and so on. The dielectric loss is responsible for conductivity and the peak temperature is characterized as the glass transition temperature by dielectric theory. The glass transition temperature of the composites increases with the increment of filler content (Figure 16.10) and the values are represented in (Table 16.2). This increment is justified by the homogeneity of dispersion of the nanofillers into PS, as revealed by SEM analysis, and by the enormous interfacial area of the nanoparticles, as the strong reinforcement between the two phases reduce the mobility of PS chains.

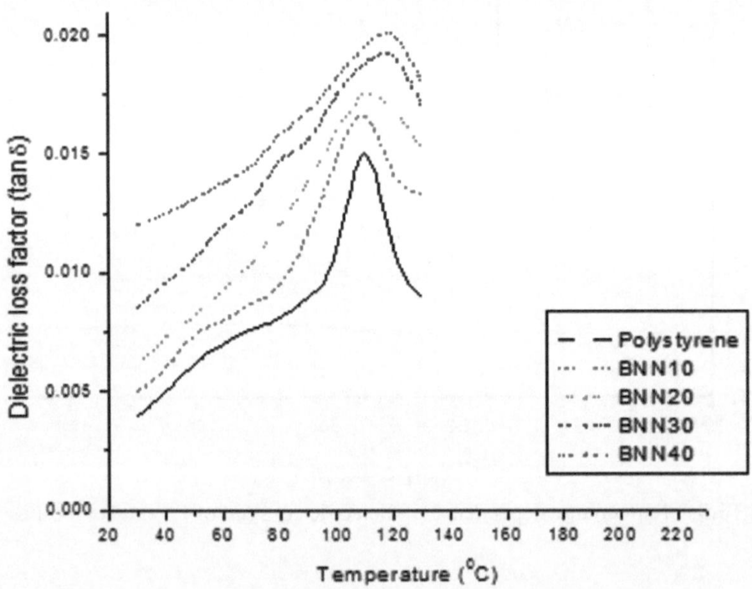

FIGURE 16.10 Temperature versus dielectric loss for PS and the composites

TABLE 16.2 Peak of dielectric losses at a frequency I MHz

Sample Name	T_g (°C)
PS	108.5
BNN10	109.5
BNN20	112.2
BNN30	116
BNN40	117.5

The magnitude of dielectric loss is an important material parameter for making capacitors. Ideally in a capacitor, the dielectric loss should be as low as possible. BNN ceramic does not show any significant dielectric losses up to megahertz frequencies. Hence the dielectric loss observed within the studied frequency region originates primarily from polymer content. It is observed that the rate of variation of dielectric loss with temperature is steeper for higher volume fraction samples. This is attributed to the internal field generated by the ceramic particles.

The increasing relaxation temperature of the composites with increasing BNN concentration may be due to an interfacial or Maxwell-Wagner-Sillars polarization (MWS) (Sillars, 1937). The pattern is schematically represented in (Figure 16.11). The relaxation arises from the fact that the free charges, which were present at the stage of processing, are now immobilized in the materials. At sufficiently high temperature, the charges can migrate in the presence of an applied electrical field. These charges are then blocked at the interface between the two media of different conductivity and dielectric constant. In BNN-PS composites, interfacial polarization is always present. Although this phenomenon is clear in a conductive filler such as metal reinforced polymer composites (Satish et al., 2005), polarizable filler-reinforced polymer composites have also been shown to exhibit the MWS effect (Saxena et al., 1991).

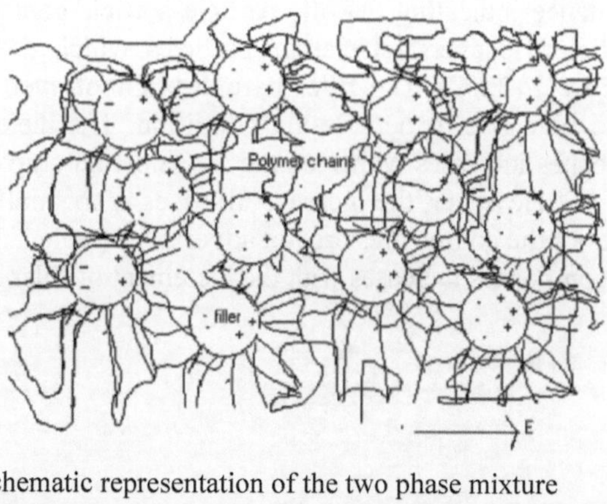

FIGURE 16.11 Schematic representation of the two phase mixture

The interpretation of the impedance, dielectric, and conductivity spectra and their electrical characteristics support the statement that these materials, are essentially insulating, although a slight increase in conductivity is observed with increasing filler content. This trace of conductivity may be attributable to an incipient tunnel effect known (Mathew et al., 2007) to allow the electrons to flow from one particle to the next through the polymer film sandwiches between the particles, thus establishing an electric current. The lesser the filler content, less likely are the electrons to leap from one particle to the next, so for BNN10 the conductivity is minimum and the composite exhibits purely capacitive behavior. As the filler content increases, chances are given for electrons to flow through it and as a consequence a drop in electrical resistivity of the polymer composite is experienced. The filler particles cannot even be brought close enough together to give rise to percolation condition. The particles are no longer in contact, but surrounded by a fine polymer film and hence infinitesimal gaps among the adjacent particles may conduct an electric current by tunneling effect. This conductivity is dictated by nearest neighbor tunneling. The percolation like behavior is observed only when the radius of the particle is superior to tunneling range (Hong et al., 2007).

16.4 CONCLUSIONS

Composites do not obey the law of physical mixtures for dielectric mixing. The logarithmic law of mixtures confirms a logarithmic dependence of the dielectric constant of the composite on the volume fraction of the filler. Dipole moment of BNN in polystyrene matrix is 55.7×10^{-30} Cm. This dipole moment of BNN particles and the induced polarization of BNN under electric field contribute for the high dielectric constant of the composites. For 10–40%, volume percentage of the filler, the composites remain in their capacitive characteristics. Dielectric permittivity variations are in accordance with Lichteneker relations and Claussis–Mossoti equations are neatly fitted for the composites. However, in the composites the impedance values are still fairly high and showed a frequency dependence indicating that the particle-particle contacts are very weak and there are thin dielectric layers between the particles, which give strong capacitive effects (dominated by $1/(2\pi fC)$. The filler particles cannot even be brought close enough together to give rise to percolation condition. The dielectric constant and losses of the composites increases with increase in temperature up to the glass transition temperature of PS. However, the increase in losses is so small and the excellent dielectric properties of the composites are not adversely affected. The glass transition temperature of the composites increases with the increment of filler content.

KEYWORDS

- **Nanocomposites**
- **Modeling**
- **Dielectric properties**
- **Claussis–Mossotti equations**
- **Polystyrene**

REFERENCE

Abraham, R.; Kuryan, S.; Isac, J.; Thomas, S. P.; Varughese, K. T.; Thomas, S. eXPress. Poly. Lett. 2009, 3, 177–193.

Abraham, R.; Kuryan, S.; Thomas, S. P.; Isac, J.; Thomas, S. J. Appl. Polym. Sci. 2010, 118, 1027–1041.

Abraham, R.; Kuryan, S.; Isac, J.; Thomas, S. J. Appl. Polym. Sci. 2011, 120, 2233–2241.

Abraham. R.; Varughese, K. T.; Isac, J.; Thomas, S .Macromol. Sym, 2012, 314, 1–14.

Blythe, A. R. Electrical Properties of Polymers; Cambridge University Press, London, 1974.

Borcia, G.; Brown, N. M. D. J. Phys. D: Appl. Phys. 2007, 40, 1927–1938.

Bottcher, C. J. F. Physics. 1942, 9, 937–942.

Brisco, B. J. Friction and Wear of Polymer Composites; R. B. Pipe Series editor Elsheveir: Amsterdam, 1986.

Devan, R. S.; Despande, S. B.; Chougule, B. K. J. Phys. D: Appl. Phys. 2007, 40, 1864–1875.

Dilip, K. P.; Choudhary, R. N.; Samantaray. P.eXPress. Poly. Lett, 2008, 2, 630–635.

Galasso, F.G. Structure Properties and Preparation of Perovskite Compound; Pergamon: Oxford, 1969.

Grossman, D. G.; Isard, J. O. J. Phys. Appl. Phys.1970, 3, 1061–1065.

Habery, F.; Wiju, H. P. Physic. Statu. Solidi. 1968, 26, 231–236.

Hong, X. K.; Hu, G. J.; Chen, J. J. Amer. Ceramic Soc. 2007, 90, 1280–1288.

Jainwen, X. U.; Wong, C. P. J. Electro. Materi. 2006, 35, 1087–1096.

Jose, L. A.; Jose, R. J. J Appl. Polym. Sci. 1955, 57, 431.

Kumar, V.; Packia, S.; Jithesh, K.; Divya, P. V. J. Phys. D: Appl. Phy. 2007, 40, 2936–2948.

Lichteneker, K. J. Appl. Phys. 1956, 27, 824–828.

Mark, G. B.; John, V. H.; Kevin, J. P.; Dan, S. P. J. Am. Ceram. Soc. 2007, 90, 1193–1199.

Mathew, G.; Nair, S. S.; Anantharaman, S. J. Phys. D: Appl. Phy. 2007, 40, 1593–1605.

Michele, T. B.;William, P. M. N.; Yurii, K. G. Nanotech. 2008, 19, 415707–415717.

Montedo, O. R. K.; Bertan, F. M.; Piccoli, R.; Oliveria, A. P. Amer. Ceram. Soc. Bull. 2008, 87, 34–54.

Myounggu, P.; Hyonny, K.; Jeffry, P. Y, Nanotech. 2008, 19, 055705–055711.

Nakano, H.; Kamegashira, S.; Urabe K. Mater. Res. Bull. 2001, 36, 57–67.

Nelson, S.; Krazewski, A.; You ,T. J. Micro. Po. Electomagnet. Ener. 1991, 26, 45–51.

Newnham, R. E.; Wolfe, R. W. Dorian. Mater. Res. Bull. 1971, 6, 1029–1033.

Ohad, L.; David, S. Phys. Rev. B. 1997, 56, 8035–8050.

Parker, R.; Elewell, D. J.; Appl. Phys. 1966, 17, 1269–1275.

Pramod, K. S.; Amreesh, C. J. Phys. D: Appl. Phys. 2003, 36, L93–98.

Satish, B.; Sridevi, K.; Vijaya, M. S. J. Phys. D: Appl. Phys. 2002, 35, 2048–2058.

Saxena, N.; Kumar, K.; Srivastava, G. P. Phys. Status. Solidi. 1991, 127, 231–238.

Sichel, E .K. Appl. Phys. Commun, 1981, 1, 83–85.

Sillars, R. W. J. Inst. Electr. Eng. 1937, 80, 378–383.

Suzhu, Y. U.; Peter, H.; Xiao, H. U. J. Appl. Phys. 2000, 88, 398.

Vemulapally, G. K. Theoretical Chemistry; Asoke Ghosh Prentice Hall; New Delhi, 2004.

Vrejoiu, J. D.; Pedarnig , M.; Dinescu. M. Appl. Phys. A. 2002, 74, 407–412.

Yang, H.; Shi, J.; Gong, M. J. Mater. Sci. 2005, 40, 6007–6015.

Yun, K.; Ricinschi, Y. D.; Kanashima, T.; Okuyamma, M. Appl. Phys. Lett. 2006, 89, 192902–192912.

CHAPTER 17

A COMPARATIVE STUDY OF METCHROMASY INDUCED BY THIAZINE DYES

R. NANDINI* and B. VISHALAKSHI

Department of Chemistry, MITE, Moodbidri, Karnataka, India

*Corresponding author: Tel: +91 9964412316

E-mail: nandinifalnir@yahoo.com

CONTENTS

ABSTRACT

The interaction between three thiazine dyes, namely Azure B, Methylene blue, and Toluidine blue with an anionic polyelectrolyte, and Sodium Carrageenate has been investigated by spectrophotometric method. The polymers induced metachromasy in the dye resulting in the shift of the absorption maximum towards shorter wavelengths. The stoichiometry and stability of the complexes formed between Thiazine dyes and the polymers are found to be dependent on the structure of the polymers. The stability of the complexes followed the order AB-NaCar>MB-NaCar>TB-NaCar. This inference was further confirmed by reversal of metachromasy by alcohols, urea, surfactants, and electrolytes. The thermodynamic parameters of interaction revealed that binding between Thiazine dyes and the polymers was found to involve both electrostatic and hydrophobic forces.

17.1 INTRODUCTION

A blue shift in the spectrum of the dye takes place when a dye is bound to the polymers. This phenomena is known as metachromasy (Bergeron and Singer, 1958; Horbin, 2002; Mitra et al., 1997; Mitra et al., 2006). Metachromasy is related to the interaction of cationic dyes with polyanions, where a single individual compound is formed by the interaction of the dye cation and the chromotrope polyanion polymer. The studies on the metachromasy of various classes of polysaccharides like polycarboxylates, polysulphates, mucopolysaccharides, and different synthetic polyanions with different cationic dyes available in literature (Pal and Schubert, 1960; Pal and Schubert, 1961; Pal and Ghosh, 1980; Smith et al., 1999). The interaction of toluidine blue with heparin has been reported in literature (Yamaoka and Tkasuki, 1974). The studies on polymer/dye interaction inducing metachromasy in different cationic dyes by synthetic polyelectolytes are also available in literature. Several physiochemical parameters, such as molecular weight of each repeating unit, stoichiometry of the polymer/dye complex, binding constant and other related thermodynamic parameters like free energy, enthalpy, and entropy change can be evaluated using polymer/dye interaction (Berret et al., 2002; Mesraos et al., 2005; Mitra and Chakraborty, 1998; Mitra et al. , 1993; Monteaux et al., 2004; Pal and Das, 1992; Tkasuki and Yamaoka, 1979; Winkleman, 1907; Yamaoka and Tkasuki, 1974). The phenomena of reversal of metachromasy by increasing the temperature of the system may be used to determine the stability of the metachromatic compounds(Chakraborty and Nath, 1989). The nature of polymer/dye interaction in the metachromatic complex formation and also the condition for the interaction between the cationic dye and the anionic site of the macromolecules has been studied by determining the thermodynamic parameters of interaction, and to study the extent of reversal of metachromasy by using alcohols, urea, surfactants, and electrolytes which is an indirect evidence for the stability of the metachromatic complex formed

(Basu et al., 1982; Chakraborty and Nath, 1989; Hugglin et al., 1986; Jiao and Liu, 1998a; Jiao and Liu, 1998b; Jiao and Liu, 1999; Jiao et al., 1999; Liu and Jiao, 1998). The objective of the present study is to compare the extent of metachromasy induced by sodium alginate in the dyes Azure B, Methylene blue and Toluidine blue, to evaluate the thermodynamic parameters of interaction, and to study the extent of reversal by using alcohols, urea, surfactants, and electrolytes which is an indirect evidence for the stability of the metachromatic complex formed.

17.2 EXPERIMENTAL SECTION

17.2.1 APPARATUS

All the absorbance measurements were recorded using a ShimadzuUV-2550 spectrophotometer (Japan).

17.2.2 REAGENTS

All chemicals and solvents used were of analytical reagents grade. Azure B was obtained from (Hi Media, India), Methylene blue, Methanol, ethanol, propanol from (Merck, India), and Toluidine Blue, Sodium Carrageenate, Sodium lauryl sulphate, Sodium dodecyl benzene sulphonate were obtained from (Loba Chemie Mumbai, India).

17.2.2.1 DETERMINATION OF STOICHIOMETRY OF POLYMER/ DYE COMPLEX

Increasing amounts of sodium carrageenate (0.0–9.0ml, 1 × 10−2M) were added to a fixed volume of dye solution (0.6 × 10−3M) in case of AB-NaCar, MB-NaCar and TB-NaCar in different sets of experiments and the total volume was made up to 10 ml by adding distilled water in each case. The absorbances were measured at 547 nm in case of AB-NaCar, 538 nm in case of MB-NaCar and at 525 nm in case of TB-NaCar, respectively.

17.2.2.2 STUDY OF REVERSAL OF METACHROMASY USING ALCOHOLS AND UREA

For measurements of the reversal of metachromasy, solutions containing polymer and the dye in the ratio 2:1 were made containing different amounts of alcohols (10–80%) or urea (1–8 M). The total volume was maintained at 10 ml in each case. The absorbances were measured at the appropriate wavelengths as mentioned earlier.

17.2.2.3 STUDY OF REVERSAL OF METACHROMASY USING SURFACTANTS AND ELECTROLYTES

For measurements of reversal of metachromasy, solutions containing polymer and in the ratio 2:1 were made containing different amounts of surfactants in all the above cases and the absorbances were measured at the wavelengths as mentioned earlier.

17.2.2.4 DETERMINATION OF THERMODYNAMIC PARAMETERS

The thermodynamic parameters were determined by measuring the absorbance of the pure dye solution at the respective monomeric band and metachromatic band in the temperature range 36–54 °C. The above experiments were repeated in presence of polymers at various polymer/dye ratios (Mitra et al., 2006; Pal and Schubert, 1961; Smith et al., 1999; Tkasuki and Yamaoka, 1979).

17.3 RESULTS AND DISCUSSION

17.3.1 EFFECT OF POLYMER CONCENTRATION ON METACHROMASY

The absorption spectra of AB, MB and TB at various concentrations are shown in Figures 17.1, 17.2 and 17.3 respectively. The absorption maxima were found to be at 645 nm, 628 nm and at 504 nm indicating the presence of a monomeric dye species in the concentration range studied. On adding increasing amounts of polymer solution, the absorption maxima shifted to 547 nm in case of AB-NaCar, 538 nm in case of MB-NaCar complex and at 525 nm in case of TB-NaCar complex. The blue shifted band in each case is attributed to the stacking of the dye molecules on the polymer backbone and this reflects high degree of co-operativity in binding (Pal and Ghosh, 1980; Pal and Schubert, 1961). The absorption spectra at various P/D ratios are shown in Figures 17.4, 17.5 and 17.6 respectively.

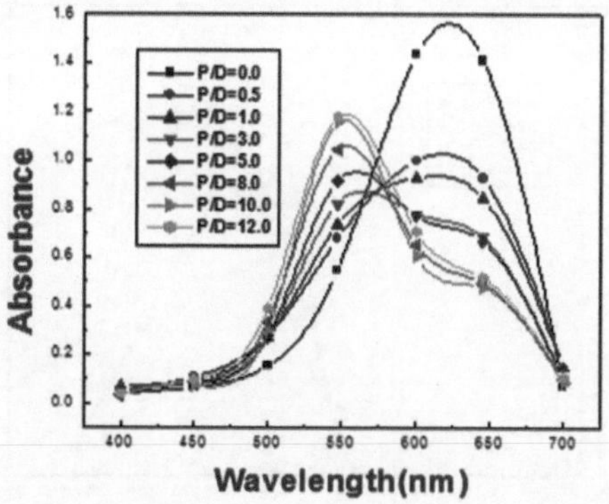

FIGURE 17.1 Absorption spectrum of AB at various concentrations

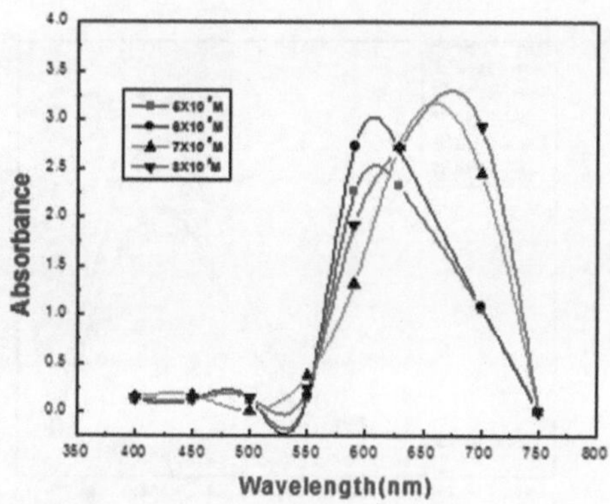

FIGURE 17.2 Absorption spectrum of MB at various concentrations

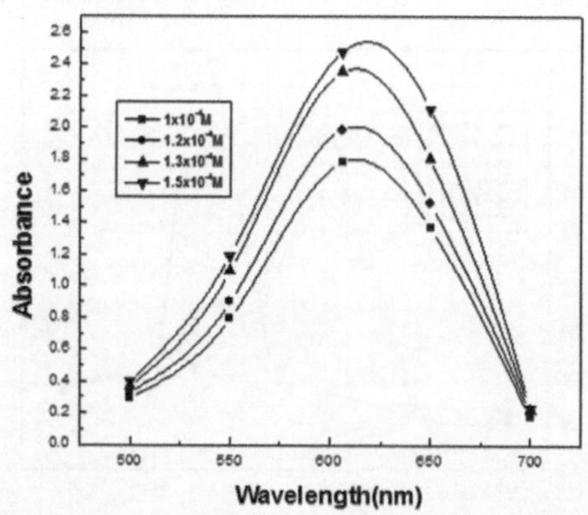

FIGURE 17.3 Absorption spectrum of TB at various concentrations

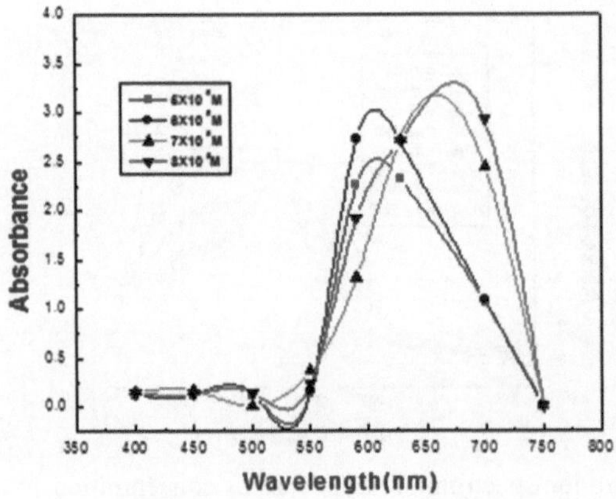

FIGURE 17.4 Absorption spectra of AB-NaCar complex at various P/D ratios

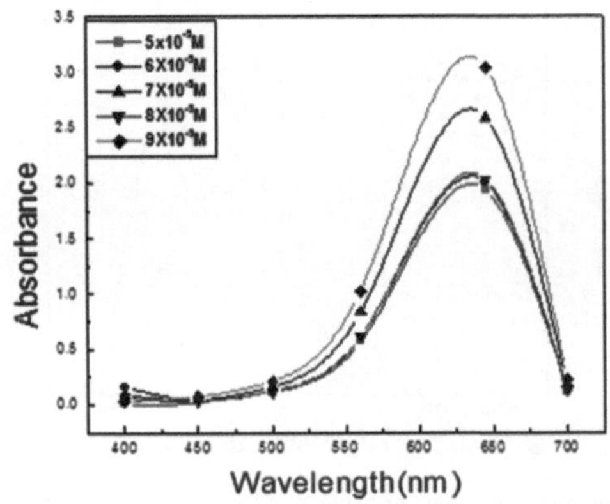

FIGURE 17.5 Absorption spectra of MB-NaCar complex at various P/D ratios

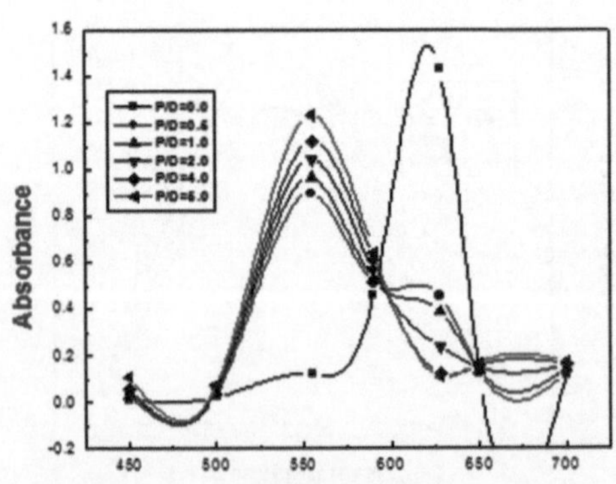

FIGURE 17.6 Absorption spectra of TB-NaCar complex at various P/D ratios

17.3.2 DETERMINATION OF STOICHIOMETRY

To determine the stoichiometry of the polymer/dye complex, a plot of AMeta/AMono versus the polmer/dye ratio was made in case of Thiazinedye-NaCar system. The stoichiometry of AB-NaCar and MB-NaCar was found to be 1:1, which indicates that binding is at adjacent anionic sites. This implies that every potential anionic site of the polyanion was associated with the dye cation and aggregation of such dye molecules was expected to lead to the formation of card pack stacking of the individual dye monomers on the molecules of the polyanion (Pal and Ghosh, 1979; Pal and Schubert, 1963). In the case of TB-NaCar complex the stoichiometry is 2:1 indicating that the binding is at alternate anionic sites. This indicates that there is lesser overcrowding of the bound dyes on the polymer chain in case of Pincyanol chloride on poly(methacrylic acid) and poly(styrene sulphonate) system. The results are given in Figure 17.7.

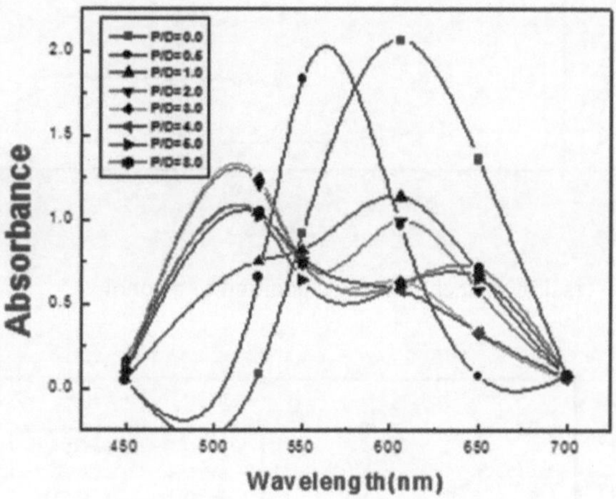

FIGURE 17.7 Stoichiometry of Thiazinedye-polymer complexes

17.3.3 REVERSAL OF METACHROMASY USING ALCOHOLS AND UREA

The metachromatic effect is presumably due to the association of the dye molecules on binding with the polyanion which may involve both electrostatic and hydrophobic interaction. The destruction of metachromatic effect may occur on addition of low molecular weight electrolytes, alcohols, or urea. The destruction of metachromasy by alcohols and urea is attributed to the involvement of hydrophobic bonding, as has already been established (Browning and Holtzer, 1961; Frank and Evans, 1945; Frank and Quist, 1961; Kauzmznn et al., 1959; Mukherjee and Ray, 1963;

Whitney and Tranford, 1962). The efficiency of alcohols in disrupting metachromasy was found to be in the order: methanol<ethanol<propanol indicating that reversal became quicker with increasing amounts of alcohols (Konradi and Ruhe, 2005; Rabinowitch and Epstein, 1941). The results are given in Figures 17.8 and 17.9 respectively and are also given in Table 17.1.

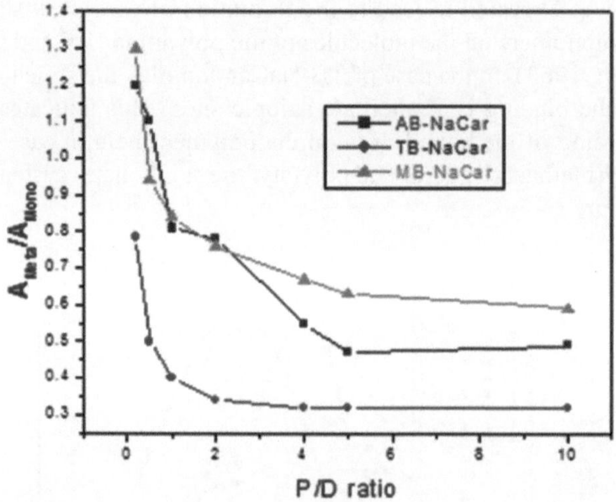

FIGURE 17.8 Reversal of metachromasy on addition of alcohol

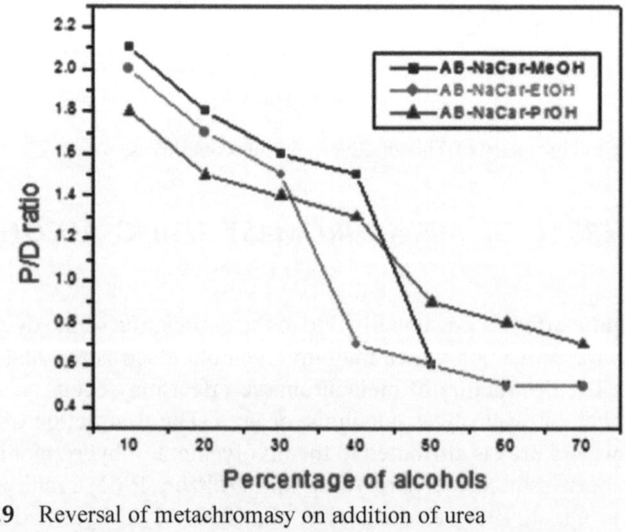

FIGURE 17.9 Reversal of metachromasy on addition of urea

TABLE 17.1 A comparative account of reversal parameters for Thiazine dyes-Polyanions

	Percentage of alcohols			Urea	Electrolytes (M)	Surfactants (M)	
System	MeOH	EtOH	PrOH	(M)	NaCl &KCl	SLS	SBS
AB-NaCar	60	50	40	4	1 × 103	1 × 10−3	1 × 10−2
MB-NaCar	65	45	35	3.5	1 × 10−6	1 × 10−6	1 × 10−4
TB-NaCar	50	40	30	4	1 × 10−4	1 × 10−4	1 × 10−5

17.3.4 EFFECT OF SURFACTANTS

The strength and nature of interaction between water soluble polyelectrolytes and oppositely charged surfactants depend on the characteristic features of both the polyelectrolyte and the surfactants (Vileti et al., 2004). The charge density, flexibility of the polyelectrolyte, hydrophobicity of the non-polar part, and the bulkiness of the polar part also play an important in the case of polysaccharide-surfactant interaction. It is known that surfactant can interact with polymers in the form of micelles(Romani et al., 2005). The molar concentration of SLS and SBS needed for reversal was found to be 1 × 10−3 M and 1 × 10−2M in case of TB-NaCar. Literature studies showed that the surfactant molecules interacted with polymers above a CAC, forming polymer supported micelles along the polymer chain thus binding, which is reinforced by hydrophobic forces (Levine and Schubert, 1958). The result is shown in Figure 17.10.

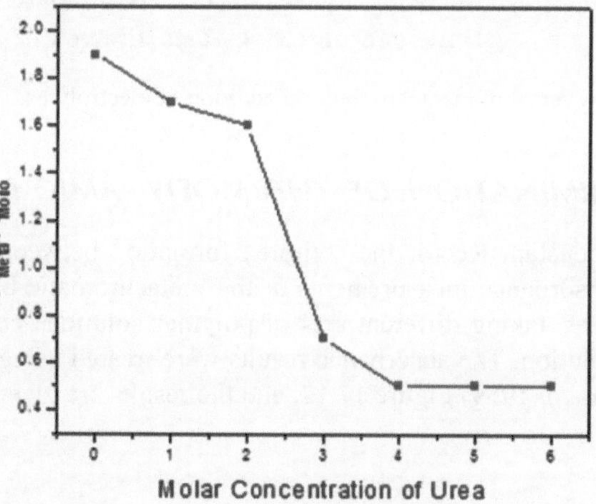

FIGURE 17.10 Reversal of metachromasy on addition of surfactants

17.3.5 EFFECT OF IONIC STRENGTH

Tan and Schneider (Tan and Schneider, 1975) have reported the disruption of the metachromatic band with variation of ionic strength. In the present study, NaCl and KCl solutions of different concentration were added to AB-Nacar, MB-NaCar and TB-NaCar complexes, as the absorbances were measured at the two wavelengths. On adding the dye, the conformation of the polycation changes to a compact coil owing to reduced repulsion due to dye binding, thus giving rise to the metachromatic band. The addition of NaCl or KCl reverses the conformational changes. From the plot of A525/A606 against molar concentration of electrolytes, the molar concentration of NaCl and KCl needed for reversal were found to be 1 × 10−4M and 1 × 10−3M respectively. The results are shown in Figure 17.11.

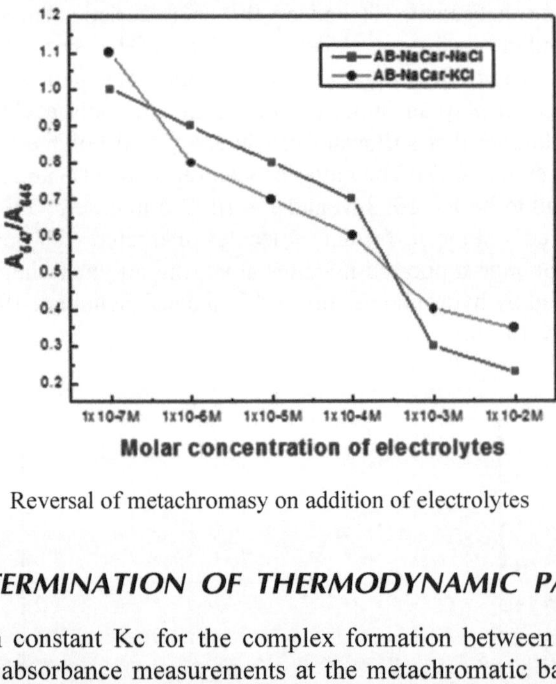

FIGURE 17.11 Reversal of metachromasy on addition of electrolytes

17.3.6 DETERMINATION OF THERMODYNAMIC PARAMETERS

The interaction constant Kc for the complex formation between AB-NaCar was determined by absorbance measurements at the metachromatic bands at four different temperatures, taking different sets of polymer solutions containing a fixed amount of dye solution. The absorbance results were treated using the Rose-Drago eqn (Rose and Drago, 1959) Figure 17.12, and the results are given in Table 17.2.

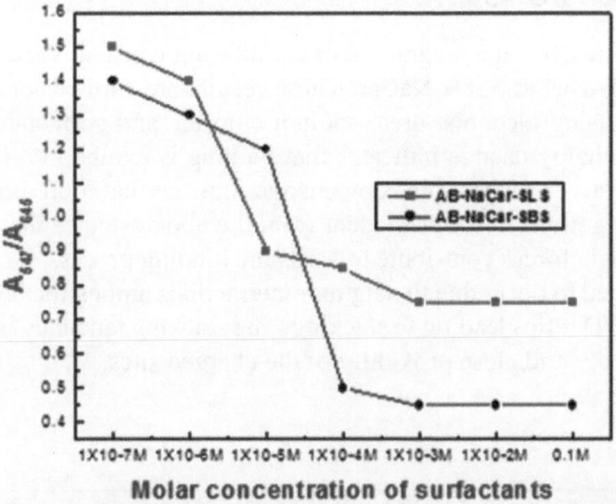

FIGURE 17.12 Plots of CD.CS/ (A–AO) Vs Cs for AB—Na Car at different temperatures

TABLE 17.2 Thermodynamic parameters for Thiazine dyes-Polyanion systems

System	Temp (K)	KC (dm3mol−1)	ΔG (kJ.mol-1)	ΔH (kJ.mol-1)	ΔS (J.mol-1K-1)
	309	10,907	-23.8		
	315	9876	-24.0		
	321	8654	-24.2		
AB-NaCar	327	7634	-24.3	-26.3	-33.3
	309	6218	-22.4		
	315	5415	–22.6		
	321	4321	–22.3		
MB-NaCar	327	3785	-23.7	-23.5	-33.3
	309	4954	-21.8		
	315	3780	-21.5		
	321	2586	-20.9		
TB-NaCar	327	1987	-20.6	-47.1	-25.0

17.4 CONCLUSIONS

In case of Thiazine dyes, the magnitude of equilibrium constant varies in the order: AB-NaCar > MB-NaCar > TB-NaCar. These results are further confirmed by the reversal studies using alcohols, urea, sodium chloride and surfactants. The negative values of enthalpy change indicates that binding is exothermic in the systems studied. The negative values of entropy change indicate that more ordered state of ions due to aggregation. It is thus evident from the above studies that both electrostatic and non-ionic forces contribute towards the binding process. Cooperativity in binding is observed to occur due to neighbor interactions among the bound dye molecules at lower P/D ratios leading to stacking. The stacking tendency is enhanced by the easy availability and close proximity of the charged sites.

KEYWORDS

- **Thiazine dyes**
- **Metachromasy**
- **Sodium carrageenate**
- **Thermodynamic parameters**

REFERENCES

Basu, S.; Gupta, A. K.; Rhotagi Mukherjee, K. K. J. Indian Chem. Soc. 1982, 59, 578.
Bergeron, J. A.; Singer, M. J. Biophys. Biochem. Cytol. 1958, 4, 433.
Berret, J. F.; Cristobal, G; Hervel, G.; Oberdisse, J.; Grillo, I. J. Eur. Phys. 2002, E9, 301.
Browning, A. Holtzer. J. Am. Chem. Soc. 1961, 83, 4865.
Chakraborty, A. K; Nath, R. K. Spectrochimica Acta. 1989, 45A, 981.
Frank, H. S.; Evans, M. W. J. Chem. Phys. 1945, 13, 493.
Frank, H. S.; Quist, A. S. J. Chem. Phys. 1961, 34, 604.
Horbin, R. W. Biochemie. Biochem. 2002, 77, 3.
Hugglin, D.; Seiffert, A.; Zimmerman, W. Histochemistry, 1986, 86, 71.
Jiao, Q. C.; Liu, Q. Anal. Letters. 1998a, 31, 1311.
Jiao, Q. C.; Liu, Q. Spectrscop Lett. 1998b, 31, 1353.
Jiao, Q. C.; Liu, Q. Spectrochimica Acta. 1999, 55A, 1667.
Jiao, Q. C; Liu, Q.; Sun, C.; He, H.; Talanta 1999, 48, 1095.
Kauzmznn, W. Advan. Protein. Chem. 1959, 14, 1.
Konradi, R.; Ruhe, J. Macromolecules, 2005, 38, 6140.
Levine, A.; Schubert, M. J. Am. Chem. Soc. 1958, 74, 5702.
Liu, Q.; Jiao, Q. C. Spectrscop .Lett. 1998, 31, 913.
Mesraos, R.; Varga, I.; Gilyani, T. J. Phys. Chem. 2005, 109, 13538.

Mitra, A.; Chakraborty, A. K.; J. Coll. Inter. A: Sci. 1998, 203, 260.

Mitra, A.; Nath, R.; Chakraborty, A. K.; Coll. Polym. Sci. 1993, 271, 1042.

Mitra, A.; Nath, R. K.; Biswas, S.; Chakraborty, A. K.; Panda, A. K. J. Photochem. Photobiol. A: Chem. 1997, 111, 157.

Mitra, A.; Nath, S.; Biswas, S.; Chakraborty, A .K.; Panda, A. K. J. Photochem. Photobiol. A: Chem. 2006, 178, 95.

Monteaux, C.; Williams, E.; Meunier, J.; Anthony, O.; Bergeron, V.; Langmuir 2004, 20, 57.

Mukherjee, P.; Ray, A.; J. Phys. Chem. 1963, 67, 190.

Pal, M. K.; Ghosh, B. K. Macromol, Chem. 1980, 181, 1459.

Pal, M. K.; Das, S. Ind. J. Biochem. Biophys. 1992, 29, 407.

Pal, M. K.; Ghosh, B. K. Macromol. Chem. 1979, 180, 959.

Pal, M. K.; Schubert, M. J. Phys. Chem. 1960, 65, 872.

Pal, M. K.; Schubert, M. J. Histochem. Cytochem. 1961, 9, 673.

Pal, M. K.; Schubert, M. J. Phys. Chem. 1963, 67, 182.

Rabinowitch, E.; Epstein, L. F. J. Am. Chem. Soc. 1941, 63, 6938.

Romani, A. P.; Gehlen, M. H.; Itri, R. Langmuir, 2005, 21, 127.

Rose, N. J.; Drago, R. S. J. Am. Chem. Soc., 1959, 81, 6138.

Smith, J. O.; Oslon, D. A; Armitage, B. A. J. Am. Chem. Soc. 1999, 121, 2686.

Tan, J. S.; Schneider, R. L. J. Phys. Chem. 1975, 79, 1380.

Tkasuki, M. Y.; Yamaoka, K .Bull. Chem. Soc. Japan, 1979, 52, 1003.

Vileti, M. A.; Borsali, R.; Crespo, J. S.; Solidi, V. O.; Fukada, K. Macromol. Chem. Phys. 2004, 205, 907.

Whitney, P. L.; Tranford, C. J. Biol. Chem. 1962, 237, 1735.

Winkleman, J. W.; Biochimicaet Biophysica Acta. 1907, 43, 140.

Yamaoka, K.; Tkasuki, M. Y. Bull. Chem. Soc. Japan, 1974, 47, 611.

CHAPTER 18

CORROSION STUDIES OF BASALT FABRIC REINFORCED EPOXY AND POLYESTER PLAIN-WEAVE LAMINATES

C. ANAND CHAIRMAN and S. P. KUMARESH BABU*

Department of Metallurgical and Materials Engineering, National Institute of Technology-Tiruchirappalli, Tamil Nadu, India

*E-mail: babu@nitt.edu

CONTENTS

ABSTRACT

The high-strength basalt fibers supplied by ASAMER TEC (Austria) are used in the present study. The laminates with 60 wt% fiber and 40 wt% resin are fabricated by hand-layup technique. The corrosion performance of basalt fabric reinforced epoxy and polyester plain-weave laminates in 3.5% NaCl solution is evaluated by electrochemical corrosion technique using ACM Gill apparatus. Based on the experimental studies it is observed that the basalt fabric reinforced epoxy (BFE) composite offers a high corrosion resistance compared to basalt fabric reinforced polyester composite (BFP).

18.1 INTRODUCTION

Basalt fibrous composites, known as a novel material has been extensively applied in construction, heating insulation, power industry, agriculture, metallurgy, aircraft and shipbuilding, mining engineering, and medical applications. For manufacturing the fiber reinforced composites, the basalt fiber is a good alternative material and has the potential to replace other high performance fibers such as carbon and glass (Bin et al., 2010, Dorigato and Pegoretti, 2014, Najafi et al., 2013). Easy processing, high strength, excellent fiber/resin adhesion suits basalt fiber to be opted in various engineering applications. It is also eco-friendly and produced from naturally occurring basalt rocks. These fibers possess chemical stability, corrosion resistance, and are nontoxic and non-combustible (Berozashvili, 2001). Because they are derived from natural minerals without any additives are hence pollution-free, and non-carcinogenic health vitreous fiber green product. For these reasons, basalt fiber has attracted wide attention to many fields as the reinforcement than the traditional ones (Xiong et al., 2005). Bin Wei et al (Wei et al., 2011) investigated the direct effects of the artificial seawater on the performance of the basalt fiber reinforced epoxy resin composites. Deák and Czigány (Deák and Czigány, 2009) compared the properties of the basalt and glass fibers and proved that continuous basalt fibers were competitive with glass fibers. But they did not mention the corrosion resistance properties. Chajes et al. (Chajes et al., 1995) studied the effect of freeze/thaw or wet/dry cycles in a calcium chloride solution of composites made with aramid, E-glass, and graphite fibers. Most of the previous studies are concentrated on the mechanical behavior and degradation of the reinforcement fibers.

Hidemitsu Hojo et al. (Hidemitsu et al., 1998) reviewed the behavior, the forms and mechanisms of corrosion of resins and glass fiber reinforced plastics under several aqueous solutions. They concluded that the concept of the corrosion rate of metals could be applied even in plastics and fiber reinforced plastics for each corrosion form. The corrosion resistance of the composites is mainly dependant on the resin's corrosion resistance properties and the corrosion crack propagation. Applications requiring corrosion resistant composite materials usually use epoxy and polyester resin as the composite matrix because these thermosets have high resistance to chemical attack. Among the various type of resins epoxies dominate the field of adhesives due to their better wetting ability, excellent mechanical properties, and high chemical and thermal resistance. Epoxy resins have the ability to bond the reinforcement, superior mechanical properties, and less degradation from water ingress. Huang Gu investigated the corrosion behaviors of the glass fiber reinforced unsaturated polyester laminates under 3.5% NaCl solutions (Gu, 2009). Knowledge of corrosion behavior of composites and a fundamental understanding of strength and material degradation is required for long-term service. In view of the above, this research article, reports a study of corrosion resistance of basalt fabric reinforced with epoxy and polyester composites in 3.5 wt% NaCl solutions.

18.2 MATERIALS AND SPECIMEN GEOMETRY

The high-strength Basalt fiber (Plain weave, 150 g/m2) which is supplied by ASA. TEC, Austria are used in the present study. The isopthalic polyester resin used for matrix system is ISO-4503 procured from Vasavibala resins (P) Ltd., Chennai (viscosity: 500–600 cps, acid value: 15–19, gel time: 15–25 min, heat deflection temperature: 95 °C). One weight percent of accelerator (cobalt Napthanate) and 1 wt% of methyl ethyl ketone peroxide (MEKP) catalyst were mixed with the polyester resin. The type of epoxy resin used here is LY 556 and the hardener HT 951 both supplied by M/s Hindustan Ciba Geigy Limited, Mumbai, India.

18.3 FABRICATION OF COMPOSITES

Basalt fiber reinforced polymer matrix composites were fabricated by using the hand layup method. Isopthalic polyester and epoxy resin were used as matrix. Twenty layers of basalt fiber mats were cut into an approximate size of 300 mm × 300 mm. These were weighed to determine the corresponding 1:1 amount of isopthalic polyester and epoxy resin respectively. The epoxy resin is mixed with the hardener in the ratio 10:1 by weight. The polyester resin was cured by incorporating one weight per cent of the methyl ethyl ketone peroxide (MEKP) catalyst. One weight per cent cobalt naphthenate (accelerator) was also added. A stirrer was used to homogenize the mixture. Then, the resin mixture was used to fabricate basalt fibers with the hand-layup technique using a roller. The samples were cured for one day at room temperature.

18.4 ELECTROCHEMISTRY CORROSION TEST OF BASALT FABRIC REINFORCED EPOXY AND POLYESTER LAMINATES

18.4.1 TEST CONDITIONS

The corrosion performance of basalt fabric reinforced epoxy and polyester plain-weave laminates in 3.5% NaCl solution was evaluated by electrochemical corrosion technique using ACM Gill apparatus. The composites to be tested are cut into square pieces of 10 mm × 10 mm × 2 mm and cleaned in acetone, alcohol, and distilled water successively. At room temperature, adopt 3.5% NaCl solution as corrosives, saturated calomel electrode (SCE) as reference electrode, and corrosion duration of 2 h.

18.5 RESULTS AND DISCUSSION

When analyzing composite properties, the corrosion resistance of the composite depends on the anti-corrosiveness of the polymeric matrix, reinforcing fiber resistance, and the diffusion-transport properties of the laminated composites. The adhesion of the matrix to the reinforcement is of prime importance for mechanical integrity and is the region of greatest importance related to corrosion. Any corrosion that takes place attacks the matrix phase first and generally more severely. Degradation in aqueous environments generally occurs by fiber/matrix debonding (Ronald, 2004). Electrochemical impedance spectroscopy was used by Wall et al. (Wall et al., 1993) to monitor the damage in graphite fiber/bismaleimide composites in contact with aluminum, steel, copper, and titanium immersed into aerated 3.5 wt% NaCl solution. Decomposition of the bismaleimide polymer was thought to occur by the action of hydroxyl ions, which break imide linkages. They concluded that the corrosion concentrated at the fiber/matrix interface was caused by cathodic polarization and was dependent upon the over potential and the cathodic reaction rate.

The polarization curves of composites in 3.5% NaCl solution are shown in Figure 18.1 (a–b). As shown in Figure 18.1(a) the current changes slightly with the increase in potential, but the current will increase with potential if the potential reaches 616.72 mV. Corrosion current density (Icorr) and corrosion potential (Ecorr) are shown in Table 18.1. Initially the rate of corrosion increases sharply for BFE composites. The corrosion rate remains stationary for BFP composites and it finally declines. The fluctuation in the corrosion rate is not an uncommon observation in composites during the initial stages.

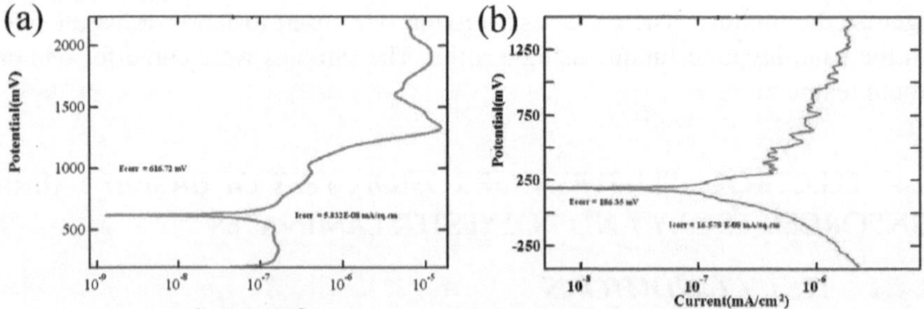

FIGURE 18.1 Polarization curves for (a) BFE composite (b) BFP composite.

The weak matrix interface leads to high corrosion in BFP composites. This is, however, not the case with BFE composites. Epoxy plays a crucial role in corrosion of BFE composites. In BFE composites the interface is strongly bonded and there is no crevice between the fiber and matrix. It is also evident from the Scanning Electron Micrograph (SEM) of an un-corroded BFE interface [Figure 18.2(a)]. During the initial exposure period, any damage to the fiber–matrix interface would lead to formation of micro crevices, resulting in an increase in the localized corrosion. Corrosion is high in BFP composites due to the discontinuity in the fiber–matrix interface, because discontinuities are the preferred site for localized corrosion due to ingress of chloride ions present in the NaCl solution. Water ingress is more in the BFP composite, which causes fiber–matrix debonding and leads to high corrosion. Since polyesters are usually made through condensation of alcohols and carboxylic acids, the presence of these ingredients in the solution is attributed to the poor bonding between the matrix and the fiber. SEM image of Figure 18.2(b) also indicated the same. Polyester resins are prone to water degradation due to the presence of hydrolysable ester groups in their molecular structures. In the case of epoxy resin, at the molecular level, the presence of polar hydroxyl and ether groups improves adhesion. Surface contact between the liquid resin and the reinforcement are not disturbed during curing because epoxies cure with low shrinkage. The result is a more homogenous bond between fibers and resin. Hence BFE have better corrosion resistance than BFP.

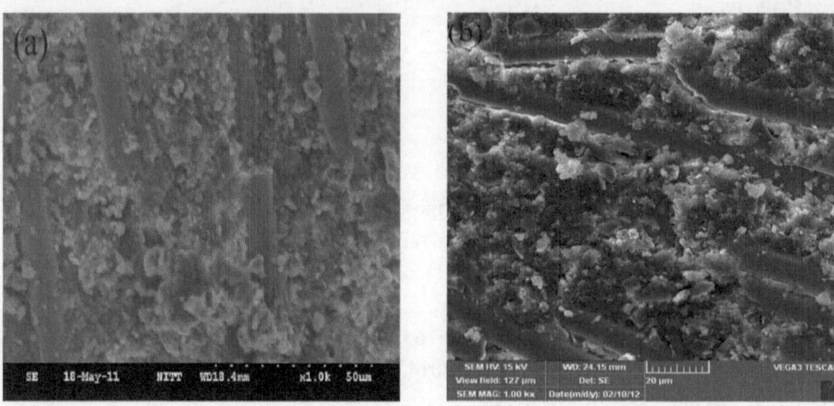

FIGURE 18.2 Scanning electron micrograph showing the (a) boundary between the basalt fibers and epoxy resin (b) boundary between the basalt fibers and polyester.

TABLE 18.1 Result of electrochemistry corrosion

Composite	Ecorr(mV)	Icorr(mA/sq.cm)
Basalt fiber-epoxy (BFE)	616.72	5.832
Basalt fiber Polyester (BFP)	186.35	16.578

18.6 CONCLUSIONS

After comparing the corrosion resistance basalt fabric reinforced epoxy (BFE) and polyester (BFP) composites certain conclusions can be drawn as follows: BFE composites possess the highest corrosion resistance than the BFP composite. The reason may be due to the better fiber–matrix interface of BFE composites than the BFP composites.

Good onding between fiber and matrix is also one of the reasons for better corrosion resistance.

KEYWORDS

- **Basalt fiber**
- **Epoxy**
- **Poly-ester**
- **Corrosion**

REFERENCES

Berozashvili, M., Adv. Mater. Com. News, Compos. Worldwide 2001, 6, 5–6.

Bin, W., Hailin C., Shenhua,. S., Mat. Sci. Engg. A 2010, 527, 4708–4715.

Bin, W., Hailin C., Shenhua, S., Corros. Sci. 2011, 53, 426–431.

Chajes, M. J., Thomson, T. A., Farschman, C. A., J. Const. Build. Mat. 1995, 9, 141–148.

Deák, T., Czigány, T., J. Tex. Res. 2009, 79, 645–651.

Dorigato, A., Pegoretti, A., J. Compos. Mater. 2014, 48, 1121–1130.

Gu, H., J. Mat. Design 2009, 30, 1337–1340.

Hojo, H., Tsuda, K., Kubouchi, M., Kim, S. D., J. Metals Mat. 1998, 4, 6, 1191–1197.

Najafi, M., Khalili, S. M. R., Eslami-Farsani, R., Int. J. Advan. Design Manufac. Technolo. 5, 2013, 4, 33–37.

Ronald A., Corrosion of Ceramic and Composite Materials; McCauley: CRC press, 2004.

Wall, F. D.; Taylor, S. R.; Cahen, G. L., STP 1174; Harris C. E., Gates, T. S., Eds.; ASTM: Philadelphia, PA, 1993; 95–113.

Xiong, S. L., Chen, Y., LI, Z. W., Shi, J. L., Fiber Glass 2005, 6, 5–11.

CHAPTER 19

STRUCTURAL INVESTIGATIONS OF PEO/PVP POLYMER BLEND SALT COMPLEXES

K. KIRAN KUMAR*, S. BHAVANI, M. RAVI, Y. PAVANI, A. K. SHARMA, and V. V. R. N. RAO

Department of Physics, Sri Venkateswara University, Tirupati-517 502, India

*Corresponding author: Tel: +91 8772242110(R)

Mobile: +91 8008528025

E-mail: kokila.12345@gmail.com

CONTENTS

ABSTRACT

The broadening and reduction in the intensity of the Bragg peaks confirm the dissolution of salt in the polymer host and the increase in amorphous nature of the films with the addition of the salt. Vibrational bands correspond to PEO and PVP confirm the miscibility of the blend. Attainment of smooth surface morphology upon the addition of salt suggests the enhancement of degree of amorphisity. A decrease in the degree of crystallinity is observed on doping, while Tg and Tm also showed similar trend. Single Tg measured as onset value indicates that the present polymer blend is miscible.

19.1 INTRODUCTION

Development and characterization of new kind of materials with uncommon properties has been drawing considerable attention in recent years. Among these new "strategic" materials, conductive polymer blends with high electric transport have reached a prominent position. The large interest in this field is due to their exciting applications such as realization of sensors, development of advanced rechargeable batteries, electrochemical cells, electro-chromic display devices and so on. The inherent merits of using blend-based polymer electrolytes are exemplified by several research groups (Kim et al., 1996). In present investigations, polyethylene oxide (PEO) based electrolyte system is prepared due to its ability to play host to various metal salt systems at a range of concentrations, to act as a binder for other phases and its excellent chemical stability. However, the semi crystalline nature of pure PEO is often regarded as a major problem in real working systems, since the ionic conductivity always predominates in amorphous portion. One of the most promising ways of enhancing the amorphous content is polymer blending. Polyvinyl pyrrolidone (PVP) is a special one among the conjugated polymers because of its high amorphous nature which can permit faster ionic mobility (Blum et al., 2011; Chen et al., 2011; Han et al., 2012; Reisfeld et al., 2011; Singh et al., 2011; Wang et al., 2010; Zhao et al., 2014). The presence of carbonyl group (C=O) in PVP, leads to form a variety of complexes with various inorganic salts and exhibits high Tg with good environmental, thermal, and mechanical stability. Sodium is cheaper and available in abundance. A few attempts have been made to develop electrolytes based on sodium complexed films (Fautex et al., 1987). The present investigations focused on the extension of these studies.

19.2 EXPERIMENTAL

PEO (4×106 mw) and PVP (1.3×105 mw) were used to prepare solid polymer blend electrolytes by solution cast technique. Films (thicknesses ~160 μm) of pure blend of PEO/PVP (70:30) and various compositions of NaF complexed films of

PEO/PVP by weight percent ratios 67.5:27.5:5, 65:25:10 and 62.5:22.5:15 were obtained. X-ray diffraction scans were recorded using SEIFERT, XRD 3003TT, X-ray diffractometer. FTIR absorption spectra of the films were recorded using "Thermo Nicolet IR200" Spectrometer. The surface morphology is observed using CARL ZEISS EVO 25 scanning electron microscope. DSC thermo grams were recorded in the present investigations using Perkin Elmer PYRIS Diamond differential scanning calorimeter. A Perkin Elmer TGA-7 (vertical furnace) system is used for the TGA.

19.3 RESULTS AND DISCUSSION

19.3.1 XRD STUDIES

XRD patterns of PEO/PVP blend films (Figure 19.1) showed peaks at around 19.2°, 22.5°, 23.6°, 25.5°, 26.2°, 27.1°, 29.7°, 32.6°, 36.6° and 39.6° and are attributed to the crystalline phase of PEO. These peaks appear to superimpose on a broad hump between 18 and 50 which could be due to the amorphous nature of PEO. A broad peak at around 13° is associated with the amorphous nature of PVP. The intensity of all crystalline peaks decreases gradually in all XRD patterns with salt concentration suggesting a decrease in the degree of crystallinity of the complex. This could be due to the disruption of the semi crystalline structure of the films by salt. No sharp peaks correspond to NaF salt were observed in all PEO/PVP complexes, indicating the complete dissolution of salt in the polymer matrix.

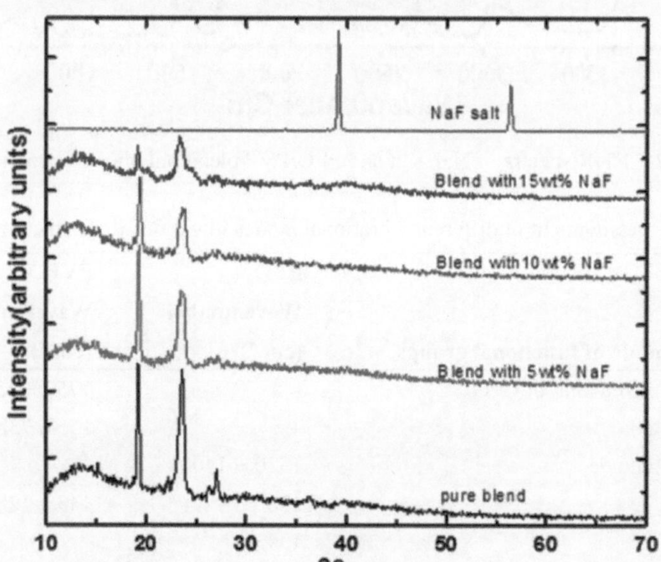

FIGURE 19.1 XRD patterns of NaF salt and PEO/PVP blend polymer electrolytes

19.3.2 FTIR STUDIES

Co-existence of well resolved bands (Figure 19.2) corresponds to the C–O–C of PEO and C=O and C-N groups of PVP (Table 19.1) indicate that PEO and PVP are miscible.

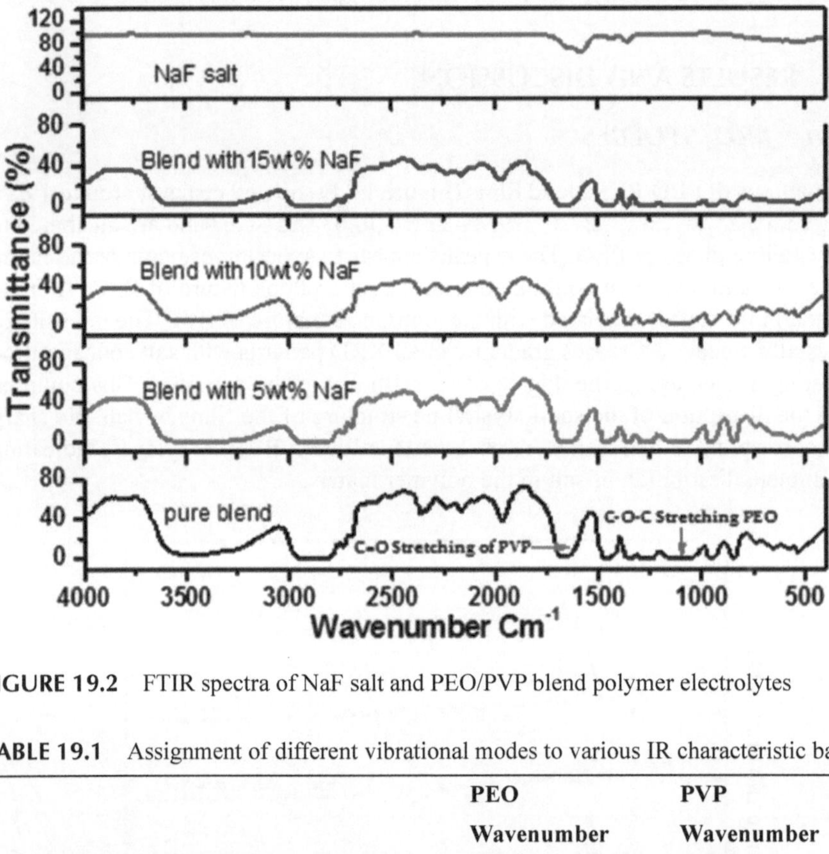

FIGURE 19.2 FTIR spectra of NaF salt and PEO/PVP blend polymer electrolytes

TABLE 19.1 Assignment of different vibrational modes to various IR characteristic bands

Vibrational mode of functional groups	PEO Wavenumber (cm^{-1})	PVP Wavenumber (cm^{-1})
Sym. & asym. stretching of C=O	–	1750 – 1550
C-N stretching	–	1557, 738
CH_2 scissoring mode	1520 – 1400	–
CH_2 wagging	1360	1433, 1283
CH_2 bending	1343	–
Swinging vibration of C-H in CH_2 group	1350	1375

Asymmetric CH_2 twisting	1282	1293
C-C stretching	1147	1166, 945
C–O–C (sym. & asym.) stretching	1100	–
C–O stretching with some CH2 asym. rocking	947	–
CH_2 rocking in PVP and with some C–O stretching in PEO	845	845
C-N bending	–	465

The following shifts (Table 19.2) in position of characteristic bands of PEO or PVP confirm the complexation of the polymer blend systems.

TABLE 19.2 Shifting of bands with salt concentration

Sl. No	Polymer electrolyte	Composition	Tg (°C)	$(\Delta H_m J/g)$	$(\chi_c\%)$	Tm (°C)
1	Pure (PEO/PVP)	(70:30)	58.49	223	100	65.25
2	PEO/PVP/NaF	(67.5:27.5:5)	54.90	134.1	60.1	62.92
3	PEO/PVP/NaF	(65:25:10)	53.69	115.5	51.8	62.21
4	PEO/PVP/NaF	(62.5:22.5:15)	52.68	95.3	42.7	61.51

19.3.3 THE SURFACE MORPHOLOGY

The pure PEO/PVP shows a rough surface morphology with lot of rumples indicating the presence of crystalline phase in pure PEO/PVP blend. The surface morphology (Figure 19.3 a–d) of the blend electrolytes changes from rough to smoother, and the reductions of rumples are also found with the increase of salt content which indicates the decrease in semi crystalline nature. As there is uniform distribution of one of the polymers, in the other, PEO and PVP form a compatible blend. Furthermore, the formation of micro voids and craters on the surfaces of some films is due to rapid evaporation of solvent from the polymer structure.

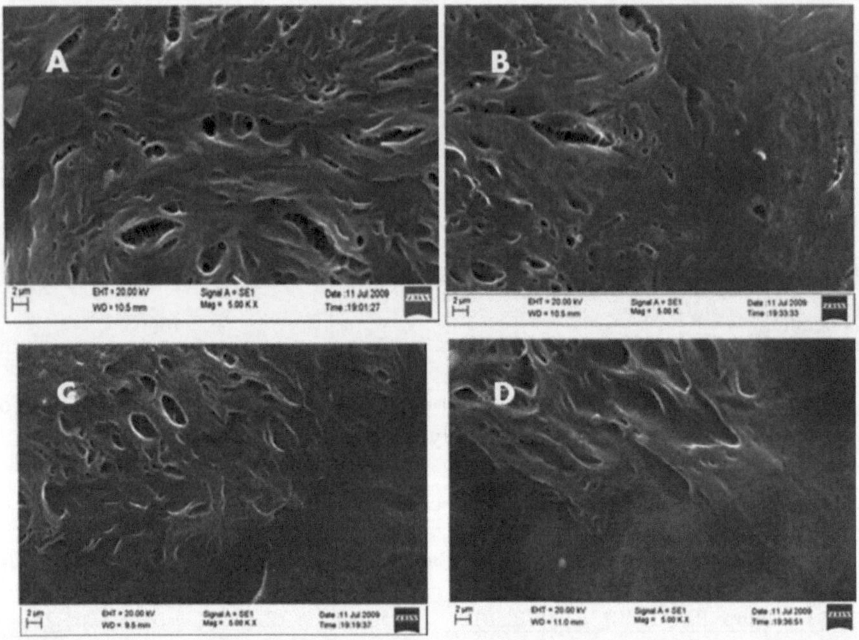

FIGURE 19.3 SEM photograph of the (a) pure PEO/PVP blend electrolyte, (b) 5 wt% NaF complexed PEO/PVP blend electrolyte, (c) 10 wt% NaF complexed PEO/PVP blend electrolyte, (d) 15 wt% NaF complexed PEO/PVP blend electrolyte

19.3.4 THERMAL ANALYSIS

The DSC thermo grams (Figure 19.4) showed a strong endothermic effect in the region between 43 °C and 75 °C is superimposed on the heat flow shift for pure and all complexed PEO/PVP blend electrolyte systems and is associated with the melting of PEO crystallites. The variation of onset glass transition temperature (Tg), melting enthalpy (ΔHm), percentage of relative crystallinity ($\chi c\% = [\Delta H_m / \Delta H_m{}^{\circ}] \times 100$, where $\Delta H_m{}^{\circ}$ is crystalline melting enthalpy of pure blend and ΔH_m, the melting enthalpy of complexed PEO/PVP films) and melting temperature (Tm) with salt concentration are given below in Table 19.3.

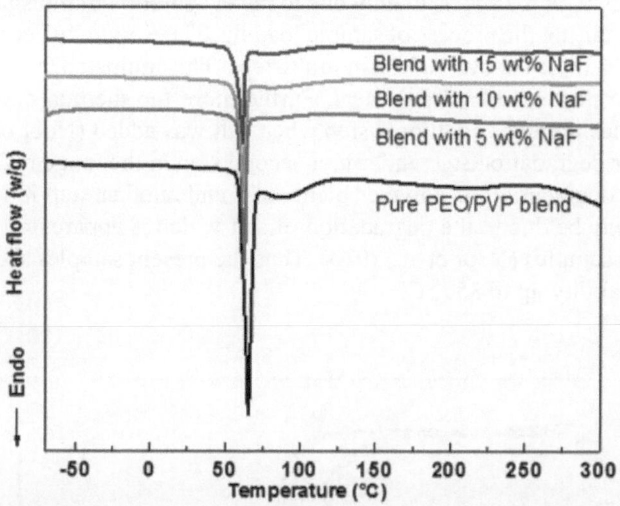

FIGURE 19.4 DSC Thermograms of PEO/PVP blend polymer electrolytes

TABLE 19.3 Thermal parameters of PEO/PVP polymer blend electrolyte system

Vibrational mode	Spectral position (Wave Number cm^{-1})			
	Pure PEO/ PVP	PEO/PVP/NaF (67.5:27.5:5)	PEO/PVP/ NaF (65:25:10)	PEO/PVP/NaF (62.5:22.5:15)
C–H stretching of CH_2	2818	2809	2796	2780
C–H bending of CH_2	1483	1479	1473	1469
CH_2 wagging	1351	1355	1359	1363
CH_2 Asym. Twisting	1282	1285	1289	1294
CH_2 Sym. Twisting	1236	1234	1231	1227
C-N stretching	1557	1555	1552	1550

The single Tg of PEO/PVP blend indicates that the present polymer blend is miscible. Furthermore the Tg and Tm are found to shift towards lower temperatures, when salt is increased in its concentration. This effect is a result of reduction in crystalline nature and increase in the amorphous content there by segmental mobility.

19.3.5 THERMO GRAVIMETRIC ANALYSIS

Typical thermo gravimetric analysis (TGA) traces of pure and NaF complexed PEO/PVP polymer blend electrolytes are depicted in Figure 19.5. The initial 1–4% weight loss between 55 °C and 85 °C is mainly due to the evaporation of moisture absorbed by the samples during the process of sample loading. The weight loss curves showed degradation of samples above 300 °C in multisteps. This multistep trend proves that the present samples are blend polymers. Furthermore the thermal degradation of polymer samples involves additional step when salt was added (Noor et al., 2010). The first major degradation step and minor second step in the range of 350–600 °C are due to the degradation of polymer blend host and another step in the range of 650–750 °C may be due to the degradation of salt which is apparent in complexed polymer blend samples (Noor et al., 2010). Thus the present samples showed excellent thermal stability up to 350 °C.

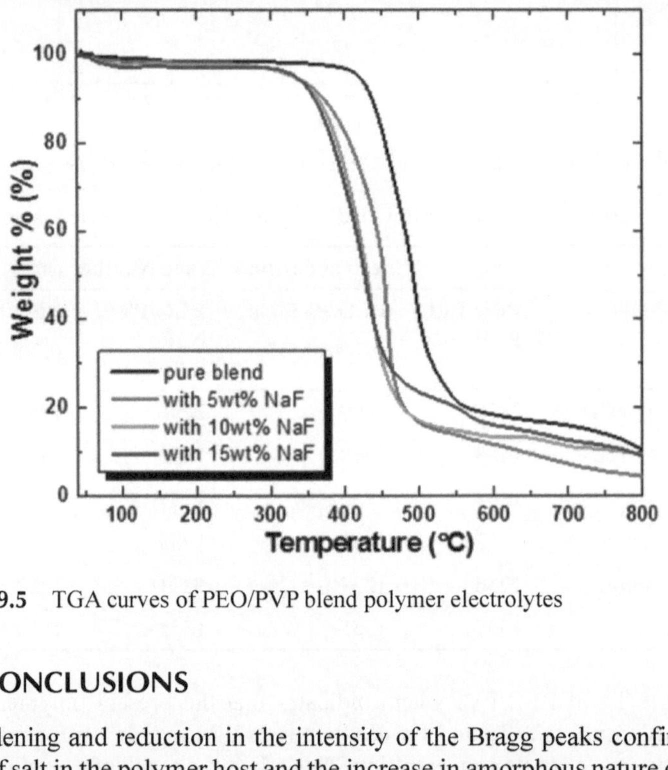

FIGURE 19.5 TGA curves of PEO/PVP blend polymer electrolytes

19.4 CONCLUSIONS

The broadening and reduction in the intensity of the Bragg peaks confirm the dissolution of salt in the polymer host and the increase in amorphous nature of the films with the addition of the salt. Vibrational bands correspond to PEO and PVP confirm the miscibility of the blend. Attainment of smooth surface morphology upon the addition of salt suggests the enhancement of degree of amorphisity. A decrease in the degree of crystallinity is observed on doping, while Tg and Tm also showed similar trend. Single Tg measured as onset value indicates that the present polymer blend is miscible. The present blend system is thermally stable up to 350 °C.

KEYWORDS

- **Polyethylene oxide**
- **Polyvinylpyrrolidone**
- **XRD studies**
- **FTIR studies**
- **Thermo gravimetric analysis**

REFERENCES

Blum, A.; Lee, L.; Eberl, D., 2011,59, 212–213.

Chen, H.; Luo, H.; Lan, Y.; Dong, T.; Hu, B.; Wang, Y., 2011,192, 44–53.

Fautex, D.; Lupien, M. D.; Robitaille, C. D. J. Electrochem. Soc. 1987, 314, 2761.

Han, H. S.; Kwak, S. W.; Kim, B. M.; Lee, T. M.; Kim, S. H.; Kim, I. Y., 2012, 22, 476–481.

Kim, D. W.; Park, J. K.; Rhee, H. W. Solid State Ionics, 1996, 83, 49.

Noor, S. A. M.; Ahmad, A.; Talib, I. A.; Rahman, Ms. Y. A. Ionics 2010, 16, 161.

Reisfeld, R.; Saraidarov, T.; Panzer, G.; Levchenko, V.; Gaft, M. Opt. Mater. 2011, 34, 351–354.

Singh, M.; Singh, A. K.; Mandal, R. K.; Sinha, I. Coll. Surf. A: Physicochem. Eng. Aspects. 2011, 390, 167–172.

Wang, J.; Tsuzuki, T.; Tang, B.; Cizek, P.; Sun, L.; Wang, X. Coll. Polym. Sci., 2010, 288, 1705–1711.

Zhao, C.; Cheng, H.; Jiang, P.; Yao, Y.; Han, J., 2014, 156, 123–128.

INDEX

C

D

K

L

M

N

T